Gene Cloning and Analysis by RT-PCR

BioTechniques Molecular Laboratory Methods Series

1. *Gene Cloning and Analysis by RT-PCR*
   Edited by P.D. Siebert and J.W. Larrick

2. *Apoptosis Detection and Assay Methods*
   Edited by L. Zhu and J.J.M. Chun

3. *Protein Staining and Identification Techniques*
   R. Allen and B. Budowle

BioTechniques Update Series

The PCR Technique: DNA Sequencing
Edited by J. Ellingboe and U. Gyllensten

The PCR Technique: DNA Sequencing II
Edited by U. Gyllensten and J. Ellingboe

The PCR Technique: Quantitative PCR
Edited by J.W. Larrick

The PCR Technique: RT-PCR
Edited by P. Siebert

Expression Genetics: Differential Display
Edited by A.B. Pardee and M. McClelland

Gene Cloning and Analysis by RT-PCR

Edited by

Paul D. Siebert
CLONTECH Laboratories
Palo Alto, CA

James W. Larrick
Palo Alto Institute of Molecular Medicine
Mountain View, CA

BioTechniques® Books
(Division of Eaton Publishing)

Paul D. Siebert
CLONTECH Laboratories
1020 East Meadow Circle
Palo Alto, CA 94303-4230

James W. Larrick
Palo Alto Institute of Molecular Medicine
2462 Wyandotte Street
Mountain View, CA 94043

Library of Congress Cataloging-in-Publication Data
Gene cloning and analysis by RT-PCR / edited by Paul D. Siebert, James W. Larrick.
    p.   cm. -- (BioTechniques molecular laboratory methods series)
   Includes bibliographical references and index.
   ISBN 1-881299-14-7
   1. Molecular cloning. 2. Polymerase chain reaction. 3. Reverse transcriptase. I. Siebert, Paul D. II. Larrick, James W. III. Series.
QH442.2.G4623 1998
572.8'645--dc21
                                                  98-28892

Copyright© 1998 Eaton Publishing

All rights reserved. No part of this publication may be reproduced, stored in a retrieval system, or transmitted, in any form or by any means, electronic, mechanical, photocopying, recording, or otherwise, without prior permission of the copyright owner.

Reproduction or translation of any part of this work beyond that permitted by Section 107 or 108 of the 1976 United States Copyright Act without the permission of the copyright owner is unlawful. Requests for permission or further information should be addressed to the Permissions Department, Eaton Publishing.

ISBN 1-881299-14-7
ISSN 1520-3182

Printed in the United States of America

9 8 7 6 5 4 3 2 1

> The cover graphic is a schematic representation of RT-PCR. Image design by Yoshimi Munch, CLONTECH Laboratories, Palo Alto, CA.

Wo ai ni XIAO JUN!

CONTENTS

SECTION I: Obtaining RNA for RT-PCR

1 RT-PCR Using Formalin-Fixed, Paraffin-Embedded (FFPE) Archival Tumor Specimens 3
   D.A. Thompson, C. Carmeci and R.J. Weigel

2 Isolation of RNA from Solid Tumors for RT-PCR 19
   J. Kitlińska and J. Wojcierowski

3 Isolating Single Cells for Analysis by RT-PCR 27
   M.C. Harbeck and P.L. Rothenberg

4 Isolation of RNA from Mammalian Cells: Application to Large mRNA 35
   D. Gruffat

5 RT-PCR without RNA Isolation 57
   R.J. Klebe, G.M. Grant, A.M. Grant, M.A. Garcia and T.A. Giambernardi

SECTION II: Quantitative RT-PCR

6 Absolute Quantification of Messenger RNA Using Multiplex RT-PCR 71
   D.E. Dostal, A.M. Kempinski, T.J. Motel and K.M. Baker

7 Standard Curve Quantitative Competitive RT-PCR (SC-QC-RT-PCR): A Simple Method to Quantify Absolute Concentration of mRNA from Limited Amounts of Sample 91
   T. Tsai and M.C. Wiltbank

8 Design and Use of an Exogenously Added Competitive RNA Multiplex Template for Measurement of mRNA Levels by RT-PCR ..103
   R.W. Tarnuzzer

9 Alternative Quantitative PCR Method 113
   A. Nicoletti and C. Sassy-Prigent

10 Quantitation of PCR Products Using DNA Hybridization Assays in Microplates 129
   C. Berndt, J. Gross and W. Henke

SECTION III: Cloning Differentially Expressed cDNAs

A. RNA Display/Fingerprinting

11 Improvements to Differential Display 145
   T. Bos and M. Hadman

12 DD/AP-PCR: Differential Display and Arbitrarily Primed PCR of Oligo(dT) cDNA 159
   C.B. Rothschild, D.W. Bowden and G.S. Shelness

13 Preferential Identification of Differentially Expressed mRNAs of
    Moderate to Low Abundance in a Microscopic System Using
    Selected Primers 171
    O.C. Ikonomov and M.H. Jacob

14 Solid-Phase Differential Display 183
    J. Lundeberg, Ø. Røsok, A. Hansen Ree and J. Odeberg

B. cDNA Subtraction

15 Isolating Differentially Expressed Genes by Representational
    Difference Analysis 193
    D. Chang and C. Denny

16 Linker-Capture Subtraction 203
    M. Yang and A.J. Sytkowski

17 Suppression Subtractive Hybridization: A Method for Generating
    Subtracted cDNA Libraries Starting from Poly(A+) or Total RNA 213
    L. Diatchenko, A. Chenchik and P.D. Siebert

18 Optimized Subtraction-Enhanced Display Technique 239
    M. Denijn, T.B.M. Hakvoort and W.H. Lamers

SECTION IV: Obtaining Full-Length cDNAs and Homologous cDNAs

19 Enriched Full-Length cDNA Expression Library by RecA-Mediated
    Affinity Capture 259
    T.B.M. Hakvoort, J.L.M. Vermeulen and W.H. Lamers

20 Gene-Capture PCR 271
    R. Mastrangeli and S. Donini

21 Improved Technique for Walking in Uncloned Genomic DNA 289
    S.S. Chen, A. Chenchik, K.A. Lukyanov and P.D. Siebert

SECTION V: Ancillary PCR Methods

22 Generation and Use of High-Quality cDNA from Small Amounts
    of Total RNA by SMART™ PCR 305
    A. Chenchik, Y.Y. Zhu, L. Diatchenko, R. Li, J. Hill and P.D. Siebert

23 Protein Production from PCR Products 321
    S.J. Winder

24 In Vitro Cloning: A Method for PCR Amplification of Individual
    Unknown DNA Fragments Suitable for Direct Sequencing 329
    K.A. Lukyanov

Index 347

CONTRIBUTORS

Kenneth M. Baker
Weis Center for Research
Henry Hood M.D. Research Program
Pennsylvania State University College of Medicine
Danville, PA, USA

Christoph Berndt
Institute of Laboratory Medicine and
 Pathobiochemistry
Medical Faculty Charité
Humboldt University
Berlin, Germany

Timothy Bos
Department of Microbiology and Immunology
Eastern Virginia Medical School
Norfolk, VA, USA

Donald W. Bowden
Department of Biochemistry
Bowman Gray School of Medicine
Wake Forest University
Winston-Salem, NC, USA

Charles Carmeci
Department of Surgery
Stanford University
Stanford, CA, USA

David Chang
Departments of Medicine and Microbiology &
 Immunology
Jonsson Comprehensive Cancer Center
UCLA School of Medicine
University of California
Los Angeles, CA, USA

Stephen S. Chen
Gene Cloning and Analysis Group
CLONTECH Laboratories
Palo Alto, CA, USA

Alex Chenchik
Gene Cloning and Analysis Group
CLONTECH Laboratories
Palo Alto, CA, USA

Luda Diatchenko
Gene Cloning and Analysis Group
CLONTECH Laboratories
Palo Alto, CA, USA

Marylene Denijn
Department of Anatomy and Embryology
Academic Medical Center
University of Amsterdam
Amsterdam, The Netherlands

Christopher Denny
Department of Pediatrics
Gwynne-Hazen Memorial Laboratories
Jonsson Comprehensive Cancer Center
UCLA School of Medicine
University of California
Los Angeles, CA, USA

Silvia Donini
Drug Discovery Department
Istituto di Ricerca Cesare Serono SpA
Ardea, Rome, Italy

David E. Dostal
Weis Center for Research
Henry Hood M.D. Research Program
Pennsylvania State University College of Medicine
Danville, PA, USA

Melissa A. Garcia
Department of Cellular and Structural Biology
University of Texas Health Science Center
San Antonio, TX, USA

Troy A. Giambernardi
Department of Cellular and Structural Biology
University of Texas Health Science Center
San Antonio, TX, USA

Anne M. Grant
Department of Cellular and Structural Biology
University of Texas Health Science Center
San Antonio, TX, USA

George M. Grant
Department of Cellular and Structural Biology
University of Texas Health Science Center
San Antonio, TX, USA

Johann Gross
Institute of Laboratory Medicine and
 Pathobiochemistry
Medical Faculty Charité
Humboldt University
Berlin, Germany

Dominique Gruffat
Laboratoire Croissance et Métabolismes des
 Herbivores
Unité Métabolismes Energétique et Lipidique
Institut National de la Recherche Agronomique
Centre de Recherches Clermont Ferrand-Theix
Saint Genès Champanelle, France

Martin Hadman
Department of Microbiology and Immunology
Eastern Virginia Medical School
Norfolk, VA, USA

Theodorus B.M. Hakvoort
Department of Anatomy and Embryology
Academic Medical Center
University of Amsterdam
Amsterdam, The Netherlands

Anne Hansen Ree
Department of Tumor Biology
Institute of Cancer Research
The Norwegian Radium Hospital
Oslo, Norway

Mark C. Harbeck
Department of Endocrinology and Metabolic Disease
Jefferson Medical College
Thomas Jefferson University
Philadelphia, PA, USA

Wolfgang Henke
Institute of Biochemistry
Medical Faculty Charité
Humboldt University
Berlin, Germany

Jason Hill
Gene Cloning and Analysis Group
CLONTECH Laboratories
Palo Alto, CA, USA

Ognian C. Ikonomov
Department of Psychiatry
Wayne State University School of Medicine
Detroit, MI, USA

Michele H. Jacob
Department of Neuroscience
Tufts University School of Medicine
Boston, MA, USA

Anna M. Kempinski
Weis Center for Research
Henry Hood M.D. Research Program
Pennsylvania State University College of Medicine
Danville, PA, USA

Joanna Kitlińska
Center for Perinatal Biology
Loma Linda University School of Medicine
Loma Linda, CA, USA

Robert J. Klebe
Department of Cellular and Structural Biology
University of Texas Health Science Center
San Antonio, TX, USA

Wouter H. Lamers
Department of Anatomy and Embryology
Academic Medical Center
University of Amsterdam
Amsterdam, The Netherlands

Roger Li
Gene Cloning and Analysis Group
CLONTECH Laboratories
Palo Alto, CA, USA

Konstantin A. Lukyanov
Shemyakin and Ovchinnikov Institute of Bioorganic Chemistry
Russian Academy of Science
Moscow, Russia

Joakim Lundeberg
Department of Biochemistry and Biotechnology
KTH, Royal Institute of Technology
Stockholm, Sweden

Renato Mastrangeli
Drug Discovery Department
Istituto di Ricerca Cesare Serono SpA
Ardea, Rome, Italy

Thomas J. Motel
Weis Center for Research
Henry Hood M.D. Research Program
Pennsylvania State University College of Medicine
Danville, PA, USA

Antonino Nicoletti
INSERM U430
Hôpital Broussais
Paris, France

Jacob Odeberg
Department of Biochemistry and Biotechnology
KTH, Royal Institute of Technology
Stockholm, Sweden

Øystein Røsok
Department of Immunology
Institute of Cancer Research
The Norwegian Radium Hospital
Oslo, Norway

Paul L. Rothenberg
Clinical Pharmacology
Novartis Pharmaceuticals
East Hanover, NJ, USA

Cynthia B. Rothschild
Department of Biochemistry
Bowman Gray School of Medicine
Wake Forest University
Winston-Salem, NC, USA

Caroline Sassy-Prigent
INSERM U430
Hôpital Broussais
Paris, France

Contributors

Gregory S. Shelness
Department of Comparative Medicine
Bowman Gray School of Medicine
Wake Forest University
Winston-Salem, NC, USA

Paul D. Siebert
Gene Cloning and Analysis Group
CLONTECH Laboratories
Palo Alto, CA, USA

Arthur J. Sytkowski
Laboratory for Cell and Molecular Biology
Division of Hematology and Oncology
Beth Israel Deaconess Medical Center
Harvard Medical School
Boston, MA, USA

Roy W. Tarnuzzer
Division of Endocrinology and Metabolism
Department of Medicine
University of Florida
Gainesville, FL, USA

Devon A. Thompson
Department of Surgery
Stanford University
Stanford, CA, USA

Shaw-Jenq Tsai
Department of Physiology
College of Medicine
National Cheng Kung University
Tainan, Taiwan

Jacqueline L.M. Vermeulen
Department of Anatomy and Embryology
Academic Medical Center
University of Amsterdam
Amsterdam, The Netherlands

Ronald J. Weigel
Department of Surgery
Stanford University
Stanford, CA, USA

Milo C. Wiltbank
Endocrinology–Reproductive Physiology
  Program and Department of Dairy Science
University of Wisconsin–Madison
Madison, WI, USA

Steven J. Winder
Institute of Cell and Molecular Biology
University of Edinburgh
Edinburgh, Scotland

Jacek Wojcierowski
Department of Medical Genetics
Medical Academy of Lublin
Lublin, Poland

Meiheng Yang
Laboratory for Cell and Molecular Biology
Division of Hematology and Oncology
Beth Israel Deaconess Medical Center
Harvard Medical School
Boston, MA, USA

York Y. Zhu
Gene Cloning and Analysis Group
CLONTECH Laboratories
Palo Alto, CA, USA

PREFACE

The principal objective of this book is to provide investigators with a contemporary RT-PCR-based methodology for the purpose of gene cloning and analysis. Since RT-PCR was first described for the detection of low-abundance mRNAs (1), the method has been extended to provide measurements of relative and absolute levels of single and multiple mRNAs. Methods for obtaining RNA for RT-PCR have also been extended to include analysis of mRNA levels in clinical samples and even in single cells.

PCR methods are available for obtaining full-length clones, homologous cDNAs and differentially expressed cDNAs. In many cases, RT-PCR is able to circumvent many of the time-consuming and technically demanding cloning steps. The use of PCR can even enable the cloning of DNA molecules without the need for biological hosts.

The authors who kindly provided chapters to this book were chosen with three criteria: *(i)* they have brought either new applications or improvements to the RT-PCR method; *(ii)* their primarily peer-reviewed manuscripts have been published during or after 1996; and *(iii)* they bring a fresh perspective to the subject. The project turned out to be an international collaboration.

The following is a brisk walk through the various chapters and sections.

Section I describes different methods for obtaining RNA for RT-PCR. Traditional methods of RNA purification are generally useful, but there are many cases in which they are not suitable, e.g., when RT-PCR is to be performed on clinical samples such as small tumor biopsies and archival specimens. The first two chapters focus on these two applications.

Following development of the long PCR technique (2), it has become important to obtain intact high-molecular-weight RNA. A chapter is included discussing methods for both the isolation and quality control of high-molecular-weight RNA. Traditional methods of RNA extraction are also not suited for examination of gene expression in single cells. One chapter describes a simple method for isolation and lysis of individual eukaryotic cells.

Section II describes several new approaches for performing quantitative RT-PCR. In two of the chapters, competitive RT-PCR is described. In one, the novel combination of multiplex and competitive RT-PCR, which is the most sophisticated and accurate method, is discussed. The other outlines a simplified method for performing competitive RT-PCR, which is useful when many experiments are to be performed. Competitive RT-PCR can be used to determine the number of mRNA molecules. It requires the generation of competititor DNA or RNA fragments. This section also contains chapters describing a simple method to generate competitor RNA fragments and a new simplified method for determining the relative levels of mRNA, which is simpler than competitive RT-PCR. Whichever method of quantitative RT-PCR is chosen, it is necessary to quantify the PCR products. The final chapter in this section describes a high-throughput method for quantifying PCR products in microplates.

Section III, which is divided into two parts, describes several PCR-based methods for cloning differentially expressed genes. The first part discusses improved methods for performing differential RNA display. The second part describes methods for performing PCR-based cDNA subtraction.

The first chapter focuses on the difficulties associated with RNA display and the improvements that have been found to reduce the level of false positives. The other three chapters include more modifications of RNA display to further reduce the level of false positives and the number of cDNA preparations required for each experiment. They also discuss quality control of cDNA to improve the likelihood of identification of low-abundance mRNAs and a method for conducting solid-phase RNA display.

The goal of the second part is to allow selective PCR amplification of individual cDNA species that differ between two populations of mRNAs, e.g., ones that are differentially expressed. The first PCR-based DNA subtraction method developed was termed representational difference analysis (RDA). It was first applied to genomic DNA and later to RNA. A chapter is provided that includes improvements of RNA RDA. The two other chapters describe recently developed PCR-based cDNA subtraction methods. One is called linker-capture subtraction and the other supression subtractive hybridization. The latter method was designed to simultaneously suppress amplification of common sequences while selectively promoting the amplification of differentially expressed sequences. The final chapter of Section III describes the use of a combination of physical cDNA subtraction and display of sequences on polyacrylamide gels, in effect an enriched differential RNA display.

Section IV describes three PCR-based methods for obtaining full-length and homologous cDNAs. The first chapter involves selective capture of homologous sequence by RecA-mediated triple helix formation with probes generated by RT-PCR. The second chapter describes the sequential hybridization and selection of hybridized cDNA molecules on magnetic beads followed by amplification. The last chapter in Section IV describes a method for rapid walking in uncloned genomic DNA. The method is very useful for obtaining promoter and other regulatory elements once a novel cDNA has been obtained.

Section V provides three ancillary RT-PCR methods. The first chapter is a very simple method for generation of high-quality full-length cDNA from small quantities of total RNA useful for cDNA subtraction, expression analysis and cDNA library construction. The second chapter provides tips that can be universally applied to the design of protein expression constructs once a cDNA has been amplified and cloned. The final ancillary method involves PCR-based cloning of individual DNA molecules without the use of biological hosts.

ACKNOWLEDGMENTS

I would like to thank Ellen Dormitzer for painstakingly correcting the format provided by the busy authors, and my wife Joy for her support throughout the years.

REFERENCES

1. **Chelly, J., J.-C. Kaplan, P. Maire, S. Gautron and A. Kahn.** 1998. Transcription of the dystrophin gene in human muscle and non-muscle tissue. Nature *333*:858-860.
2. **Barnes, W.M.** 1994. PCR amplification of up to 35-kb DNA with high fidelity and high yield from delta bacteriophage template. Proc. Natl. Acad. Sci. USA *91*:2216-2220.

Paul Siebert, PhD
CLONTECH Laboratories
Palo Alto, CA

Section I
Obtaining RNA for RT-PCR

1

RT-PCR Using Formalin-Fixed, Paraffin-Embedded (FFPE) Archival Tumor Specimens

Devon A. Thompson, Charles Carmeci and Ronald J. Weigel
Department of Surgery, Stanford University, Stanford, CA, USA

OVERVIEW

As the size of breast tumors continues to decrease, it has become more difficult to obtain adequate tumor tissue for molecular studies. We have used the estrogen receptor (ER) gene as a model to study the ability to perform a quantitative analysis of ER mRNA extracted from formalin-fixed, paraffin-embedded (FFPE) archival breast carcinoma specimens using reverse transcription polymerase chain reaction (RT-PCR). These data demonstrate that archival breast tumor specimens can be characterized for ER mRNA abundance and that this technique can be designed to provide a quantitative analysis of gene expression for any gene of interest using archival tumor specimens.

BACKGROUND

The amount of ER protein is determined routinely in all breast biopsy specimens by an enzyme immunoassay method. The level of ER determined is used both as a prognostic marker to categorize the tumor phenotype and in clinical decision making regarding the efficacy of anti-estrogen therapy. Shortly after the ER gene was cloned (4,5), a number of studies were published that examined breast carcinomas for ER mRNA expression (1,3,6,7) by Northern blot analysis or RNase protection using RNA extracted from fresh-frozen tumor specimens. These studies established a correlation between ER mRNA expression and ER status as determined by ligand binding. The percentage of ER-negative tumors characterized by binding assay that were found to lack ER mRNA varied from 29% to 78%. This variability was likely due to differences in sensitivity of RNA detection achieved in these assays. Previous studies were limited in their clinical utility because the assays required relatively large quantities of high-quality RNA. The techniques used in these assays are also labor-intensive and not readily applicable as a routine clinical assay. These technical obstacles have limited the ability to examine large series of tumors for ER mRNA expression. Many of the new findings related to breast carcinoma oncogenesis and progression are based on studies in human cell lines or animal models. The clinical relevance of these discoveries requires extending these observations to an examination of primary breast tumors. As breast cancer detection methods improve, biopsy specimens are often quite small, and consequently it has become more difficult to obtain adequate amounts of primary tumor specimens for study. Most investigations are necessarily limited to large tumors or metastatic lesions. In addition, special arrangements are required to ensure rapid processing of the specimens. In this study, we have examined primary breast carcinomas for ER mRNA expression using a sensitive RT-PCR assay. Based on previous reports, we expected to confirm an association between ER mRNA expression and ER protein level as determined by enzyme immunoassay. The available data on ER status for each tumor specimen provide an excellent correlate for our analysis of ER gene expression that would not be available for any other common neoplasms. These studies using the ER gene as a model have been designed to determine if a reliable analysis of tumor RNA could be accomplished using archival material.

PROTOCOLS

Tissue Samples and Clinical Data

Collect tumor tissue samples from archival FFPE cell blocks. The hematoxylin/eosin-stained slides corresponding to each sample must also be obtained for visual localization of the tumor.

For this study, FFPE archival cell-block tumor samples were collected from patients with breast cancer treated at Stanford University Hospital from 1990 to 1995. Tissue was collected from 142 cell-block specimens representing 120 patients. Medical records were searched to obtain clinical data on all patients from whom tumor tissue was analyzed. Data collected included patient age, surgical procedure, tumor size, tumor histology, Bloom-Richardson tumor grade and ER and progesterone receptor (PR) status.

Protocol 1: RNA Isolation—FFPE Archival Cell-Block Tumor Samples

Materials and Reagents

- FFPE archival cell-block tumor samples
- Xylene
- Lysis buffer: 20 mM Tris-HCl, pH 8.0, 20 mM EDTA, pH 8.0, 2% sodium dodecyl sulfate (SDS) (wt/vol), 500 µg/mL proteinase K
- Phenol (water-saturated)
- 3 M sodium acetate, pH 5.2
- Isopropanol
- Diethyl pyrocarbonate (DEPC)-treated sterile, distilled water

Procedure

Isolate total RNA from FFPE archival cell-block tumor samples using the method of Weizsäcker et al. (9).

1. Compare cell blocks to the hematoxylin/eosin-stained slides. Remove 5–15 mg shaving from an area determined to be composed of invasive tumor and place in a sterile 1.5-mL microcentrifuge tube.
2. Remove the paraffin by incubating the tissue slice in 500 µL xylene at 55°C for 5 min.
3. Add a fresh 500-µL aliquot of xylene and continue incubating the tissue slice at 55°C for 5 min.
4. Wash the tissue to remove remaining solvent by adding 500 µL of 100% ethanol. Centrifuge at $12\,000\times g$ for 5 min and discard the supernatant.
5. Add a fresh 500-µL aliquot of 100% ethanol to the tissue. Centrifuge at $12\,000\times g$ for 5 min and discard the supernatant.
6. Extract the nucleic acids from the tissue section by incubating overnight at 60°C in 0.5 mL of lysis buffer (20 mM Tris-HCl, pH 8.0, 20 mM EDTA, pH 8.0, 2% SDS [wt/vol], 500 µg/mL proteinase K).
7. Extract the sample with 0.5 mL of water-saturated phenol, mix by inversion and then centrifuge at $12\,000\times g$ for 3 min. Remove the aqueous layer to a new microcentrifuge tube.
8. Repeat step 7.
9. Add 0.1 vol (50 µL) of 3 M sodium acetate (pH 5.2) and 0.6 vol (330 µL) isopropanol to precipitate the RNA. Place at 4°C for 5 min and then centrifuge at $12\,000\times g$ for 10 min at 4°C.
10. Carefully decant the isopropanol and discard. Air dry the pellet for approximately 10 min. Resuspend the pellet in 15 µL of DEPC-treated sterile, distilled water.
11. Remove 2 µL and dilute in 198 µL of DEPC-treated sterile, distilled water to determine concentration by reading the optical density $(OD)_{260}$ with a spectrophotometer. Store the remaining 13 µL at -80°C until required.

Protocol 2: RNA Isolation—Positive Control Tissue

Materials and Reagents

- Cell line or source of tissue
- TRIzol® Reagent (Life Technologies, Gaithersburg, MD, USA) (for total RNA isolation)

Procedure

As a positive control, a source of RNA that definitely contains the gene of interest should be included in the RT-PCR and slot-blotting steps.

For this study, the ER-positive breast adenocarcinoma cell line MCF7 was used as a positive control, and the ER-negative breast adenocarcinoma cell line HBL-100 was used as a negative control. Total RNA was isolated from cell lines using TRIzol reagent according to the manufacturer's instructions.

Protocol 3: RT-PCR—Standard Conditions

Materials and Reagents

- GeneAmp® RNA PCR Kit (Perkin-Elmer, Norwalk, CT, USA): 25 mM $MgCl_2$, 10× PCR Buffer II (500 mM KCl, 100 mM Tris-HCl, pH 8.3), 10 mM each dNTP, 50 µM random hexamers, 20 U/µL RNase inhibitor, 50 U/µL Moloney murine leukemia virus (MMLV) reverse transcriptase, 5 U/µL AmpliTaq® DNA Polymerase
- DEPC-treated sterile water
- Thin-walled 200-µL PCR tubes (Perkin-Elmer)
- Gene-specific primers, mixture of 25 µM each (20–24-mers, T_m >60°C; designed across intron/exon junctions)

Procedure

1. Reverse transcribe 2 µL of RNA from each sample (control and FFPE archival cell-block tumor samples) using random hexamers with the GeneAmp RNA PCR Kit according to the manufacturer's recommendations as detailed below.
   a. Separate into aliquots 2 µL of RNA per tube for each reaction.
   b. To each sample add:

		Final concentration
DEPC-treated sterile water	4.0 µL	
25 mM $MgCl_2$	8.0 µL	5 mM
10× PCR Buffer II:	4.0 µL	1× reaction buffer:
500 mM KCl		50 mM
100 mM Tris-HCl, pH 8.3		10 mM
dNTP mixture, 2.5 mM each	16.0 µL	1 mM each dNTP
50 µM random hexamers	2.0 µL	2.5 µM
RNase inhibitor (20 U/µL)	2.0 µL	1 U/µL
MMLV reverse transcriptase (50 U/µL)	2.0 µL	2.5 U/µL
Final volume	38.0 µL	

   Total volume of the RT mixture is 40 µL.
   c. Incubate the reactions at 42°C for 30 min.
   d. Heat the samples to 99°C for 5 min.
   e. Cool the tubes to 4°C for 5 min and store at -20°C until required.
2. Label PCR tubes (200-µL, thin-walled; Perkin-Elmer) for each sample (tumor and controls) to be analyzed.
3. Set up the PCRs using the GeneAmp RNA PCR Kit as described by the manufacturer (and detailed below) and gene-specific primers. The primers should be 20–24-mer oligonucleotides with melting temperatures (T_m) >60°C, designed to span intron/exon junctions of the gene of interest. In addition, when using FFPE samples, we have found that the gene fragments amplified should generally be less than 250 bp; thus, the primers should be designed accordingly.

   Note: In addition to the gene of interest to be examined, the level of expression of a control gene such as β-actin, glyceraldehyde-3-phosphate dehydrogenase

RT-PCR Using FFPE Archival Tumor Specimens

(*GAPD*) or ubiquitin must also be examined. Inclusion of such a control will allow data to be expressed with respect to the control gene, thus serving to normalize the results. As such, the control gene should be one that is normally expressed in similar amounts across the tissue samples being examined.

In this study, each sample was analyzed for the expression of ER and GAPD messages.

a. Divide the RT samples into two 20-µL aliquots for the two PCRs and place in appropriately labeled PCR tubes.

b. To each 20-µL RT sample add:

		Final concentration
DEPC-treated sterile water	65.5 µL	
25 mM $MgCl_2$	4.0 µL	2 mM (including RT $MgCl_2$)
10× PCR Buffer II:	8.0 µL	1× reaction buffer:
500 mM KCl		50 mM
100 mM Tris-HCl, pH 8.3		10 mM
primer mixture (25 µM each)	2.0 µL	0.5 µM each primer
AmpliTaq DNA Polymerase (5 U/µL)	0.5 µL	0.025 U/µL
Final volume	80 µL	

Total volume of the PCR mixture is 100 µL.

c. Amplify the genes using PCR as follows:
- one cycle at: 94°C for 1 min
  60°C for 2 min
  72°C for 1 min
- 25 cycles at: 95°C for 1 min
  60°C for 30 s
  72°C for 1 min
- incubate at 72°C for 10 min.

d. Store the reaction mixtures at -20°C until analyzed.

For these studies, the ER gene was amplified using the primers 5′-TACCTGGA-GAACGAGCCCAGCGGC-3′ and 5′-GTCATTGGTACTGGCCAATCTTTC-3′ that generate a 120-bp DNA fragment. The *GAPD* gene was amplified using the primers 5′-CACATCGCTCAGACACCATG-3′ and 5′-GCCATGGAATTTGCCATGGG-3′ that generate a DNA fragment of 191 bp. Amplification was performed in a GeneAmp PCR System 9600 (Perkin-Elmer).

Protocol 4: RT-PCR—Alternate Conditions

Materials and Reagents

- Advantage™ RT-for-PCR Kit (CLONTECH Laboratories, Palo Alto, CA, USA): DEPC-treated water, 20 µM random hexamers, 5× reaction buffer (250 mM Tris-HCl, pH 8.3, 375 mM KCl, 15 mM $MgCl_2$), dNTP mixture, 10 mM each, 20 U/µL recombinant RNase inhibitor, 200 U/µL MMLV reverse transcriptase

- Thin-walled 200-μL PCR tubes (Perkin-Elmer)
- Advantage cDNA PCR Kit with Advantage KlenTaq Polymerase Mix (both from CLONTECH): 10× KlenTaq PCR buffer [400 mM Tricine-KOH, pH 9.2, 150 mM KOAc, 35 mM Mg(OAc)$_2$, 750 μg/mL bovine serum albumin (BSA)], 50× dNTP mixture (10 mM each), 50× Advantage KlenTaq Polymerase Mix
- PCR-grade water
- Gene-specific primers, mixture of 10 μM each (30–34-mers, T_m >70°C; designed across intron/exon junctions)

Procedure

The PCR conditions used to produce Figures 3–6 were generated using the standard conditions for RT-PCR outlined above and were originally published in Carmeci et al. (2). However, since publication of this manuscript (2), we have ascertained that the alternate RT-PCR conditions (described as follows) have proven to be more sensitive, requiring less template, and thus have been adopted for routine analysis in our laboratory.

1. Reverse transcribe approximately 1 μg of RNA from each sample (control and FFPE archival cell-block tumor samples) using random hexamers with the Advantage RT-for-PCR Kit according to the manufacturer's recommendations as detailed below.
   a. Dilute 1 μg of RNA in DEPC-treated sterile water to a volume of 12.5 μL.
   b. Add 1 μL of 20 μM random hexamers.
   c. Heat at 70°C for 2 min.
   d. Transfer the samples to ice rapidly and incubate for 10 min.
   e. To each sample add:

		Final concentration
5× reaction buffer:	4.0 μL	1× reaction buffer:
250 mM Tris-HCl, pH 8.3		50 mM
375 mM KCl		75 mM
15 mM MgCl$_2$		3 mM
dNTP mixture (10 mM each)	1.0 μL	500 μM each dNTP
Recombinant RNase inhibitor (20 U/μL)	0.5 μL	0.5 U/μL
MMLV reverse transcriptase (200 U/μL)	<u>1.0 μL</u>	10 U/μL
Final volume	6.5 μL	

   Total volume of the RT mixture is 20 μL.
   f. Incubate the reaction mixtures at 42°C for 1 h.
   g. Heat the samples to 94°C for 5 min.
   h. Dilute each 20-μL RT sample to 100 μL with sterile, distilled water and store at -80°C until required.
2. Label PCR tubes (200-μL, thin-walled; Perkin-Elmer) for each sample (tumor and controls) to be analyzed.
3. Set up the PCRs using the Advantage cDNA PCR Kit with Advantage KlenTaq Polymerase Mix as described by the manufacturer (as detailed below) and gene-specific primers. The primers should be 30–34-mer oligonucleotides with T_m >70°C, designed to span intron/exon junctions of the gene of interest. In

addition, when using FFPE samples, we have found that the gene fragments amplified should generally be less than 250 bp; thus the primers should be designed accordingly.

a. Remove 2 µL of each RT sample per PCR to be performed and place in appropriately labeled PCR tubes.
b. To each 2-µL RT sample add:

		Final concentration
PCR-grade water	18.5 µL	
10× KlenTaq PCR buffer:	2.5 µL	1× reaction buffer:
400 mM Tricine-KOH, pH 9.2		40 mM
150 mM KOAc		15 mM
35 mM Mg(OAc)$_2$		3.5 mM
750 µg/mL BSA		75 µg/mL
Primer mixture, 10 µM each	1.0 µL	0.4 µM each primer
50× dNTP mixture, 10 mM each	0.5 µL	200 µM each dNTP
50× Advantage KlenTaq Polymerase Mix	0.5 µL	1×
Final volume	23 µL	

Total volume of the PCR mixture is 25 µL.

c. Amplify the genes using two-step PCR as follows:
  - incubate at 94°C for 1 min
  - perform 30 cycles at: 94°C for 30 s
    68°C for 3 min
  - incubate at 68°C for 3 min.
d. Store the reaction mixtures at -20°C until analyzed.

Protocol 5: Visualization of PCR Products

Materials and Reagents

- 30% acrylamide/bisacrylamide solution (29:1) (Bio-Rad, Hercules, CA, USA)
- 0.5× TBE buffer: 45 mM Tris-borate, 1 mM EDTA, pH 8.0
- Gel-loading buffer (30% glycerol, 0.025% bromophenol blue, 0.025% xylene cyanol FF)
- Ethidium bromide

Procedure

To visualize the gene fragments, analyze PCR samples by electrophoresis on an 8% nondenaturing polyacrylamide gel.

1. Pour a vertical 8% nondenaturing polyacrylamide gel in 0.5× TBE using 0.8-mm spacers.

   Note: Do not use 1.5-mm spacers because the PCR fragments are more difficult to visualize, and if electroblotting is required for visualization (see the note at the end of this section), DNA does not transfer adequately from 1.5-mm-thick gels.

2. Place 10 μL of the 50-μL PCR mixture (RT-PCR standard conditions) or 5 μL of the 25-μL PCR mixture and 5 μL of water (RT-PCR standard conditions) into a new microcentrifuge tube.
3. Add 2 μL of gel-loading buffer (30% glycerol, 0.025% bromophenol blue, 0.025% xylene cyanol) to each sample.
4. Load sample on the gel and electrophorese at 200 V in 0.5× TBE until bromophenol blue runs off the bottom of the gel.
5. Remove the polyacrylamide gel from the glass plates and stain in 0.5 μg/mL ethidium bromide with gentle rocking for 30 min, followed by 20 min destaining in water.
6. Photograph the gel while illuminated by a UV transilluminator to visualize the PCR gene fragments.

    Note: If the amplified fragments are not readily visible, as often is the case with rare transcripts, the DNA can be transferred from the gel to 0.2-μm Nytran® nylon membranes using a Pronto™ Semi-Dry Electroblotter (Schleicher & Schuell, Keene, NH, USA). The blots can then be hybridized with a ^{32}P-labeled internal oligonucleotide probe (according to the conditions described in Protocol 6) and the fragments visualized by autoradiography.

Protocol 6: Slot Blotting, Hybridization and Phosphor Imaging

Materials and Reagents

- 0.2-μm Nytran nylon membrane (Schleicher & Schuell)
- Minifold® II slot blotter (Schleicher & Schuell)
- 20× SSPE: 3 M NaCl, 200 mM NaH_2PO_4, 20 mM EDTA, pH 7.4
- 6× SSPE: 0.9 M NaCl, 60 mM NaH_2PO_4, 6 mM EDTA, pH 7.4
- Gel-loading buffer
- Gene-specific oligonucleotides (designed internal to the primers used for the PCRs)
- Hybridization solution (20% formamide, 5× Denhardt's, 5× SSPE, 1% SDS [wt/vol], 100 μg/mL denatured salmon sperm DNA)
- [γ-^{32}P]ATP (3000 Ci/mmol; Amersham Pharmacia Biotech, Piscataway, NJ, USA)
- T4 polynucleotide kinase (10 000 U/mL; New England Biolabs, Beverly, MA, USA)
- T4 polynucleotide kinase 10× reaction buffer (New England Biolabs): 700 mM Tris-HCl, pH 7.6, 100 mM $MgCl_2$, 50 mM dithiothreitol
- Wash solution: 2× standard saline citrate (SSC), 0.1% SDS (wt/vol)
- Kodak Phosphor screen (Molecular Dynamics, Sunnyvale, CA, USA)
- PhosphorImager® (Molecular Dynamics)
- ImageQuant® software (Molecular Dynamics) to quantitate image signal

Procedure

1. Following confirmation by visualization, directly transfer an aliquot of the PCR samples onto 0.2-μm Nytran nylon membranes using a Minifold II slot blotter.

a. In a 1.5-mL microcentrifuge tube, combine 5 µL of the PCR sample with 130 µL of water, 60 µL of 20× SSPE and 5 µL of gel-loading buffer.
b. Denature the DNA by incubating the sample at 100°C for 10 min and then place on ice for 5 min.
c. Spin the samples for 30 s to collect all liquid.
d. Presoak the 0.2-µm Nytran membrane in water for 5 min and then in 6× SSPE for 15 min.
e. Using vacuum suction, load the 200-µL samples onto the presoaked 0.2-µm Nytran nylon membranes using a Minifold II slot blotter.

Note: Slot blots should be organized so that all the PCR fragments corresponding to each gene amplified are loaded together on one Nytran membrane, thus facilitating hybridizations. For example, in this study PCR products corresponding to ER were loaded together onto one sheet of Nytran, and PCR products corresponding to *GAPD* were loaded together onto a separate sheet of Nytran.

2. Hybridize the slot-blot membranes with gene-specific oligonucleotide probes designed internal to the primers used for the PCRs.
   a. Prehybridize the slot-blot membranes in 10 mL of hybridization solution (20% formamide, 5× Denhardt's, 5× SSPE, 1% SDS [wt/vol] and 100 µg/mL denatured salmon sperm DNA) at 42°C for 3 h, rotating in a hybridization oven.
   b. End label 10 pmol of each of the internal gene-specific oligonucleotides (8) as follows:
      - 1 µL of oligonucleotide at 10 pmol/µL
      - 2 µL of sterile, distilled water
      - 5 µL (50 µCi) of [γ-^{32}P]ATP (3000 Ci/mmol)
      - 1 µL of T4 polynucleotide kinase 10× reaction buffer (700 mM Tris-HCl, pH 7.6, 100 mM MgCl$_2$, 50 mM dithiothreitol)
      - 1 µL (10 U) of T4 polynucleotide kinase (10 000 U/mL).

   Incubate at 37°C for 30 min. Remove the unincorporated mixture by spinning through a Bio-Spin® 6 column (Bio-Rad).

   The ER-specific probe was 5'-GGCATTCTACAGGCCAAATT-3', designed to hybridize to the 120-bp ER PCR product, and the *GAPD*-specific probe was 5'-CATGTAGTTGAGGTCAATGA-3', designed to hybridize to the 191-bp GAPD PCR product.

   c. Add each ^{32}P-labeled oligonucleotide to a fresh 10-mL aliquot of hybridization solution (20% formamide, 5× Denhardt's, 5× SSPE, 1% SDS [wt/vol] and 100 µg/mL denatured salmon sperm DNA). Mix gently by swirling and then add to the hybridization tube containing the corresponding slot blot.
   d. Continue incubating at 42°C in a rotating hybridization oven for a minimum of 6 h.

   Note: For convenience, the hybridization can be continued overnight.

   e. Following hybridization, remove the ^{32}P-labeled oligonucleotide probe and store or dispose of as outlined by radiation guidelines.
   f. To wash the slot blots, add 150 mL of 2× SSC, 0.1% SDS (wt/vol) and continue incubating at 42°C for 20 min.

g. Discard the first wash as outlined by radiation guidelines. Place the slot-blot membranes in a container with 250 mL of 2× SSC, 0.1% SDS (wt/vol) and incubate in a shaking water bath at 45°C for 20 min.

h. Discard the second wash, add another 250 mL of 2× SSC, 0.1% SDS (wt/vol) and continue incubating in a shaking water bath at 45°C for an additional 20 min.

i. Remove the slot blots, wrap in plastic wrap and expose to X-ray film with an intensifying screen for 10 min to obtain an autoradiograph for visualization.

3. Phosphor image the slot blots to obtain quantitative values for each sample.

a. Place ^{32}P-labeled slot blots in a cassette on a Kodak phosphor screen for 1 h.

b. Obtain a quantitative value for each PCR sample using a PhosphorImager in conjunction with ImageQuant software.

c. Calculate the level of expression for the gene of interest normalized against the level of expression for the control gene (usually β-actin, *GAPD* or ubiquitin).

RESULTS AND DISCUSSION

RT-PCR was used to analyze ER mRNA recovered from FFPE cell blocks of primary breast tumors. Figure 1 depicts localization of the invasive tumor in a cell block by comparison to the hematoxylin/eosin-stained slide of the specimen. Primers were chosen to amplify a region of the ER gene, as shown in Figure 2A. These primers amplify across a splice site of the ER gene to avoid inaccurate data that would arise from amplification of genomic DNA. The PCR products were detected using an internal oligonucleotide probe (also shown in Figure 2A). The use of an internal probe increased the sensitivity and specificity of the assay. To control for variation in quality and quantity of RNA obtained from tumor specimens, cDNA was also amplified with primers specific for *GAPD*, as shown in Figure 1B. Because *GAPD* mRNA was used as a positive control, these primers were chosen to yield a larger PCR product to ensure that RNA degradation

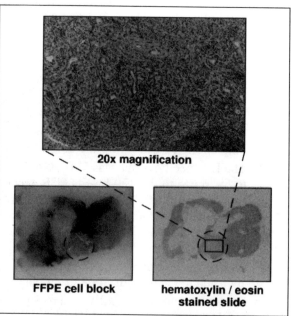

Figure 1. Example of a cell block and hematoxylin/eosin-stained slide of an archival tumor sample. An FFPE cell block of a breast carcinoma specimen is shown on the left. The corresponding hematoxylin/eosin-stained slide for this FFPE cell block is depicted on the right. The dashed circle indicates the region of the tumor sample that contained the highest concentration of tumor cells, as determined by microscopic examination. The top picture shows the invasive tumor cells at 20× magnification.

would have a greater effect on the positive control. All PCR products were visualized by electrophoresis; Figure 3 shows a representative analysis of ten tumor specimens with MCF7 RNA as a positive control. PCR products of the appropriate size were obtained using both the ER (120 bp) and *GAPD* (191 bp) primer pairs.

Various PCR cycles were tested to optimize the assay. Twenty-five was chosen as the smallest number of cycles that reliably gave a signal from ER-positive tumor specimens. Serial dilutions of MCF7 RNA were used to test the reproducibility of the assay and to establish a linear range of RNA concentration over which the assay

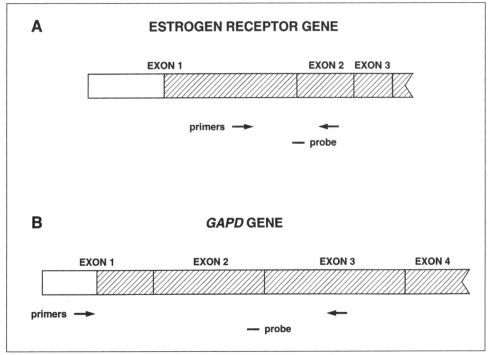

Figure 2. **A schematic map of the ER and *GAPD* genes.** (A) The 5' end of the ER gene is shown with the approximate locations of the primers used in RT-PCR analysis of ER mRNA. The location of the internal oligonucleotide probe is also shown. Primers were chosen to yield an RT-PCR product of 120 bp. (B) Map of *GAPD* gene with locations of primers and probe. Primers were chosen to yield an RT-PCR product of 191 bp. Stippled regions of both genes correspond to coding regions.

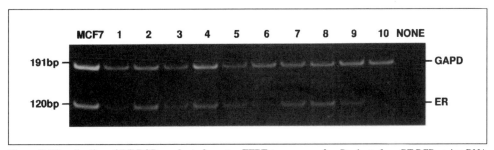

Figure 3. **Visualization of RT-PCR products from ten FFPE tumor samples.** Products from RT-PCRs using RNA from MCF7 cells and ten tumor specimens were analyzed by electrophoresis. Last lane is a PCR lacking template RNA. PCR amplification was performed using either *GAPD* or ER primers. Aliquots of PCR products from these reactions were combined and loaded on 8% polyacrylamide gels. After electrophoresis, gels were stained with ethidium bromide, and DNA was visualized by UV illumination.

could be considered semiquantitative. Figure 4 shows the results of these experiments. In the range of MCF7 mRNA from 10 pg to 1 ng, the phosphor imaging data varied proportionally to the quantity of RNA added. At the upper end, increasing the amount of RNA template did not change the amount of PCR product.

RNA extracted from archival specimens was analyzed using slot blotting and hybridization and quantitated using a PhosphorImager. Figure 5 shows the representative data for 142 tumor samples analyzed for ER and *GAPD*. By comparing the signal with MCF7 mRNA, the tumor specimens can be seen to fall within the linear range of the assay. The data in Figure 5 were used to determine the relative amount of ER mRNA present in each tumor. The 142 specimens were obtained from 120 patients.

Quantitative phosphor imaging data for ER mRNA were corrected for sample variability by dividing the ER signal by the corresponding *GAPD* signal. Values were normalized by dividing each corrected value by the average for the entire group. Normalized values presented are the average for two or three separate PCRs; 74% of the samples were analyzed three times, and 26% were analyzed twice. To test the hypothesis that the normalized ER mRNA value was associated with tumor biology, the ER mRNA data were used to segregate the patients into three groups represented by low, medium and high mRNA levels. The data were analyzed for an association between ER mRNA levels and ER and PR protein levels as determined by enzyme immunoassay. (For a full explanation of statistical analysis, see Reference 2.) Figure 6 is a graph of normalized ER mRNA signal for each of the 120 tumors. A lower level of RNA expression was established, above which ER-positive tumors were common. This group comprises approximately 20% of the tumors. An upper group corresponding to 23% of the tumors was chosen with RNA expression greater than twice the median value. The middle group, comprising 57% of the tumors, was determined to have intermediate RNA expression. Based on these divisions, the tumors were characterized as having high, medium or low ER mRNA expression. When characterized by these parameters, MCF7 cells fell into the medium ER mRNA group, and HBL-100 cells fell into the low ER mRNA group. These RNA data were compared to known characteristics of the tumors, including standard ER and PR protein levels and histopathologic grade. Tumors with low levels of ER transcription demonstrate an ER-/PR-negative phenotype with a high histologic grade compared to tumors with abundant ER mRNA characterized by an

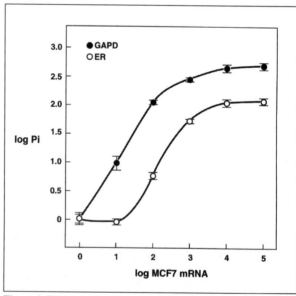

Figure 4. Titration of RNA from the MCF7 breast carcinoma cell line. Tenfold serial dilutions of MCF7 poly(A+) RNA were amplified using RT-PCR with primers for *GAPD* or ER. The log of MCF7 mRNA (pg) is plotted vs. the log of phosphor imaging value (P_i); log P_i = log P_{sample} - log$_{control}$ (no RNA). Values are the average of five separate RT-PCRs with standard deviations shown by bars.

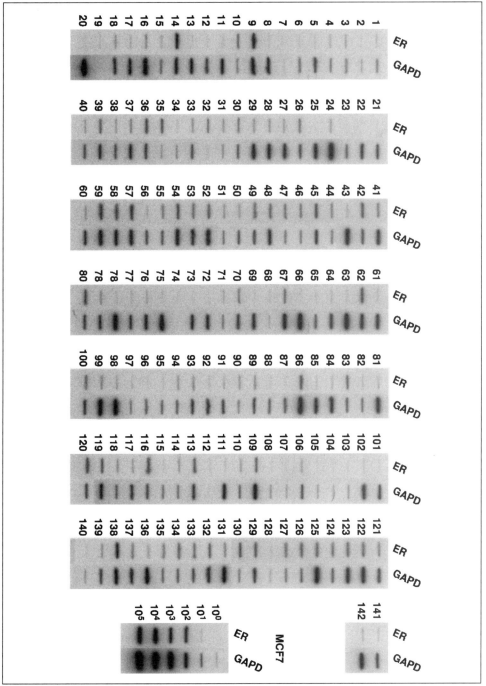

Figure 5. Example of slot-blot hybridization for each FFPE tumor sample. The RT-PCR products using RNA from 142 tissue samples amplified using ER or *GAPD* primers were immobilized on a Nytran membrane using a slot-blotting manifold. Subsequently, the slot blots were hybridized with probes specific to ER or *GAPD* (Figure 2). The corresponding autoradiographs are shown. Also depicted is a serial dilution of MCF7 mRNA from 1 to 10^5 pg of RNA. A quantitative value of hybridization signal was obtained using a PhosphorImager. Samples 60 and 139 have signals similar to MCF7 cells that map to the middle ER mRNA group. Samples 9 and 14 are examples of high ER mRNA tumors, and sample 20 is a low ER mRNA tumor. (See Figure 6 and text for description of high, medium and low ER mRNA groups.)

ER-/PR-positive phenotype with a significantly lower histologic grade. Invasive lobular carcinomas were uniformly found to transcribe the ER gene because this variety of breast carcinoma was not found in the low ER mRNA group. These data demonstrate that a quantitative evaluation of tumor ER mRNA content (based on a measurement of steady-state ER mRNA levels) can predict breast carcinoma biologic phenotype.

Breast carcinomas display a wide variation in the abundance of ER mRNA with differences between high and low expression ranging from two to three orders of magnitude. There is an association between ER mRNA expression and the ER phenotype as determined by standard quantitative assays. These points of agreement with our results validate the use of RT-PCR as a semiquantitative assay for evaluation of ER mRNA in primary tumors. In addition, our results extend previous studies of this subject in several significant aspects. The sensitivity of the RT-PCR assay allows a semiquantitative analysis of tumors with low levels of ER mRNA and demonstrates that gene expression can be analyzed from archival tumor specimens. The importance of this accomplishment is enhanced by several technical considerations. First, because of the instability of RNA, previous quantitative studies required the use of fresh-frozen tumor specimens. In addition, as breast carcinoma screening methods have improved, tumor size at presentation has continued to decrease. This smaller tumor size reduces the availability of adequate amounts of tumor tissue. The use of archival material does not require special processing arrangements and can be readily performed from as few as 5 mg of tissue.

There are several particular advantages in using archival tumor specimens for analysis of gene expression. First, archival specimens embedded in paraffin blocks are available in abundance and easily obtainable. Also, using archival specimens allows for selection of tumors correlating to patients for whom there is a long clinical

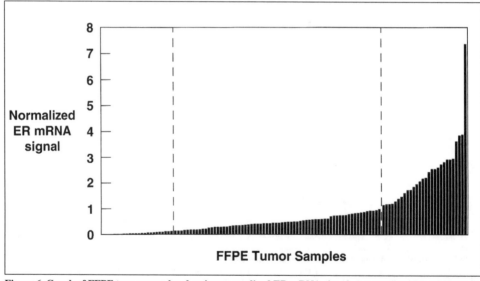

Figure 6. Graph of FFPE tumor samples showing normalized ER mRNA signal. A normalized ER mRNA value was calculated for RNA extracted from tumors representative of 120 patients. Data were normalized by dividing each value by the average value for the entire group. Value represented by bar graph was the average of two or three separate experiments. Data are presented from lowest to highest normalized value. Dotted lines separate patients into three groups: the bottom 20% (low ER mRNA group), the middle 57% (medium ER mRNA group) and the top 23% (high ER mRNA group). See text for details.

follow-up. This unique feature allows for true determination of the in vivo biological activity of the tumor under study. Studies that use fresh primary tumor samples are often undermined by a lack of sufficient samples combined with a short clinical follow-up. Another important advantage of using archival FFPE samples is the ability to minimize the nonhomogeneity inherent in tumor tissue samples. Generally, when biopsy specimens are examined histologically, they comprise a mixture of tumor cells, infiltrating lymphocytes and surrounding fibro-adipose tissue. With FFPE samples, the regions of the tissue biopsy that contain carcinoma cells predominantly are carefully excised for RNA extraction and analysis of gene expression. This histological specificity is not available for tumor samples that are frozen, pulverized and subsequently used for analysis of gene expression. RT-PCR using FFPE archival specimens should be equally applicable in other, non-oncologic, pathological states for which paraffin-embedded samples are available.

ACKNOWLEDGMENTS

D. Thompson, Ph.D. and C. Carmeci, M.D. are funded by National Institutes of Health National Research Service Award Training Grants Nos. F32 CA69751 and F32 CA69715, respectively. R.J. Weigel, M.D., Ph.D. is supported in part by a fellowship from the American Surgical Association and by a Clowes Career Development Award from the American College of Surgeons. This work was supported in part by NIH Grant No. R29 CA63251 and the United States Department of the Army Medical Research and Development Command Grant No. DAMD 17-94-J-4353. We would like to thank Ciarán N. Cronin, Ph.D. for critical reading of the manuscript.

REFERENCES

1. **Barrett-Lee, P.J., M.T. Travers, R.A. McClelland, Y. Luqmani and R.C. Coombes.** 1987. Characterization of estrogen receptor messenger RNA in human breast cancer. Cancer Res. *47*:6653-6659.
2. **Carmeci, C., E.C. deConinck, T. Lawton, D.A. Bloch and R.J. Weigel.** 1997. Analysis of estrogen receptor messenger RNA in breast carcinomas from archival specimens is predictive of tumor biology. Am. J. Pathol. *150*:1563-1570.
3. **Garcia, T., S. Lehrer, W.D. Bloomer and B. Schachter.** 1988. A variant estrogen receptor messenger ribonucleic acid is associated with reduced levels of estrogen binding in human mammary tumors. Mol. Endocrinol. *2*:785-791.
4. **Green, S., P. Walter, V. Kumar, A. Krust, J.-M. Bornert, P. Argos and P. Chambon.** 1986. Human oestrogen receptor cDNA: sequence, expression and homology to v-*erb*-A. Nature *320*:134-139.
5. **Greene, G., P. Gilna, M. Waterfield, A. Baker, Y. Hort and J. Shine.** 1986. Sequence and expression of human estrogen receptor complementary DNA. Science *231*:1150-1154.
6. **Henry, J.A., S. Nicholson, J.R. Farndon, B.R. Westley and F.E.B. May.** 1988. Measurement of oestrogen receptor mRNA levels in human breast tumors. Br. J. Cancer *58*:600-605.
7. **May, E., H. Mouriesse, F. May-Levin, G. Contesso and J.-C. Delarue.** 1989. A new approach allowing an early prognosis in breast cancer: the ratio of estrogen receptor (ER) ligand binding activity to ER-specific mRNA level. Oncogene *4*:1037-1042.
8. **Sambrook, J., E.F. Fritsch and T. Maniatis.** 1989. Molecular Cloning: A Laboratory Manual, 2nd ed., p. 10.59-10.67, 11.31-11.33. CSHL Press, Cold Spring Harbor, NY.
9. **Weizsäcker, F.V., S. Labeit, H.K. Koch, W. Oehlert, W. Gerok and H.E. Blum.** 1991. A simple and rapid method for the detection of RNA in formalin-fixed, paraffin-embedded tissues by PCR amplification. Biochem. Biophys. Res. Commun. *174*:176-180.

2 Isolation of RNA from Solid Tumors for RT-PCR

Joanna Kitlińska and Jacek Wojcierowski
Department of Medical Genetics, Medical Academy of Lublin, Lublin, Poland

OVERVIEW

Elaboration of a simple and safe method of high-molecular-weight RNA isolation remains attractive and needful. Especially important is the ability to obtain DNA-free preparations in connection with widespread application of reverse transcription polymerase chain reaction (RT-PCR). Moreover, the RNA isolation method should be suitable for the kind of sample that is analyzed. Neoplastic tissue processing seems to be particularly difficult because of diversity, compact consistency and frequent necrotic changes of the samples.

We present two methods of RNA isolation from solid tumors. Both of them use high salt concentration and phenol extraction in acidic pH. They allow intact RNA to be obtained with high efficiency. Because of final DNase treatment, the preparations are DNA-free and ready to use in RT-PCR applications. Furthermore, we give some advice related to neoplastic tissue selection, sampling and storage.

BACKGROUND

Efficient extraction of the total RNA from neoplastic tissue can sometimes be difficult because of the extremely compact consistency of the tumor. Suitable sample of the tissue can also determine the quality of the RNA obtained.

There are some technical problems in the isolation of RNA. The most important of these is the inhibition of the contaminating RNases released especially during the cell disruption. Because of this problem, the lysis buffer should contain strong RNase inhibitors such as guanidinium thiocyanate, 2-mercaptoethanol or sodium dodecyl sulfate (SDS). Moreover, the prevention of degradation by ribonucleases is dependent on the efficiency of the homogenization. Therefore, a suitable choice of the lysis buffer and tissue disruption manner is very important, especially in the case of compact solid tumors.

Table 1. RNA and DNA Contents in Particular Fractions after Extraction Procedure with Ammonium Sulfosalicylate

Fraction	RNA Range	RNA Mean	RNA %	DNA Range	DNA Mean	DNA %	% DNA in RNA	A_{260}/A_{280} Ratio Range	A_{260}/A_{280} Ratio Mean
I–III phenol phase	115–270	178	31	198–281	248	94.5			
Phenol phases after DNase treatment	27–58	46	8	3–9	5	2			
Final RNA pellet	332–444	350	61	5–13	6.5	2.5	1.8	1.61–1.8	1.77

Values from five experiments expressed in micrograms per 100 mg of analyzed tumor tissue.

The RNA purification from DNA contamination is another essential problem. This can be achieved by lithium chloride precipitation (3), ultracentrifugation through a cesium chloride cushion (1,15) or phenol extraction in acidic pH (5,8). However, if isolated RNA is used in RT-PCR, the DNase treatment might be necessary because of the high sensitivity of this method (even small traces of DNA can cause false-positive results) (4,10,11).

We propose two methods of RNA isolation. Both of them are efficient and safe. One of them uses a high concentration of ammonium sulfosalicylate (5). The second one is based on the procedure of Chomczynski and Sacchi modified by additional DNase treatment (2,10).

PROTOCOLS

Creating a Ribonuclease-Free Environment

To obtain good preparations of RNA, it is necessary to avoid an accidental introduction of RNases during or after the isolation procedure. Whenever possible, sterile disposable plasticware should be used. Nondisposable plasticware should be treated with 0.1% diethyl pyrocarbonate (DEPC) overnight at room temperature and then autoclaved for 20 min to remove DEPC traces. The last step is necessary because of the DEPC ability to modify the RNA. Glassware should be baked at 180°C for 2 h. All solutions except Tris buffers (DEPC reacts with amines) should be treated with 0.1% DEPC and then autoclaved for 20 min. Gloves must be worn during the preparation of materials and the isolation procedure.

Apart from these precautions, all manipulations involving RNA should be done on ice except steps in which the RNA is dissolved in buffers containing strong RNase inhibitors such as guanidinium thiocyanate.

Tissue Sampling

Close cooperation with surgeons is necessary to obtain good results in work with tumors. A suitable choice of the tissue sample determines the RNA quality and reliability of the research.

The sample should be derived from a homogeneous and representative region of tumor. It is recommended to sample the tissue from the border of the tumor to avoid

Table 2. RNA and DNA Contents in Particular Fractions after Extraction Procedure with Guanidinium Thiocyanate

Fraction	RNA Range	RNA Mean	RNA %	DNA Range	DNA Mean	DNA %	% DNA in RNA	A_{260}/A_{280} Range	A_{260}/A_{280} Mean
I phenol phase	131–197	160	28	182–289	254	96.5			
Phenol phases after DNase treatment	39–72	52	9	2–6	4	1.5			
Final RNA pellet	290–405	362	63	3–11	5	2	1.3	1.65–1.9	1.82

Values from five experiments expressed in micrograms per 100 mg of analyzed tumor tissue.

the sampling of a necrosis. The specimen should be free from fat and connective tissue if possible. The neoplastic tissue from patients after chemotherapy is often necrotic, and RNA preparations obtained from such samples are generally degraded.

The samples to be extracted should be rinsed in sterile phosphate-buffered saline (PBS) and processed as soon as possible. During transportation from an operating room, the tube with the sample should be placed on ice. However, if it is not possible to isolate RNA immediately after obtaining the sample of interest, the tissue can be frozen with liquid nitrogen and stored at -70°C. We have also obtained good results with tissue stored at -20°C.

Tissue Disruption

Freshly harvested tissue should be weighed, cut into small pieces and placed directly into homogenization buffer. High-speed homogenizers such as Polytron® (Brinkmann Instruments, Westbury, CT, USA) or a Dounce glass-Teflon® homogenizer is recommended. Alternatively, the tissue can be disrupted in a glass homogenizer because of the strong lytic properties of the buffers used. This is more time-consuming and inefficient, but the yield and quality of the preparations obtained are sufficient to use in all procedures involving RNA.

Tissues frozen in liquid nitrogen should be powdered in a chilled homogenizer or mortar. However, the homogenization buffer should be added to powdered sample while it is still in the frozen state. It is very important because freezing and thawing of the tissue results in RNA degradation. So we recommend instead to weigh and cut the sample before freezing, transfer the frozen tissue directly to the homogenizer and then immediately add the lysis buffer and start the homogenization. This procedure minimizes the possibility of RNA degradation.

Determination of RNA Yield and Quality

To check the efficiency of the RNA isolation method, quantitative estimations of RNA and DNA were performed. All phenol fractions were precipitated with 2 vol of 96% ethanol. RNA and DNA in these fractions (as well as in the final pellet) were separated according to the Schmidt and Tanhauser procedure (12). Quantitative estimation of RNA and DNA contents were carried out spectrophotometrically (13). The purity and protein content were also detected based on the absorbance ratio (A_{260}/A_{280}).

The integrity of the RNA preparations obtained was verified by electrophoresis on 2% agarose, 6% formaldehyde gels. The RT-PCR was performed to check the presence of poly(A) mRNA. The reaction was carried out using primers for the β-actin gene (sense 5′-AACGGCTCCGGCATGTGCAA-3′ and antisense 5′-CTTCTGAC-CCATGCCCACCA-3′). By overlapping one of the introns using these primers, it is possible to distinguish between amplification products obtained from cDNA (107 bp) and DNA (245 bp) and to check the DNA contamination of the RNA sample (7,9).

Protocol for RNA Isolation with Ammonium Sulfosalicylate

Materials and Reagents

- Tissue homogenizer (Polytron, Dounce, alternative glass)
- 10-mL polypropylene tubes
- 1.5-mL microcentrifuge tubes
- 37°C water bath
- Ice bath
- Lysis buffer: 2 M ammonium sulfosalicylate, 50 mM EDTA, 0.3% SDS, 20 µL/mL 2-mercaptoethanol, pH 4.8
- Phenol saturated with water
- Chloroform
- Phenol–chloroform–isoamyl alcohol (PCI) (25:24:1)
- Ice-cold 96% and 80% ethanol
- RNase-free DNase
- 1× TE buffer (10 mM Tris-HCl, 1 mM EDTA, pH 7.6) or sterile water
- 3 M sodium acetate, pH 7.4

Procedure

All manipulations should be done on ice.
1. Homogenize the tumor tissue (0.5 g) with 3 mL of lysis buffer.
2. Transfer the homogenate into a 10-mL polypropylene tube, add 1 mL of water-saturated phenol and shake for 5 min. Then add 2 mL of chloroform and shake again for 10 min.
3. Centrifuge at 6000× g for 15 min at 4°C.
4. Transfer the aqueous phase to a fresh cold tube. Repeat steps 2–5 three times.
5. Add 0.9 vol of cold 96% ethanol to the last aqueous phase. Mix gently and precipitate for 1 h at 20°C.
6. Centrifuge at 10 000× g for 20 min at 4°C.
7. Wash the pellet with 80% ethanol and centrifuge at 10 000× g for 10 min at 4°C.
8. Dissolve the obtained pellet in 300 µL of DNase buffer. Add 30 µL of RNase-free DNase (10 U/µL). Incubate at 37°C for 20 min.
9. Add 1 vol of PCI (25:24:1) and shake for 15 s.
10. Centrifuge at 3000× g for 10 min at 4°C.
11. Transfer the aqueous phase to a fresh tube, add 1 vol of chloroform and

centrifuge as above.
12. Precipitate the aqueous phase with 0.1 vol of 3 M sodium acetate (pH 7.4) and 2 vol of ethanol for 1 h at 20°C.
13. Centrifuge at 13 000× g for 30 min at 4°C.
14. Dry the pellet under vacuum and dissolve in sterile water or TE buffer.

Protocol for RNA Isolation with Guanidinium Thiocyanate

Materials and Reagents

- Tissue homogenizer
- 10-mL polypropylene tubes
- 1.5-mL microcentrifuge tubes
- 37°C water bath
- Ice bath
- Lysis buffer: 4 M guanidinium thiocyanate, 25 mM sodium citrate, pH 7.0, 0.5% N-lauroylsarcosine, 0.1 M 2-mercaptoethanol
- 2 M sodium acetate, pH 4.0
- 3 M sodium acetate, pH 7.4
- Phenol saturated with water
- Chloroform–isoamyl alcohol (49:1)
- Chloroform
- PCI (25:24:1)
- Isopropanol
- Ice-cold 70% ethanol
- RNase-free DNase

Procedure

1. Homogenize 100 mg of the tumor tissue with 1 mL of the lysis buffer at room temperature.
2. Transfer the homogenate to the fresh tube and sequentially add 0.1 mL of 2 M sodium acetate, pH 4.0, 1 mL of water-saturated phenol and 0.2 mL of chloroform. Mix gently after addition of each reagent. Finally, shake vigorously for 15 s. Keep on ice for 15 min.
3. Centrifuge at 6500× g for 20 min at 4°C.
4. Transfer the upper aqueous phase to a fresh tube, add 1 vol of ice-cold isopropanol and precipitate for at least 1 h at -20°C.
5. Centrifuge at 6500× g for 20 min at 4°C.
6. Dissolve the RNA pellet in 300 µL of DNase buffer. Add 30 µL RNase-free DNase (10 U/µL) and incubate for 20 min at 37°C.
7. Add 1 vol of PCI mixture and shake vigorously for 15 s.
8. Centrifuge at 3000× g for 10 min at 4°C.
9. Transfer the aqueous phase to a fresh tube, add 1 vol of chloroform and shake for 15 min.
10. Centrifuge as in step 8.

Table 3. Troubleshooting

Problem	Reason	Recommendation
Low yield	Incomplete homogenization of the sample	Do not discard the phenol phases before the end of procedure. It is possible to extract them again or repeat the homogenization step.
	Incomplete solution of the final RNA pellet	Add more water or TE buffer. The washing of the final pellet with TE buffer before the solubilization step makes it easier. Moreover, it is possible to use 0.5% SDS to dissolve the RNA pellet.
	The low quality of the sample used for RNA isolation—necrotic tissue, the tumor after chemotherapy	
RNA degradation	The low quality of the sample—see above	
	The tissue has not been processed or frozen immediately after the resection	Try to process tissue as soon as possible. Keep it on ice before starting the procedure.
	The tissue has been frozen and thawed	Do not permit the tissue to thaw until it is in lysis buffer with RNase inhibitors.
	RNase contamination	Sterilize plasticware, glassware and buffers properly. Use gloves during all manipulations with RNA.
	Tissue disruption is too slow	If the glass homogenizer is used, it is useful to resign maximum tissue disruption and to shorten the homogenization step.
DNA contamination	Too small volume of lysis buffer used for sample homogenization	The sample should be weighed, and an adequate volume of the buffer should be used.
	Insufficient DNase treatment	Repeat DNase treatment.
Low A_{260}/A_{280}	The protein contamination often due to contamination of aqueous phase with phenol phase	Perform an additional phenol–chloroform extraction on the purified RNA.

11. Collect the aqueous phase and precipitate with 0.1 vol of 3 M sodium acetate (pH 7.4) and 2 vol of ethanol for 1 h at -20°C.
12. Centrifuge at 13 000× g for 30 min at 4°C.
13. Wash the pellet with 80% ethanol.
14. Centrifuge at 13 000× g for 10 min at 4°C.

15. Dissolve the pellet in sterile water or TE buffer.

While the RNA is not dissolved in lysis buffer, all manipulations should be done on ice.

RESULTS

Different methods of RNA isolation and different manners of tissue disruption have been compared. The samples derived from various kinds of tumors (colorectal, breast, ovarian cancers) have been used. After application of SDS/EDTA lysis buffer, the obtained preparations were often depolymerized (6,10). The modification of the Chomczynski and Sacchi method by combining the first three steps of the protocol into a single one (the guanidinium thiocyanate, phenol and sodium citrate were mixed together) results in worse tissue disruption than in the classic method. Moreover, the A_{260}/A_{280} was a little lower (about 1.6) (14). However, good results were obtained using some commercial reagents such as TRI Reagent® (Molecular Research Center, Cincinnati, OH, USA) based on the modified Chomczynski and Sacchi method.

Both of the methods described here seem to be efficient, safe and simple. They do not require time-consuming procedures such as ultracentrifugation through the cesium chloride cushion. These methods allow intact RNA to be obtained (Figure 1). The disruption of the compact solid tumor tissue was comparable in both of them but easier than in others. The yield was high (up to 65%) (Tables 1 and 2). DNA

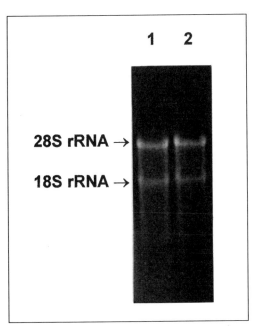

Figure 1. Control of the integrity of RNA isolated from colon adenocarcinoma by two described methods. RNA (6 µg) was separated on a 2% agarose, 6% formaldehyde gel and stained with ethidium bromide (1 µg/mL). Lane 1: RNA isolated with ammonium sulfosalicylate. Lane 2: RNA isolated with guanidinium thiocyanate.

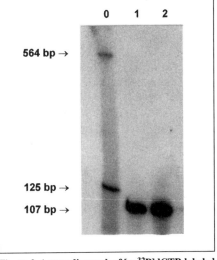

Figure 2. Autoradiograph of [α-^{32}P]dCTP-labeled RT-PCR amplification products of β-actin mRNA isolated by two described methods. RT-PCR products were separated on a 5% polyacrylamide gel with 8 M urea. Only one kind of DNA fragment (107 bp) corresponding to the cDNA amplification product (245 bp) was found. Lane 0: Molecular weight standard λ DNA HindIII-digested. Lane 1: RNA isolated with ammonium sulfosalicylate. Lane 2: RNA isolated with guanidinium thiocyanate.

contamination appeared to be minimal because of DNase treatment. There were no DNA amplification products (245 bp) after RT-PCR using cDNA derived from the investigated RNA preparations and primers for β-actin (Figure 2). Moreover, only traces of protein were found in the final RNA pellet (A_{260}/A_{280} = 1.7–1.9).

Both the procedures appear to be very useful in RNA isolation from solid tumors. The method with ammonium sulfosalicylate is slightly more time-consuming but does not require harmful substances such as guanidinium salts. Moreover, in the presence of ammonium sulfosalicylate, it is possible to use SDS as a disrupting agent and ribonuclease inhibitor, while in high concentrations of the other salts, it precipitates.

In addition to testing different methods of RNA isolation, different manners of tissue disruption were compared. The most efficient appears to be mechanical disruption using high-speed homogenizers like Polytron. Use of the glass homogenizer decreases the efficiency a little but is also sufficient. The powdering of the frozen tissue does not increase the yield but can cause the RNA degradation if the sample is allowed to thaw. Therefore, we recommend using fresh tissue if possible.

REFERENCES

1. **Chirgwin, J.M., A.E. Przybyla, R.J. MacDonald and W.J. Rutter.** 1979. Isolation of biologically active ribonucleic acid from sources enriched in ribonuclease. Biochemistry *18*:5294-5299.
2. **Chomczynski, P. and N. Sacchi.** 1987. Single-step method of RNA isolation by acid guanidinium thiocyanate-phenol-chloroform extraction. Anal. Biochem. *162*:156-159.
3. **Dahle, C.E. and D.E. Macfarlane.** 1993. Isolation of RNA from cells in culture using Catrimox14™ cationic surfactant. BioTechniques *15*:1102-1105.
4. **Gruffat, D., C. Piot, D. Durand and D. Bauchart.** 1996. Comparison of four methods for isolating large mRNA: apolipoprotein B mRNA in bovine and rat livers. Anal. Biochem. *242*:77-83.
5. **Kitlińska, J. and J. Wojcierowski.** 1995. RNA isolation from solid tumor tissue. Anal. Biochem. *228*:170-172.
6. **Majumdar, D., Y.J. Avissar and J.H. Wyche.** 1991. Simultaneous and rapid isolation of bacterial and eucaryotic DNA and RNA: a new approach for isolating DNA. BioTechniques *11*:94-101.
7. **Miyake, K., K. Inokuchi, K. Dan and T. Nomura.** 1993. Alternations in the deleted in colorectal carcinoma gene in human primary leukemia. Blood *82*:927-930.
8. **Monstein, H.-J., A.-G. Nylander and D. Chen.** 1995. RNA extraction from gastrointestinal tract and pancreas by a modified Chomczynski and Sacchi method. BioTechniques *19*:339-343.
9. **Nakijima-Iijima, S., H. Hamada, P. Reddy and T. Kakunga.** 1985. Molecular structure of the human cytoplasmic β-actin gene: interspecies homology of sequences in the introns. Proc. Natl. Acad. Sci. USA *82*:6133-6137.
10. **Peppel, K. and C. Baglioni.** 1991. A simple and fast method to extract RNA from tissue culture cells. BioTechniques *9*:711-713.
11. **Sambrook, J., E.F. Fritsch and T. Maniatis.** 1989. Molecular Cloning: A Laboratory Manual, 2nd ed. CSHL Press, Cold Spring Harbor, NY.
12. **Schmidt, G. and S.J. Tanhauser.** 1941. A method for the determination of deoxyribonucleic acid, ribonucleic acid and phosphoproteins in animal tissue. J. Biol. Chem. *83*:1961.
13. **Tsanev, R. and G.G. Markow.** 1960. Substances interfering with spectrophotometric estimation of nucleic acids and their elimination by the two-wavelength method. Biochim. Biophys. Acta *42*:442.
14. **Xie, W. and L.I. Rothblum.** 1991. Rapid, small-scale RNA isolation from tissue culture cells. BioTechniques *11*:325-327.
15. **Zarlenga, D.S. and H.R. Gamble.** 1987. Simultaneous isolation of preparative amounts of RNA and DNA from *Trichinella spiralis* by cesium trifluoroacetate isopycnic centrifugation. Anal. Biochem. *162*:569-574.

3 Isolating Single Cells for Analysis by RT-PCR

Mark C. Harbeck[1] and Paul L. Rothenberg[1,2]
[1]Department of Endocrinology and Metabolic Disease, Jefferson Medical College, Thomas Jefferson University, Philadelphia, PA and [1,2]Clinical Pharmacology, Novartis Pharmaceuticals, East Hanover, NJ, USA

OVERVIEW

We describe a simple, convenient and inexpensive microcapillary tube method for isolating single cells for mRNA analysis, which does not involve the use of a micromanipulator. Individual cells isolated by this method are transferred to a polymerase chain reaction (PCR) tube containing lysis buffer. Both reverse transcription (RT) and PCR amplification are carried out in the same tube, and amplification with nested PCR primers results in a product yield sufficient for visualization on conventional ethidium bromide-stained agarose gels. Complete protocols for constructing the microcapillary tube apparatus, isolating single cells and subsequent RT-PCR amplification are described, using amplification of the low-abundance insulin-receptor mRNA from H35 hepatoma cells as an example.

BACKGROUND

The amplification of mRNA from a single cell, or single-cell RT-PCR, has become an increasingly important method for characterizing expression patterns of genes in specific cell types. A major difficulty of this technique lies in the required isolation and physical manipulation of an individual cell. Traditional methods to isolate single cells include serial dilutions of a cell suspension (6) or the use of a micromanipulator-mounted capillary pipet (2,10). The serial dilution approach suffers from a lack of certainty that any particular reaction indeed contains only a single cell, which can lead to difficulty in the interpretation of experimental results, particularly in a heterogeneous population of cells. Although the use of a micromanipulator permits the unambiguous isolation of single cells, such expensive equipment is not generally

Gene Cloning and Analysis by RT-PCR
Edited by Paul Siebert and James Larrick
© 1998 BioTechniques Books, Natick, MA

available in every laboratory and requires a trained and experienced operator. The technique we describe here provides a simple and inexpensive alternative to a micromanipulator. Our technique uses readily available materials and allows the isolation of individual cells. The utility of this method is demonstrated here by the amplification of the low-abundance insulin-receptor mRNA (9) by nested RT-PCR from single rat H35 hepatoma cells (8). A suspension of disaggregated H35 cells is treated with ribonuclease (RNase A) to digest any extracellular mRNA released from cells damaged during the cell isolation protocol. Individual cells are isolated, and each is transferred to a 0.5-mL PCR tube and lysed in a buffer containing RNase inhibitors to protect the released intracellular mRNA. Both RT and PCR are carried out in the same tube, reducing the potential for contamination of the minute amount of cDNA generated from a single cell. Amplification with nested PCR primers results in a product yield sufficient for visualization on conventional ethidium bromide-stained agarose gels following the second round of PCR. The described procedure should be generally applicable to the amplification of mRNA from specific cells isolated from heterogeneous cell mixtures.

PROTOCOLS

Protocol for Construction of Microcapillary Pipets

Materials and Reagents

- 1-μL Microcaps®, 32 mm long × 0.1 mm inside diameter (i.d.) (Drummond Scientific, Broomall, PA, USA). These pipets, commonly used for spotting thin-layer chromatography (TLC) plates, are available from Fisher Scientific (Pittsburgh, PA, USA)
- Dimethyldichlorosilane (Sigma Chemical, St. Louis, MO, USA)
- Toluene, reagent grade
- Methanol, reagent grade
- P-2 or P-10 adjustable Pipetman® (Rainin Instrument, Woburn, MA, USA)
- Disposable plastic pipet tips, 0–10-μL size for P-2 or P-10 Pipetman

Procedure

1. Silanize glass microcapillary pipets by immersion in dimethyldichlorosilane for 1 h in a ventilated safety hood. It is convenient to silanize 100–200 pipets at a time.
2. To remove excess silane, wash the pipets in toluene followed by methanol. This is rapidly and efficiently accomplished by flame bending a Pasteur pipet to a right angle approximately 1 cm from the narrow end, so that vacuum aspiration can be used to pick up individual microcapillary pipets. Then aspirate toluene and methanol through the internal bore of each microcapillary pipet.
3. Dry the rinsed, silanized pipets at 110°C for 2 h. Cool to room temperature.
4. Insert each microcapillary glass pipet into the narrow end of a disposable plastic pipet tip (Figure 1a). Insert the microcapillary tube only partially into the plastic tip, thus leaving enough internal clearance for the barrel of the

micropipettor. It might be necessary to trim the end of the plastic pipet tip and widen its orifice so that the microcapillary glass pipet can slide in while maintaining a tight fit.

5. Place the completed microcapillary pipet tips in an appropriately sized pipet rack and autoclave (storage racks used for long gel-loading pipet tips work well).

Protocol for Isolation of Single Cells

The following protocol describes the isolation of single cells from a cell suspension using the microcapillary pipets described above. The specific method used to obtain a cell suspension depends on the tissue or cell type being examined. The example given below is for cultured H35 cells, which grow adherent to plastic tissue culture flasks (3). The method can be adapted for obtaining single cells from animal organs and tissues (4). For other cell types, consult the appropriate sources for protocols on obtaining cell suspensions.

Materials and Reagents

- Lysis buffer: 0.8% (wt/vol) Nonidet® P-40 (NP40; Pierce Chemical, Rockland, ME, USA), 100 µg/mL yeast tRNA (Life Technologies, Gaithersburg, MD, USA), 8 U/mL RNasin® (Promega, Madison, WI, USA) and 10 mM dithiothreitol (DTT; Boehringer Mannheim, Indianapolis, IN, USA)
- Dulbecco's modified Eagle medium (D-MEM; Life Technologies)
- 0.05% trypsin (Life Technologies)
- Dulbecco's phosphate-buffered saline (D-PBS; Life Technologies)
- RNase A (Catalog No. E70194; Amersham Pharmacia Biotech, Piscataway, NJ, USA)
- 0.5-mL thin-walled PCR tubes (Perkin-Elmer, Norwalk, CT, USA)
- Hemacytometer
- Inverted phase-contrast microscope

Procedure

1. Detach H35 cells by a 5-min incubation in 0.05% trypsin at 37°C, resuspend and wash once by centrifugation at 100× g at 22°C for 4 min in D-MEM tissue culture medium, followed by two washes in D-PBS containing 1 mM glucose. Resuspend the cells in ice-cold (0°–4°C) D-PBS with glucose.
2. Determine the cell concentration by counting in a standard hemacytometer.
3. Dilute the cell suspension to a concentration of 10–20 cells per microliter so that 0.1–0.2 µL will, on average, contain 1 cell.
4. To remove extracellular mRNA from the cell suspension, incubate the diluted cells with 100 ng/mL RNase A for 10 min at 22°C, then place on ice.
5. A sterile microcapillary pipet attached to the end of an adjustable 0–10-µL micropipettor is used to draw up 0.1–0.2 µL from the cell suspension (a 0.1–0.2-µL aliquot fills the glass microcapillary pipet to a height of 3–6 mm). After drawing the cell suspension into the microcapillary pipet, snap the distal end of the glass tube containing the cell suspension with a sterile forceps into a

sterile petri dish, cover with the dish lid and determine the number of trapped cells within the capillary using an inverted phase-contrast microscope (Figure 1b). Initially, some practice focusing and adjusting the contrast might be required to readily identify cells.

6. After confirming the presence of one or more cells, aseptically transfer the glass tube to a 0.5-mL PCR tube containing 5 µL of lysis buffer. Expel the trapped cells into the lysis buffer by centrifugation at $10\,000\times g$ for 10 s at 22°C and store on ice.

7. As a control to ensure that any amplified mRNA is derived from isolated and intact single cells and not from extracellular mRNA, centrifuge the cell suspension at $2000\times g$ for 30 s, add 0.1 µL of the cell-free supernatant to 5 µL of lysis buffer and process in parallel as described below.

Note: The cell lysis buffer described above contains RNasin ribonuclease inhibitor to inhibit the endogenous RNase A added in step 4. When using RNasin, it is necessary to maintain a temperature below 50°C to prevent the release of active ribonuclease from the RNasin:RNase complex.

Protocol for RT and PCR Amplification

mRNA is reverse-transcribed using both the oligo(dT) and 3′ gene-specific outer PCR primer. The resulting cDNA is amplified using both outer gene-specific PCR primers. The product from the first round of PCR amplification is then subjected to a second round of PCR amplification using the inner, or nested, set of gene-specific PCR primers.

Materials and Reagents

- Oligo(dT)$_{15}$ (Promega)
- Gene-specific PCR primers [nested PCR primers for insulin-receptor amplification (5) synthesized by Integrated DNA Technologies (Coralville, IA, USA) and Life Technologies]
- 1.82× RT reaction mixture: 91 mM Tris-HCl, pH 8.3, 137 mM KCl, 5.5 mM MgCl$_2$, 18 mM DTT, 910 µM dNTPs (Promega) and 200 U SUPERSCRIPT™ II Reverse Transcriptase (Life Technologies) per reaction (use 11 µL of RT reaction mixture per sample)
- PCR mixture A: 20 mM Tris-HCl, pH 8.4, 50 mM KCl, 1.5 mM MgCl$_2$, 1.25 µM outer upstream (5′) insulin-receptor primer and 5 U of *Taq* DNA polymerase (Life Technologies) per 80 µL (use 80 µL PCR mixture A per sample)
- PCR mixture B: 20 mM Tris-HCl, pH 8.4, 50 mM KCl, 200 µM dNTPs, 1.5 mM MgCl$_2$, 1 µM of each inner insulin-receptor primer and 2.5 U of *Taq* DNA polymerase per 48 µL (use 48 µL PCR mixture B per sample)

Procedure

Reverse transcription

1. To the 5.1 µL lysate from a single cell, add 50 pmol of oligo(dT)$_{15}$ and 100 pmol of outer downstream insulin-receptor primer in 4 µL double-distilled (dd)H$_2$O. Incubate at 42°C for 10 min to anneal the primers.

2. Add 11 μL of 1.82× RT reaction mixture, yielding a final concentration of 50 mM Tris-HCl, pH 8.3, 75 mM KCl, 3 mM $MgCl_2$, 10 mM DTT, 500 μM dNTPs and 200 U reverse transcriptase in a final volume of 20 μL.
3. Incubate for 50 min at 42°C, followed by 10 min at 95°C to inactivate the reverse transcriptase.
4. To confirm that the RT-PCR product is due to mRNA and not genomic DNA, a negative control reaction is performed identically but without the addition of reverse transcriptase.

Polymerase chain reaction

5. Add 80 μL of PCR mixture A to the 0.5-mL PCR tube containing the RT reaction (step 3 above) for a final volume of 100 μL.
6. Overlay the reaction mixtures with mineral oil, cap tubes and commence thermal cycling using the following parameters: 60 cycles of denaturation at 94°C for 1 min, annealing at 45°C for 1 min and extension at 72°C for 3 min. In the first cycle, the denaturation step is extended 2 min, and the last extension step is 7 min.
7. Two microliters of the product from the first-round PCR are added to 48 μL of PCR mixture B for a second round of PCR with initial denaturation for 1 min at 94°C, 50 cycles for 1 min at 94°C, 1 min at 61°C and 1.5 min at 72°C, followed by a final extension for 7 min at 72°C.

Note: The thermal cycling parameters described above were optimized for PCR using the DNA Thermal Cycler 480 (Perkin-Elmer) and PCR primers designed for the nested PCR amplification of the insulin-receptor cDNA. The use of different thermal cyclers and the amplification of other cDNA templates might require different PCR cycling conditions. Therefore, the optimal conditions for the specific cycler and template should be determined using purified mRNA before attempting to amplify cDNA obtained from a single cell.

The identity of the PCR products can be confirmed by digestion with restriction enzymes that cut at sites present in the amplified cDNA or by Southern blot analysis.

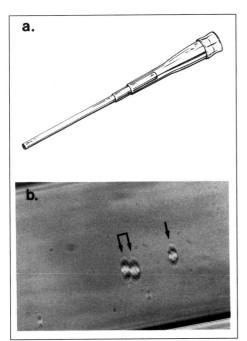

Figure 1. Microcapillary pipet isolation of single cells. (a) Illustration of the microcapillary pipet tip, demonstrating the placement of the silanized microcapillary tube in the disposable plastic pipet tip. (b) A cell suspension was aspirated into the silanized microcapillary pipet. The trapped cells are visualized using an inverted phase-contrast microscope at 200× magnification. A single cell is indicated by the single arrow and a cell doublet by the double arrow.

RESULTS

Figure 1b shows that a single cell (single arrow) is readily visible by phase-contrast microscopy inside the barrel of the microcapillary tube and is easily distinguished from a cell doublet (double arrow). The liquid volume in the capillary tube, including the cell, is expelled by a 10-s centrifugation of the capillary tube inside a microcentrifuge tube also containing 5 µL of lysis buffer.

The apparatus shown in Figure 1 was used to isolate single H35 rat hepatoma cells, which express the plasma membrane insulin receptor (9). Figure 2 shows the products generated from H35 cells by RT-PCR using nested insulin-receptor PCR primers (5). The RT-PCR products from purified total H35 RNA (100 ng or ca. 10^4 cell equivalents) (Figure 2, lane 1) and the products obtained from a single, isolated H35 cell (Figure 2, lane 2) both contain the predicted 331-bp product from insulin-receptor mRNA. No product was observed in control reactions that lacked either reverse transcriptase or template mRNA (Figure 2, lanes 3 and 4, respectively). The absence of product in the RT-PCR performed using a 0.1-µL aliquot of the cell-free supernatant from the H35 cell suspension (obtained after pelleting the RNase A-treated cells at 2000× g; Figure 2, lane 5) indicates that the 331-bp product results from RT-PCR amplification of the mRNA from the isolated cell rather than extracellular mRNA that can be released from some damaged cells during random cell lysis. When we omitted ribonuclease digestion of the cell suspension from our protocol, a specific 331-bp band was frequently observed in such cell-free supernatant control reactions (Figure 2, lane 6).

In our experiments, nested PCR was required for amplification of the insulin-receptor cDNA to levels detectable on conventional ethidium bromide-stained agarose gels. No insulin-receptor cDNA was identified from first-round PCR, even after 60 cycles of amplification. Detection of more abundant transcripts might not require a second round of PCR amplification.

DISCUSSION

In many investigations, it is necessary to amplify rare transcripts that are present at the level of only a few molecules per cell. We have used nested primers and two rounds of PCR to amplify a low-abundance mRNA (9) from a single lysed cell. However, the extreme sensitivity of nested RT-PCR increases the opportunity for false-positive results originating from trace amounts of contaminating mRNA [e.g., a single molecule of viral RNA can be detected by RT-PCR (7)]. Treatments required to produce a cell suspension might damage some fraction of the cell population, causing

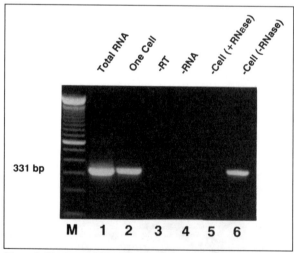

Figure 2. Single-cell RT-PCR of insulin-receptor mRNA. 100 ng of total H35 RNA (lane 1) or a single H35 cell (lane 2; representative product from 9 separate reactions) were subjected to RT-PCR as described in the text. The products were analyzed by electrophoresis on 1.6% agarose and stained with ethidium bromide. The predicted product size for the insulin-receptor mRNA is 331 bp. Negative controls used 20 cells but without reverse transcriptase (lane 3; representative of 3 separate reactions) or RNA template (lane 4), and with cell-free supernatant (lane 5; representative of 6 separate reactions). Lane 6 shows the products generated from the cell-free supernatant when the RNase treatment of the cell suspension is omitted. Lane M is 1 µg of a 100-bp DNA ladder.

leakage of intracellular mRNA into the extracellular medium. Contaminating endogenous ribonucleases should not be assumed to adequately digest all extracellular mRNAs, because cytosolic extracts of eukaryotic cells have been shown to possess ribonuclease inhibitors (1), and the modifications of eukaryotic mRNAs, the 5′ cap structure and 3′ polyadenylation, can render these RNAs resistant to degradation by endogenous ribonucleases. To analyze exclusively the mRNA of only a single cell that has been selected from a cell suspension, 100 ng/mL ribonuclease (RNase A) are added to the cell suspension to digest any extracellular mRNA. Omission of this treatment resulted in the generation of the 331-bp insulin-receptor RT-PCR product from the cell-free supernatant in our experiments. Thus, inclusion of a ribonuclease treatment might be essential for certain applications of this method. This is an especially important factor to consider when amplifying mRNA from a heterogeneous population of cells, in which mRNA from cells of a type different from that being analyzed could generate misleading results.

The use of oligo(dT), random hexamers and/or gene-specific primers in the RT step permits the amplification of multiple mRNAs from a single RT reaction. It is therefore possible, by amplification of mRNA for a characteristic marker protein(s), to identify the type of cell being subjected to RT. We have successfully used this protocol to amplify specific mRNA transcripts in single insulin-secreting beta cells derived from enzymatically disaggregated islets of Langerhans, a "microorgan" composed of several distinct cell types (4).

Although the single-cell RT-PCR protocol described here provides a very high degree of sensitivity, as yet it is limited by the qualitative nature of the results. Because of the use of nested primers and a total of 110 rounds of amplification, even large differences in starting template concentration would not yield significantly different final PCR product band intensities. Moreover, because of the very limited amounts of RNA present in a single cell, small differences in RT efficiency could further mask variations in template levels.

REFERENCES

1. Coutts, M., A. Krowczynska and G. Brawerman. 1993. Protection of mRNA against nucleases in cytoplasmic extracts of mouse sarcoma ascites cells. Biochim. Biophys. Acta *1173*:49-56.
2. Eberwine, J., H. Yeh, K. Miyashiro, Y. Cao, S. Nair, R. Finnell, M. Zettel and P. Coleman. 1992. Analysis of gene expression in single live neurons. Proc. Natl. Acad. Sci. USA *89*:3010-3014.
3. Harada, S., E.G. Loten, R.M. Smith and L. Jarett. 1992. Nonreceptor mediated nuclear accumulation of insulin in H35 rat hepatoma cells. J. Cell. Physiol. *153*:607-613.
4. Harbeck, M.C., D.C. Louie, J. Howland, B.A. Wolf and P.L. Rothenberg. 1996. Expression of insulin receptor mRNA and insulin substrate 1 in pancreatic islet β-cells. Diabetes *45*:711-717.
5. Harbeck, M.C. and P.L. Rothenberg. 1995. A technique for isolating single cells for analysis by reverse transcription polymerase chain reaction. Anal. Biochem. *230*:193-196.
6. Molesh, D.A. and J.M. Hall. 1994. Quantitative analysis of CD34+ stem cells using RT-PCR on whole cells. PCR Methods Appl. *3*:278-284.
7. Nakamura, S., S. Katamine, T. Yamaoto, S.K.H. Foung, T. Kurata, Y. Hirabayashi, K. Shimadu, S. Hino and T. Miyamoto. 1993. Amplification and detection of a single molecule of human immunodeficiency virus RNA. Virus Genes *7*:325-338.
8. Taub, R., A. Roy, R. Dieter and J. Koontz. 1987. Insulin as a growth factor in rat hepatoma cells. Stimulation of proto-oncogene expression. J. Biol. Chem. *262*:10893-10897.
9. Tewari, M., D.S. Tewari and R. Taub. 1991. Posttranscriptional mechanisms account for differences in steady state levels of insulin receptor messenger RNA in different cells. Mol. Endocrinol. *5*:653-660.
10. Trumper, L.H., G. Brady, A. Bagg, D. Gray, S.L. Loke, H. Griesser, R. Wagman, R. Brazeil et al. 1993. Single-cell analysis of Hodgkin and Reed-Sternberg cells: molecular heterogeneity of gene expression and p53 mutations. Blood *81*:3097-3115.

4. Isolation of RNA from Mammalian Cells: Application to Large mRNA

Dominique Gruffat
Laboratoire Croissance et Métabolismes des Herbivores, Unité Métabolismes Energétique et Lipidique, Institut National de la Recherche Agronomique, Centre de Recherches Clermont Ferrand-Theix, Saint Genès Champanelle, France

OVERVIEW

A typical mammalian cell contains about 10 pg of RNA, 80%–85% of which is rRNA (chiefly 28S, 18S and 5S). Most of the remaining RNA (15%–20%) consists of a variety of low-molecular-weight (LMW) species of RNA such as tRNAs, small nuclear (sn)RNAs etc. Messenger RNA, which represents 1%–5% of the total cellular RNA is heterogeneous in both size (from a few hundred bases to many kilobases) and base sequence. The resulting heterogeneous population of molecules collectively virtually encodes for all of the polypeptides synthesized by the cell.

Characterization of isolated mRNA species and their translation products has largely contributed to understanding the regulation of gene expression at different levels within the cell. These studies have required development of procedures for obtaining undegraded, biologically active mRNA from a variety of sources. There are many different procedures described in the literature to extract eukaryotic mRNA because of the large diversity of cells and tissues studied.

This chapter first describes techniques involved in isolation of RNA from eukaryotic cells and then gives several examples of large mRNA isolation.

BACKGROUND

Isolation of Eukaryotic RNA

Several general techniques are commonly used for isolation of RNA. Each of these methods might require modifications to be applicable to a particular system. However, they have certain features and requirements in common: *(i)* the absolute need to prevent ribonuclease activity and *(ii)* the need to separate RNA from proteins and DNA in a quantitative manner.

Ribonuclease Inactivation

A major technical problem in isolation of RNA is the control of ribonuclease activity. A variety of ribonucleases are present in all tissues and occur as contaminants in many reagents derived from biological sources. Inhibition of ribonuclease (liberated during cell lysis) using RNase inhibitors or methods that disrupt cells and inactive RNases simultaneously is essential for succeeding in RNA isolation. Because quantity, type and specificity of ribonuclease vary widely among tissues, procedures that are adequate in a given system can be totally ineffective in other systems. A combination of methods is usually necessary.

Heparin

This sulphated polysaccharide is widely used as a ribonuclease inhibitor (49). It adsorbs the nucleases and competitively inhibits them. Addition of heparin (ca. 1 mg/mL) to a buffer can substantially improve yield and translational activity of purified mRNA. However, heparin must be removed from RNA (with which it co-purifies) before initiation of translational assays because it is a potential inhibitor of polypeptide chain initiation.

Diethyl pyrocarbonate (DEPC)

DEPC is an effective inhibitor of ribonucleases (49). It is often proposed to clean glassware and plasticware. At intermediate pH values, DEPC attacks ribonucleases as well as other cell proteins by reacting especially with the imidazole nitrogens of histidine residues and of free amino groups, leading to a loss of enzyme activity. However, it also reacts with single-stranded nucleic acids, modifying purine residues of RNA by carboxymethylation. Carboxymethylated RNA is translated with a very low efficiency in cell-free systems. However, its ability to form DNA–RNA or RNA–RNA hybrids is not deeply affected unless a large fraction of purine residues has been modified. The kinetics of a reaction of this substance are apparently slower with nucleic acids than with proteins. Moreover, DEPC is suspected to be a carcinogenic agent and consequently must be handled with care.

Proteinase K

This proteolytic enzyme rapidly and preferentially inactivates nucleases from many sources (33). Proteinase K, which is active at pH 7.0 in the presence of sodium dodecyl sulfate (SDS) (27), is added to the buffer just prior to homogenization. This enzyme is also advantageous in reducing the amount of proteins at the interface after phenol extraction.

Detergents

Many anionic detergents, particularly SDS and sodium N-lauroylsarcosine, are strong denaturants of proteins and thereby can act as ribonuclease inhibitors (49). They dissociate nucleoprotein complexes and further inhibit ribonuclease activity.

Endogenous ribonuclease inhibitors

Some tissues, such as rat liver or human placenta, can synthesize proteins that function as ribonuclease inhibitors. These proteins bind ribonuclease, forming noncovalent complexes and reversibly inhibiting RNase activity. Rat liver protein is a sulfhydryl-containing protein that can be stabilized by low levels of reducing agents such as dithiothreitol (DTT). Use of this RNase inhibitor is not advised when denaturing agents are added in lysis solution in initial stages of extraction of RNA. However, the inhibitor should be included when more gentle methods of lysis are used and should be present at all stages during the subsequent purification of RNA. It should be added several times during the purification procedure because it is removed by extraction with phenol. The inhibitor does not interfere with reverse transcription or cell-free translation of mRNA.

High pH and ionic strength of extraction buffers

Generally, mammalian tissues are homogenized in buffered salt solutions, sometimes containing 0.35 M sucrose to prevent disruption of lysosomes and subsequent release of ribonucleases. Large volumes of medium (10–20 times the equivalent volume of tissue weight) are used to dilute out endogenous ribonuclease in conditions of high pH (pH 8.5–9.0) and of high ionic strength (0.2–0.5 M salt) to further inhibit their activity.

Extraction of Total Cellular RNA

An additional problem arising in isolation of cellular RNA is the protein removal from cell homogenates in such a way that RNA keeps total integrity. Because cellular RNA (including mRNA) is tightly associated with specific proteins (histones) as a ribonucleoprotein complex, purification of RNA requires specific reagents that will completely disrupt this complex and allow further extraction of undegraded RNA.

Many different methods are proposed in different cell systems and types of RNA, but no systematic comparisons are made, and individual requirements of experimental conditions vary greatly.

Phenol

Phenol, a protein-denaturing agent, is first proposed for extracting RNA from mammalian tissues (30,31). Addition of chloroform facilitates transfer of RNA from the mixture to the aqueous phase (45). The general procedure (Protocol 1) consists first of treating tissue homogenate with phenol–chloroform (1:1, vol/vol). Then, aqueous and organic phases of the mixture are separated by centrifugation, with denatured proteins compacted at the interface. Nucleic acids were mainly located in the aqueous phase, but some nucleic acids can be trapped with the interface material. Nucleic acids of the interface and phenol phase can be re-extracted with an excess of buffer, and aqueous phases are then combined. Total nucleic acids in aqueous phase are precipitated by 2.5 vol of cold (-20°C) ethanol in the presence of 0.2 M sodium (or potassium) acetate at pH 5.0 (minimizing chemical hydrolysis of RNA). However, these salt solutions also solubilize numerous contaminants such as DNA, tRNA, 5S rRNA, glycogen and heparin (30).

Precipitation of proteins with phenol alone depends on experimental conditions such as temperature, levels of salts and pH. When phenol treatment was carried out in cold conditions, alkaline buffers (about pH 9.0) are requisite for the transfer of most of the extracted RNA to the aqueous phase. At neutral pH, RNA is mainly found at the interface containing denatured proteins (49). However, following RNA extraction at room temperature in the presence of an equal volume of chloroform, RNA is found in the aqueous phase (45). Other additives, such as isoamyl alcohol (an antifoaming agent), EDTA (a chelating agent that disrupts RNA–protein interactions) and *m*-cresol [an antifreeze that potentiates protein denaturation (30)] are often included in the organic extraction mixture.

RNA extraction with phenol has several disadvantages (49): *(i)* ribonuclease can still be active following RNA extraction treatments; *(ii)* phenol is contaminating by various impurities; and *(iii)* the sequences of shacking and transfer are difficult to make completely reproducible and contribute to variable and often substantial losses. Moreover, phenol extraction results in aggregation of RNA, and rather specific aggregates can be formed between mRNA and other RNA species such as rRNA (24). Aggregates appear to form in any modifications of the phenol extraction procedures and are apparently not due to ethanol precipitation. Treatments of extracted RNA with strong denaturants such as formamide or dimethyl sulfoxide (DMSO) or by heating RNA to about 65°C in a dilute buffer at neutral pH for 10–15 min followed by a rapid cooling (0°C) appear to dissociate phenol-induced aggregates efficiently (24).

A number of alternative methods that avoid phenol have been developed proposing concentrated salt solutions to dissociate ribonucleoprotein complexes as described below.

Lithium chloride–urea

RNA can be dissociated from proteins and precipitated by concentrated salt solutions such as 2 M LiCl (Protocol 2). Barlow et al. (6), Kruh (32) and Schimke et al. (44) have reported a simple method to isolate undegraded high-molecular-weight (HMW) RNA from rabbit reticulocytes by two successive treatments of homogenates with 2 M LiCl. This method allows preparation of pure RNA containing less than 1% proteins but needs to treat high amounts of tissue or tissue rich in RNA (500 µg/mL or greater) to obtain a quantitative precipitation of RNA.

LiCl treatment separates HMW RNA from proteins, heparin and LMW RNA, in contrast to phenol treatment, which does not remove heparin and LMW RNA (42).

More recently, Auffrey and Rougeon (5) have described a method for isolating mouse RNA from total myeloma tumor by precipitation with 3 M LiCl and 6 M urea (Protocol 3). Because the half-life of RNase averages 3 min in 8 M urea (7), these conditions favor *(i)* inhibition of RNase by high concentrations of salt and urea, *(ii)* selective precipitation of RNA with LiCl without contaminants such as DNA (sheared to small fragments during homogenization), polysaccharides and proteins (denatured by urea) that still remain in aqueous solution, *(iii)* prevention of RNA aggregation by phenol and *(iv)* extraction in a single step of a high yield of RNA, allowing isolation of translationally active mRNA.

Cesium chloride centrifugation

An alternative method for isolation of RNA consists of RNA sedimentation by CsCl equilibrium density gradient ultracentrifugation (22) (Protocol 4). Because this method was often recommended for isolation of DNA, it is apparent that RNA can be isolated as well. However, RNA can be obtained only as a pellet because there is no attainable CsCl concentration at which it will band. Usually, pellets can be contaminated by other components of the mixture, but because RNA is the most dense component, the pellet will represent highly purified RNA.

This method can efficiently eliminate nucleases, other proteins and DNA (migrating in the opposite direction) without any additional treatments by RNase and DNase.

Compared to phenol extraction, CsCl ultracentrifugation has the advantages of ease of use, short processing time, a high percentage of recovery, good integrity of structure and biological activity of resulting RNA.

Guanidinium salts

Proteins are readily solubilized in concentrated solution (4 M or more) of guanidinium chloride and become biologically inactive by alteration of secondary structure. The half-life of RNase was 10 s in 4 M guanidinium hydrochloride (37). In contrast, secondary structure of nucleic acids is not affected to the same extent (16).

Guanidinium thiocyanate combines the strong denaturing characteristics of guanidinium with the chaotropic action of thiocyanate while, like guanidinium chloride, the guanidinium cation is only chaotropic and hence active in denaturation. Consequently, guanidinium thiocyanate can be considered among the most effective denaturants of proteins (38).

Following dispersion of RNA in guanidinium solution (in which endogenous nucleases are inhibited), final separation of RNA from other macromolecular components of homogenates is achieved either by a selective precipitation of RNA (in ethanol or LiCl) or by a selective sedimentation by CsCl density gradient ultracentrifugation (36).

Guanidinium thiocyanate–CsCl

Several authors have proposed to combine a denaturing treatment of protein by guanidinium thiocyanate followed by a selective isolation of RNA by CsCl centrifugation for RNA extraction from tissues rich in ribonucleases such as the pancreas and the liver (14,41,50).

Thus, the method of Raymond and Shore (41) (Protocol 5) consists of preparing total RNA from rat liver by homogenizing tissue samples in 4 M guanidinium thiocyanate, 0.1 M 2-mercaptoethanol, 25 mM sodium acetate (pH 7.0), followed by centrifugation of the mixture in 5.7 M CsCl and 0.1 M EDTA (pH 7.0). Efficiency of this method is controlled by measurement of the translational activity of rat liver mRNA, which is generally 5 times higher than that of mRNA purified by phenol extraction.

Table 1. Procedural Anecdotes Provided by Chirgwin et al. (14)

- The prevention of degradation by ribonuclease is dependent on the efficiency of the initial seconds of the homogenization (the use of high-speed Tissumizer™ [Tekmar-Dohrmann, Cincinnatti, OH, USA] or Polytron® [Brinkmann Instruments] is recommended).
- The use of a conventional blender and homogenization of tissue that has been frozen and thawed or pulverized in liquid nitrogen results in degradation of RNA (diminution of the 28S peak height and concomitant appearance of lower-molecular-weight species on electrophoresis in denaturing gels).
- The use of detergent sodium N-lauroylsarcosine is not essential but gives a cleaner initial precipitate of RNA and can accelerate the initial dissolution of tissue.
- The 2-mercaptoethanol is essential for tissue containing RNase, but increasing concentrations beyond 0.1 M final concentration have no effect.
- DTT can be used with guanidinium hydrochloride stock as a disulfide bond reductant, but it undergoes a chemical reaction with the thiocyanate anion to produce hydrogen sulfide and a green color.
- The initial precipitation of RNA uses 0.75 vol of ethanol relative to guanidinium thiocyanate stock; this precipitates some DNA (eliminated by re-precipitations) as well as RNA but is necessary to prevent guanidinium thiocyanate from crystallizing out of solution at -20°C.
- The re-precipitation of the RNA helps to eliminate already-denatured ribonuclease from the nucleic acid pellets.
- It is convenient to decrease the volumes of the successive precipitations to concentrate RNA.

Guanidinium thiocyanate and hydrochloride–ethanol

Chirgwin et al. (14) proposed a method (Protocol 6) to prepare intact RNA from rat pancreas (rich in ribonuclease) by efficient homogenization of tissue samples in 4 M guanidinium thiocyanate containing 0.1 M 2-mercaptoethanol to disrupt protein disulfide bonds. RNA is isolated free of proteins by precipitation with ethanol acidified with 0.1 M acetic acid. The RNA pellet is dissolved in 7.5 M guanidinium hydrochloride, and RNA is re-precipitated with ethanol. Operations are repeated a second time, and RNA is washed again with ethanol. Different anecdotes of this procedure are given in Table 1. Because of high concentrations of both RNase and RNA in rat pancreas, polyanionic competitive inhibitors of RNase (heparin, protein inhibitors of RNase) cannot be practically used because of the high levels required for good efficiency. Compared with a method using phenol (plus SDS), this method appeared more efficient to denature RNase structure, therefore preventing a massive degradation of pancreatic RNA during its extraction.

Han et al. (25) proposed to improve the original method of Chirgwin et al. (14), which, even in near-saturating concentrations, does not completely inhibit RNase activity (Protocol 7). Rat pancreas tissue is quickly homogenized in guanidinium thiocyanate solution at 0°C, and RNA is rapidly precipitated by ethanol at -10°C. This treatment eliminates the bulk of RNase activity and makes the remaining steps less critical. Efficiency of this method is compared to that of guanidinium thiocyanate methods including CsCl centrifugation (14), ethanol precipitation (14) and LiCl precipitation (12). All methods are satisfactory for the isolation of RNA from tissues that exhibit moderate RNase activity (e.g., spleen). However, the ultracentrifugation method is deficient when RNA is isolated from tissues exibiting high or extremely high RNase activity (e.g., rat embryonic and adult pancreas). Following purification of mRNA by chromatography on oligothymidylic acid cellulose, mRNA is mainly lost with the method of RNA precipitation by ethanol but is still present with the LiCl method and especially with the low-temperature method.

Guanidinium thiocyanate–LiCl

Cathala et al. (12) have described a method for RNA preparation that combines advantages of direct precipitation by LiCl and homogenization of tissue in guanidinium thiocyanate (Protocol 8). Homogenizing treatment by guanidinium thiocyanate allows prevention of RNA against degradation, which otherwise can occur within the first seconds of homogenization. This ensures complete cell or tissue solubilization that minimizes both DNA contamination and RNA losses. RNA is precipitated with 4 M LiCl, and remaining denatured proteins and contaminating DNA are removed by further washes in 3 M LiCl and 2 M LiCl, 4 M urea.

This method is especially suitable for simultaneous preparations of several RNA samples from small quantities of tissue or cell samples. It is a useful method for isolation of large RNA species (up to 10 kb). However, RNA precipitation by LiCl results in losses of small-size RNA species (less than 300 nucleotides) such as tRNA, 5S RNA and snRNA.

Efficiency of RNA extraction by the guanidinium thiocyanate–LiCl method is similar to that of methods using guanidinium thiocyanate–CsCl or guanidinium thiocyanate–ethanol. However, precipitation of RNA by LiCl yielded more poly(A+) RNA (12).

Guanidinium thiocyanate–phenol–chloroform

A single-step method for RNA isolation by extraction with guanidinium thiocyanate–phenol–chloroform is proposed by Chomczynski and Sacchi (15) (Protocol 9). Addition of 4 M guanidinium thiocyanate until the last step of extraction prevents RNA degradation by inhibiting ribonuclease. This method can be used for RNA preparations from both small (3 mg tissue or 10^6 cells) and large (30 g tissue) quantities of tissue samples. It provides high yields of RNA with a high degree of purity (only traces of contaminating proteins, undetectable DNA) and nondegraded as shown for HMW thyroglobulin mRNA (8.5 kb) isolated from rat FRTL-5 cells (15). Because of its simplicity and rapidity, this method allows simultaneous processing of a large number of samples. Moreover, intensity of degradation and losses of RNA are minimized by limited handlings of this method.

Comparison of methods of RNA isolation by guanidinium thiocyanate–phenol–chloroform (15) and by ultracentrifugation of guanidinium lysate through a CsCl cushion (14) showed less contamination of RNA by proteins, higher yield of RNA recovery and a greater amount of LMW RNA with the first method than with the guanidinium–CsCl method.

Advantages and disadvantages of the different methods proposed for RNA extraction from mammalian tissues described here are summarized in Table 2. Selection of the efficient method will depend also on the precise requirements of the experiment (representation and size of RNA, number of samples etc.) and on the characteristics of the biological material (quantity and origin of tissue samples, tissue RNase activity etc.).

Extraction and Analysis of Large mRNA

Thyroglobulin mRNA

Thyroglobulin (Tg) is the major protein produced by the thyroid gland as a precursor for thyroid hormone (T3 and T4) synthesis. It is a homodimeric glycoprotein with a molecular weight of 660 kDa, and its corresponding mRNA (8.5 kb) has a sedimentation coefficient of 33 s. Abnormal splicing of Tg mRNA is described in several thyroid diseases such as Grave's disease and thyroid carcinomas (9). Purification and molecular weight determination of Tg mRNA in bovine (51) and ovine animals (13) have been reported following isolation of RNA from thyroid polysomes using SDS–phenol–chloroform treatment. In these conditions, size distribution and molecular weight of Tg mRNA have been determined by polyacrylamide gel electrophoresis (PAGE) (33 s; 8.5 kb).

Tg gene expression and its cell regulation have been extensively studied in human thyroid tissue (9,18) and in rat thyroid follicular cell line FRTL-5 (19,48) to understand molecular mechanisms involved in carcinomas or hyperthyroidism in humans (1). All of these studies used the method of Chomczynski and Sacchi (15) to extract total thyroid RNA and demonstrated by Northern blot analysis (Figure 1) complete integrity of extracted Tg mRNA. Only a few studies on hormonal control (11) or methimazole regulation (therapy used in Grave's hyperthyroidism) (29) of Tg gene expression have proposed alternatively the method of Chirgwin et al. (14) to extract total RNA from FRTL-5 cells. In the same way, Northern blot analysis confirmed integrity of Tg mRNA thus extracted (Figure 1).

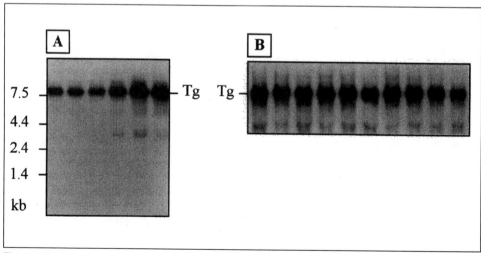

Figure 1. Northern blot analysis of thyroglobulin mRNA extracted using the method of Chirgwin et al. (14) (A) or Chomczynski and Sacchi (15) (B). (A) From Isozaki et al. (29). (B) From Tang et al. (48).

Table 2. Advantages and Disadvantages of Different Methods of RNA Extraction

RNA Extraction	Advantages	Disadvantages
Phenol	• rapid procedure	• impurities in phenol • numerous steps of shacking and transfer • aggregation of RNA
LiCl–Urea	• simple procedure • removal of heparin • strong denaturation of proteins and polysaccharides by urea • low contamination by proteins • no aggregation of RNA	• necessity of high RNA amounts • loss of low-molecular-weight RNA
CsCl	• small amount of starting material • very effective for eliminating nucleases, other proteins and DNA without use of RNase inhibitors and DNase	• time-consuming
Guanidinium Thiocyanate–Ethanol	• denaturation of RNase within the first seconds of homogenization • action of guanidinium thiocyanate enhanced by N-lauroylsarcosine and 2-mercaptoethanol	• time-consuming
Guanidinium Thiocyanate–CsCl	• small amounts of starting material • very effective for eliminating nucleases, other proteins and DNA without use of RNase inhibitors and DNase	• time-consuming
Guanidinium Thiocyanate–LiCl	• treatment of a large number of samples • small amounts of tissue or cells • high-molecular-weight RNA detectable	• loss of low-molecular-weight RNA species
Guanidinium Thiocyanate–Phenol–Chloroform	• treatment of a large number of samples • small and large amounts of starting tissue or cells • no detection of DNA • detection of high-molecular-weight RNA	• impurities in phenol • aggregation of RNA

Fatty Acid Synthase mRNA

Fatty acid synthase (FAS) is a multifunctional enzyme that catalyzes the seven reactions implicated in biosynthesis of long-chain fatty acids from acetyl coenzyme A (CoA) and malonyl CoA. FAS is present in animal tissues as a homodimer (average mol wt 500 kDa) produced by an 18-kb gene. In the rat, two equally abundant FAS mRNAs (8.1–8.8 and 8.8–9.5 kb) are produced from the FAS gene (3). Hormonal and nutritional regulations of FAS mRNA are studied mostly in adipose tissues (20), liver (39) and lung (40,52) of rats. In all cases cited above, tissue total RNA is extracted according to the method of Chomczynski and Sacchi (15). However, in a few studies, FAS mRNA is extracted from rat liver (4) or from liver or adipose tissues of mice (46) by the method of LiCl precipitation (5) or CsCl ultracentrifugation (14). Whatever method of RNA extraction was used in these studies, the degree of integrity of FAS mRNA checked by Northern blot analysis (Figure 2) is correct.

Insulin-Like Growth Factor I mRNA

Human insulin-like growth factor I (IGF-1), a polypeptide of 77 amino acids, shows a structural homology with proinsulin and shares several biological activities with insulin. IGF-1, a potent mitogen involved in growth of skeletal muscle, plays a fundamental stimulating role in postnatal growth. Liver is the major source of IGF-1. In rat liver, IGF-1 mRNA exists as two predominant size classes of 7.5–7.0 and 1.2–0.9 kb (26). The complete nucleotide sequence of the HMW human IGF-1 mRNA has been reported by Steenberg et al. (47) with RNA extracted from liver and uterus according to the method of Chirgwin et al. (14). The half-life of IGF-1 mRNA in rat liver (4 h for the 7.5–7.0-kb mRNA and 14 h for the 1.2–0.9-kb mRNA) was determined by Hepler et al. (26) using the method of Chirgwin et al. (14) for extracting total RNA. Studies on the regulation of IGF-1 gene expression are realized with tissue RNA isolated with various methods of RNA extraction. Indeed, in studies on rat or mouse osteoblasts or muscles, the methods of Chomczynski and Sacchi (10,17,43), Chirgwin et al. (28,34) and Cathala et al. (35) are used. In every case, an apparent integrity of mRNA is noted by Northern blot analysis (Figure 3). However, no direct comparison of the integrity of IGF-1 mRNA extracted by these different methods is done, preventing any recommendations.

Apolipoprotein B mRNA

Apolipoprotein B (apoB) represents a major structural component of plasma triglyceride-rich lipoproteins (chylomicrons, very low-density lipoproteins [VLDLs]). It plays a crucial role in the assembly and secretion of VLDLs by small intestine and liver, and in tissue uptake of low-density lipoproteins (LDLs) as a ligand for the LDL receptor (21). ApoB is encoded by a single-copy gene in the genome of mammals leading to synthesis of a 14.1-kb apoB mRNA. Most of the studies on the regulation of the liver biosynthesis of apoB are performed on rat hepatocytes or human hepatoma cell line HepG2 in culture as cell models that exhibit a high capacity for apoB synthesis (2). In ruminant animals, plasma

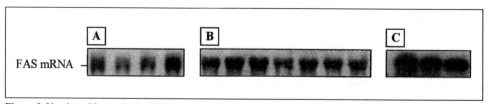

Figure 2. Northern blot analysis of fatty acid synthase mRNA extracted using the method of Chomczynski and Sacchi (15) (A), Auffrey and Rougeon (5) (B) or Chirgwin et al. (14) (C). (A) From Foufelle et al. (20). (B) From Asins et al. (4). (C) From Soncini et al. (46).

concentrations of apoB-containing lipoproteins are low because of the low level of lipids in the diet and of the low capacity of the liver to synthesize and secrete apoB as a part of VLDL particles (8).

Recently, 4 methods for isolating large mRNA encoding for hepatic bovine apoB were compared for effectiveness (23). Total RNA of liver (positive control) and lung (negative control) samples taken from bovine and rat animals were extracted simultaneously according to the methods of Chirgwin et al. (14), Chomczynski and Sacchi (15), Cathala et al. (12) and RNAzol™ (distributed by Bioprobe Systems, Montreuil-sous-Bois, France). The method of Chomczynski and Sacchi (15) represents the classic method frequently used for the extraction of numerous types of tissue mRNA. The RNAzol method is proposed because of its rapidity. In contrast, the method of Cathala et al. (12) is frequently used for isolating HMW RNAs. Finally, the method of Chirgwin et al. (14) is recommended for delicate extractions of RNA from tissues rich in ribonucleases.

Although electrophoresis on a 1% agarose gel and staining with ethidium bromide demonstrate the integrity of total RNA extracted by the four methods, yield of total RNA extraction is 2- to 4-fold higher for all samples with the methods of Chirgwin et al. (14) and RNAzol than with the two other methods (Table 3). These results show the importance of the choice of RNA isolation procedure in the case of low-represented mRNA. Indeed, a low yield of extraction can induce up to the total loss of the mRNA of interest.

By Northern blot analysis, a single band corresponding to apoB mRNA (14 kb) is shown in both bovine and rat livers but only with the method of Chirgwin et al. (14) (Figure 4). This method appears to be the more appropriate technique, probably because of the introduction of inhibitors of cellular RNases at high levels in the different solutions used for RNA extraction. Moreover, it should not be excluded that purification of RNA is improved by mild treatment with ethanol in the presence of sodium acetate. The RNAzol technique is frequently used for the isolation of hepatic apoB mRNA, which is abundant in rats and humans. However, this method probably leads to deep damage of the mRNA structure of bovine apoB because such mRNA is only visualized in the rat liver. Indeed, in rats and humans, the apoB gene is probably expressed more intensively than in the bovine species, and slight damage of rat apoB mRNA is not sufficient to suppress the presence of

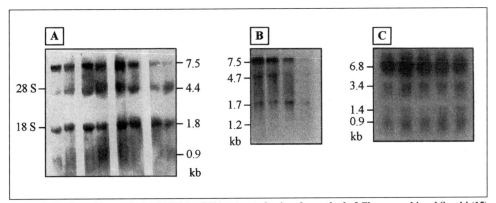

Figure 3. Northern blot analysis of IGF-1 mRNA extracted using the method of Chomczynski and Sacchi (15) (A), Chirgwin et al. (14) (B) or Cathala et al. (12) (C). (A) From Scharla et al. (43). (B) From Lund et al. (34). (C) From MacCarthy et al. (35).

Table 3. Total RNA Levels from Liver and Lung Tissues in Preruminant Calves and Adult Rats Extracted by Four Different Methods

	RNA Extraction Efficiency			
	Chirgwin et al.	Chomczynski and Sacchi	Cathala et al.	RNAzol
	mg/g fresh tissue			
Calf liver	2.48 ± 0.18	0.96 ± 0.19	1.43 ± 0.06	2.46 ± 0.17
Calf lung	1.31 ± 0.11	1.05 ± 0.84	0.55 ± 0.42	1.58 ± 0.16
Rat liver	3.90 ± 0.68	0.43 ± 0.14	0.55 ± 0.44	4.15 ± 0.48
Rat lung	1.22 ± 0.27	0.39 ± 0.09	0.32 ± 0.08	1.80 ± 0.02

From Gruffat et al. (23).
Each RNA level is the mean of 3 extractions ± standard error.

apoB mRNA on nylon membranes used in Northern blot analysis.

To verify whether results obtained by Northern blot analysis with the cDNA probe of a large-size mRNA (apoB mRNA, 14 kb) might be applied to the majority of mRNAs (which are of medium size, <3 kb), Northern blot membranes are

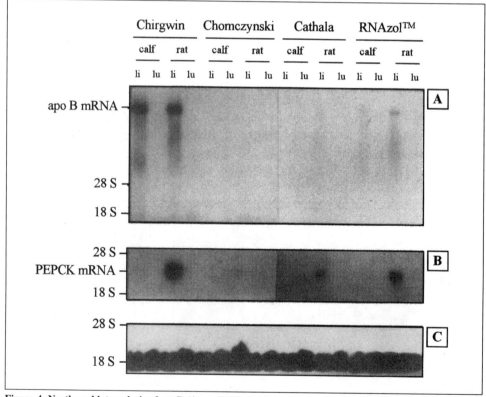

Figure 4. Northern blot analysis of apoB (A) and PEPCK (B) mRNA and 18S RNA (C) extracted using the methods of Chirgwin et al. (14), Chomczynski and Sacchi (15), Cathala et al. (12) and RNAzol. From Gruffat et al. (23). li: liver; lu: lung.

also hybridized with a cDNA probe of rat phosphoenolpyruvate carboxykinase (PEPCK) mRNA (Figure 4). This probe cross-reacts with a 2.6-kb mRNA corresponding to the length of PEPCK mRNA only in rat liver. Despite the relatively short length of this mRNA, Northern blot analysis reveals a band with a high intensity for rat liver RNA only with the methods of Chirgwin et al. (14) and RNAzol. These results indicate that whatever the length of the mRNA studied, the choice of the method of RNA extraction appears to be crucial if a sensitive method is needed. Indeed, although the four methods are suitable for PEPCK mRNA analysis, they differ essentially by their relative sensitivity.

PROTOCOLS

Protocol 1: RNA Extraction by Phenol–Chloroform Procedure (45)

Materials and Reagents

- 0.35 M sucrose, 50 mM KCl, 10 mM magnesium acetate, 1.5% Triton® X-100, 0.2 M Tris-acetate (pH 8.5), 6 mM 2-mercaptoethanol
- SDS
- EDTA
- Phenol–chloroform (1:1, vol/vol)
- 0.1 M sodium acetate, 2 mM EDTA, 0.1 M Tris-acetate (pH 9.0)
- Potassium acetate (pH 5.5)
- Ethanol
- 3 M sodium acetate (pH 6.0)
- 0.1 M sodium acetate

Procedure

1. Homogenize the tissue sample in 15 vol of cold buffer containing 0.35 M sucrose, 50 mM KCl, 10 mM magnesium acetate, 1.5% Triton X-100, 0.2 M Tris-acetate (pH 8.5) and 6 mM 2-mercaptoethanol.
2. Centrifuge at 2000× g for 10 min to pellet nuclei.
3. Collect the supernatant, add SDS to 1% and EDTA to 2 mM and bring to room temperature.
4. Shake for 10 min with 2 vol of phenol–chloroform (1:1, vol/vol).
5. Centrifuge at 10 000× g for 10 min at 20°C.
6. Collect the aqueous phase and store on ice (do not contaminate with interface material).
7. Shake the phenol phase and interface materials for 10 min with an equal volume of 0.1 M sodium acetate, 2 mM EDTA and 0.1 M Tris-acetate (pH 9.0).
8. Centrifuge as in step 5.
9. Combine the aqueous layer with the previous aqueous phase (from step 6) and re-extract them by shaking with an equal volume of phenol–chloroform (1:1, vol/vol).
10. Centrifuge as in step 5.
11. Collect the aqueous phase. Add potassium acetate (pH 5.5) to 0.2 M and

precipitate RNA by adding 2.5 vol of cold (-20°C) ethanol. Mix and leave at -20°C for 16 h.
12. Wash precipitates two or three times with 3 M sodium acetate (pH 6.0) followed by 70% ethanol containing 0.1 M sodium acetate.
13. Decant the supernatant and dry the pellet in a gentle stream of nitrogen.
14. Dissolve RNA in water or buffer.

Protocol 2: RNA Extraction by LiCl Procedure (6)

Materials and Reagents

- 1 mM $MgCl_2$
- 4 M LiCl
- 2 M LiCl

Procedure

1. Lyse cells with 3 vol of 1 mM $MgCl_2$.
2. Add an equal volume of 4 M LiCl.
3. Store the mixture at 4°C for 16 h to allow the formation of a white gelatinous precipitate.
4. Centrifuge at 5000× g for 5 min.
5. Wash twice by resuspension in 2 M LiCl followed by recentrifugation.

Protocol 3: RNA Extraction by LiCl–Urea Procedure (5)

Materials and Reagents

- 3 M LiCl, 6 M urea
- 10 mM Tris-HCl (pH 7.6), 0.5% SDS
- Chloroform–isoamyl alcohol (24:1, vol/vol)
- 2 M potassium acetate, pH 5.0
- Ethanol

Procedure

1. Homogenize the tissue sample (1 g) in 10 mL of 3 M LiCl, 6 M urea for 2 min at full speed in a homogenizer.
2. Keep the homogenate overnight at 0°C.
3. Centrifuge at 16 000× g for 20 min at 0°C and discard the supernatant.
4. Dissolve the pellet in 0.5% SDS, 10 mM Tris-HCl (pH 7.6) (10 mL/g of tissue).
5. Mix this solution by shaking it for 10 min with an equal volume of chloroform–isoamyl alcohol (24:1, vol/vol).
6. Centrifuge the mixture at 10 000× g for 10 min at 18°C in glass tubes using a

Isolation of RNA from Mammalian Cells

swing-out rotor.
7. Collect the aqueous phase.
8. Add 0.1 vol of 2 M potassium acetate (pH 5.0) and 2 vol of ethanol, mix and leave overnight at -20°C.
9. Centrifuge at 13 000× g for 20 min at -10°C.
10. Discard the supernatant, resuspend the pellet in 95% ethanol and recentrifuge as above.
11. Dry the pellet and dissolve it in water or buffer.

Protocol 4: RNA Extraction by CsCl Ultracentrifugation Procedure (22)

Materials and Reagents

- 0.1 M Tris-HCl (pH 8.0), 4% *N*-lauroylsarcosine (wt/vol)
- CsCl solid
- 5.7 M CsCl, 0.1 M EDTA
- 4% *N*-lauroylsarcosine
- 80% ethanol, 0.2 M NaCl

Procedure

1. Homogenize the cell samples at 0°C with a homogenizer in 3–10 vol of 0.1 M Tris-HCl (pH 8.0) and 4% *N*-lauroylsarcosine (wt/vol).
2. Keep the homogenate in an ice bath during intermittent homogenization for 10–15 min.
3. Add 1 g/mL of solid CsCl and mix until complete dissolution.
4. Layer the homogenate (4 mL) onto 1.2 mL of a 5.7 M CsCl and 0.1 M EDTA cushion (density = 1.707 g/L) in a cellulose nitrate centrifuge tube and layer on the top with 4% *N*-lauroylsarcosine.
5. Centrifuge in a Model SW 50.1 rotor (Beckman Instruments, Fullerton, CA, USA) at 35 000 rpm for 12 h at 25°C.
6. Remove the DNA banding at the interface of the two CsCl solutions with a Pasteur pipet.
7. Invert the tube to obtain the clear pellet corresponding to RNA.
8. Dissolve the pellet in buffer or precipitate it with 2.5 vol of 80% ethanol, 0.2 M NaCl.

Protocol 5: RNA Extraction by Guanidinium Thiocyanate–CsCl Procedure (41)

Materials and Reagents

- 4 M guanidinium thiocyanate, 0.1 M 2-mercaptoethanol, 25 mM sodium citrate (pH 7.0)
- 5.7 M CsCl, 0.1 M EDTA (pH 7.0)
- 2 M potassium acetate
- Ethanol

Procedure

1. Homogenize the tissue sample (1 g) at room temperature in 22 mL of 4 M guanidinium thiocyanate, 0.1 M 2-mercaptoethanol, 25 mM sodium citrate (pH 7.0).
2. Centrifuge at 25 000× g for 10 min at 20°C.
3. Layer a 3.6-mL aliquot of supernatant over 1.2 mL of 5.7 M CsCl, 0.1 M EDTA (pH 7.0).
4. Centrifuge for 19–22 h at 40 000 rpm and at 20°C in a Beckman SW 50.1 rotor.
5. Resuspend the pellet in 40 mL of sterile water and mix well at 55°C for several minutes.
6. Precipitate RNA by adding 1/10 vol of 2 M potassium acetate and 2.5 vol of ethanol and leave overnight at -20°C.

Protocol 6: RNA Extraction by Guanidinium Thiocyanate–Ethanol Procedure (14)

Materials and Reagents

- 4 M guanidinium thiocyanate, 0.5% sodium *N*-lauroylsarcosine (wt/vol), 0.1 M 2-mercaptoethanol, 0.1% Antifoam A (vol/vol) (Sigma Chemical, St. Louis, MO, USA), 25 mM sodium citrate (pH 7.0)
- 1 M acetic acid
- Ethanol
- 7.5 M guanidinium chloride, 5 mM dithiothreitol (DTT), 25 mM sodium citrate (pH 7.0)

Procedure

1. Homogenize the tissue sample (1 g) rapidly and thoroughly for 30–60 s at 20°C using a high-speed motor-driven pestle in 16 mL of 4 M guanidinium thiocyanate, 0.5% sodium *N*-lauroylsarcosine (wt/vol), 0.1 M 2-mercaptoethanol, 0.1% Antifoam A (vol/vol) and 25 mM sodium citrate (pH 7.0).
2. Centrifuge at 8500× g for 10 min at 10°C.
3. Collect the supernatant, add 0.025 vol of 1 M acetic acid (to lower pH from 7.0 to 5.0) and 0.75 vol of ethanol; shake thoroughly and leave at -20°C overnight.
4. Centrifuge at 5000× g for 10 min at -10°C.
5. Drain off the supernatant and resuspend the pellet vigorously by homogenizing in 0.5 vol of 7.5 M guanidinium chloride, 5 mM DTT and 25 mM sodium citrate (pH 7.0) (brief warming to 68°C might help).
6. Re-precipitate the RNA by adding 0.025 vol of 1 M acetic acid and 0.5 vol ethanol. Leave the mixture at -20°C for at least 3 h.
7. Centrifuge at 5000× g for 10 min at -10°C.
8. Repeat steps 5–7 using half the volumes previously used.
9. Remove the supernatant, resuspend the pellet thoroughly in ethanol at room temperature and centrifuge at 5000× g for 5 min at -10°C.
10. Drain the pellet and dry it under a gentle stream of nitrogen.
11. Dissolve the RNA in cold water (1 mL/g of tissue) and clarify the solution by

centrifugation at 22 500× g for 10 min at 10°C.

12. Save the supernatant and resuspend the pellet in 0.5 mL of water per gram of tissue, centrifuge at 22 500× g for 10 min at 10°C and recover the supernatant.
13. Combine the supernatants from step 12, add 0.1 vol of 2 M potassium acetate (pH 5.0) and 2 vol of ethanol, mix and then leave overnight at -20°C.
14. Centrifuge the mixture at 13 000× g for 20 min at -10°C.
15. Discard the supernatant, resuspend the pellet in 95% ethanol and re-centrifuge at 13 000× g for 20 min at -10°C.
16. Dry the pellet with a stream of nitrogen and dissolve it in water or buffer.

Protocol 7: RNA Extraction by Guanidinium Thiocyanate–Ethanol (25)

Materials and Reagents

- 5 M guanidinium thiocyanate, 50 mM Tris-HCl (pH 7.5), 25 mM EDTA, 8% 2-mercaptoethanol (added just prior to use)
- Ethanol
- 1 M acetic acid
- 6 M guanidinium hydrochloride, 25 mM EDTA (pH 7.0), 10 mM 2-mercaptoethanol (added just prior to use)
- 2 M sodium acetate
- DEPC-treated water

Procedure

1. Homogenize the tissue sample (1 g) in 20 mL of 5 M guanidinium thiocyanate, 50 mM Tris-HCl (pH 7.5), 25 mM EDTA and 8% 2-mercaptoethanol (precooled on ice for 5 min) in a precooled polypropylene tube with an ice-chilled Polytron® (Brinkmann Instruments, Westbury, CT, USA) for 20 s at maximum speed.
2. Add 0.3 vol of cold ethanol (-20°C), mix and immediately centrifuge in an HB4 rotor (Sorvall, Newtown, CT, USA) (precooled to -10°C) at 10 000 rpm for 5 min.
3. Remove the protein film at the top of the fluid and discard the remaining supernatant by aspiration while the tube is maintained at -10°C.
4. Add 10 mL of ice-cold 5 M guanidinium thiocyanate, 50 mM Tris-HCl (pH 7.5), 25 mM EDTA and 8% 2-mercaptoethanol to the pellet and homogenize for 10 s as above.
5. Centrifuge the suspension at 10 000 rpm in the HB4 rotor (3 min, 0°C) to remove contaminating tissue debris.
6. Mix the clear supernatant with 0.025 vol of 1 M acetic acid and 0.75 vol of cold ethanol (-20°C), incubate at -20°C for longer than 3 h and centrifuge at 6000 rpm in the HB4 rotor for 10 min at -10°C.
7. Discard the supernatant and resuspend the pellet in 2 mL of cold 6 M guanidinium hydrochloride (-20°C), 25 mM EDTA and 10 mM 2-mercaptoethanol.
8. Adjust the volume to 10 mL with 6 M guanidinium hydrochloride, 25 mM EDTA and 10 mM 2-mercaptoethanol, transfer to polypropylene tubes and

vortex mix.

9. Add 0.25 mL of 1 M acetic acid and 0.5 vol of cold ethanol (-20°C), mix, incubate at -20°C for more than 3 h and centrifuge at 6000 rpm in the HB4 rotor for 10 min at -10°C.
10. Repeat steps 7–9 twice and resuspend the RNA pellet in progressively decreasing volumes of guanidinium hydrochloride (finally 5 mL).
11. Dissolve the final RNA pellet in 5 mL DEPC-treated water, mix with 0.1 vol of 2 M sodium acetate and 2.5 vol of ethanol.

Protocol 8: RNA Extraction by Guanidinium Thiocyanate–LiCl Procedure (12)

Materials and Reagents

- 5 M guanidinium thiocyanate, 10 mM EDTA, 50 mM Tris-HCl (pH 7.5), 8% 2-mercaptoethanol (vol/vol)
- 4 M LiCl
- 3 M LiCl
- 2 M LiCl, 4 M urea
- 0.1% SDS, 1 mM EDTA, 10 mM Tris-HCl (pH 7.5)
- Phenol
- Chloroform
- Ammonium acetate (saturated)
- Ethanol

Procedure

1. Homogenize the tissue (1 g) or cell (1 mL) samples in 7 mL of 5 M guanidinium thiocyanate, 10 mM EDTA, 50 mM Tris-HCl (pH 7.5), 8% 2-mercaptoethanol (vol/vol) (added just before use) by vortex mixing 3 times for 10 s (for cells) or with a Polytron blender (for tissue).
2. Add 7 vol of 4 M LiCl and leave the mixture for 15–20 h at 4°C.
3. Centrifuge at 11 000× g at 4°C for 90 min in a swing-out rotor.
4. Discard the supernatant and resuspend RNA pellet with vortex mixing in 3 M LiCl.
5. Centrifuge at 11 000× g at 4°C for 1 h in a swing-out rotor.
6. Wash the pellet with 2 M LiCl and 4 M urea and centrifuge at 11 000× g at 4°C for 1 h.
7. Resuspend RNA pellet in 0.1% SDS, 1 mM EDTA, 10 mM Tris-HCl (pH 7.5) for 45 min with vortex mixing for 20 s every 10 min.
8. Extract RNA with phenol–chloroform (1:1, vol/vol).
9. Precipitate RNA by adding 0.05 vol of saturated ammonium acetate and 2 vol of ethanol at -20°C.

Protocol 9: RNA Extraction by Guanidinium Thiocyanate–Phenol–Chloroform Procedure (15)

Materials and Reagents

- 4 M guanidinium thiocyanate, 25 mM sodium citrate (pH 7.0), 0.5% *N*-lauroyl-

sarcosine, 0.1 M 2-mercaptoethanol
- 2 M sodium acetate (pH 4.0)
- Phenol (water-saturated)
- Chloroform–isoamyl alcohol mixture (49:1)
- Isopropanol
- Ethanol
- 0.5% SDS

Procedure

1. Homogenize the tissue or cell samples (100 mg) with 1 mL of 4 M guanidinium thiocyanate, 25 mM sodium citrate (pH 7.0), 0.5% *N*-lauroylsarcosine, 0.1 M 2-mercaptoethanol in a glass Teflon® homogenizer at room temperature.
2. Add sequentially 0.1 mL of 2 M sodium acetate (pH 4.0), 1 mL of phenol and 0.2 mL of chloroform–isoamyl alcohol mixture (49:1) with thorough mixing by inversion after the addition of each reagent. Shake vigorously the final suspension for 10 s and cool on ice for 15 min.
3. Centrifuge at $10\,000 \times g$ for 20 min at 4°C.
4. Transfer the aqueous phase to a fresh tube, mix it with 1 mL of isopropanol and then place at -20°C for at least 1 h to precipitate RNA.
5. Centrifuge at $10\,000 \times g$ for 20 min.
6. Dissolve the resulting RNA pellet in 0.3 mL of denaturing solution and precipitate with 1 vol of isopropanol at -20°C for 1 h.
7. Centrifuge at $10\,000 \times g$ for 10 min at 4°C.
8. Resuspend the RNA pellet in 75% ethanol. Centrifuge at $10\,000 \times g$ for 10 min at 6°C. Vacuum dry the pellet for 15 min and dissolve in 50 µL of 0.5% SDS at 65°C for 10 min or precipitate with 2 vol of ethanol.

RESULTS AND DISCUSSION

Numerous methods of RNA isolation have been proposed in the past two decades, but the methods that are the most currently used are those of Chomzcynski and Sacchi (guanidinium thiocyanate–phenol–chloroform), Chirgwin et al. and Han et al. (guanidinium thiocyanate–ethanol) and, to a lesser extent, those of Auffrey and Rougeon and Cathala et al. (guanidinium thiocyanate–LiCl). However, research workers frequently used these different methods without specifying the reasons for their choice.

Several suggestions can be proposed to help choose the RNA isolation method best adapted to their case.

Before the first assay of RNA isolation, one must take into account: *(i)* the size of the mRNA to study, *(ii)* the predicted representation of this mRNA according to the animal species and the tissue, *(iii)* the quantity of tissue available, *(iv)* the potential RNase activity of tissue and *(v)* the number of samples to extract.

The answers to these different criteria and the comparison with the advantages and disadvantages of the different techniques proposed could help in the choice of a few methods that seem to be the most adequate (Table 2).

Before making a final and judicious choice, these methods must be tested on the same tissue samples to verify: *(i)* the integrity of total RNA extracted, *(ii)* the yield of RNA extraction and *(iii)* the integrity of the mRNA studied.

ACKNOWLEDGMENTS

I would like to thank Dr. D. Bauchart and B. Graulet for critical reading of the manuscript.

REFERENCES

1. **Acebron, A., P. Aza-Blanc, D.L. Rossi, L. Lamas and P. Santisteban.** 1995. Congenital human thyroglobulin defect due to low expression of the thyroid-specific transcription factor TTF-1. J. Clin. Invest. *96*:781-785.
2. **Adeli, K., A. Mohammadi and J. Macri.** 1995. Regulation of apolipoprotein B biogenesis in human hepatocytes: posttranscriptional control mechanisms that determine the hepatic production of apolipoprotein B-containing lipoproteins. Clin. Biochem. *28*:123-130.
3. **Amy, C.M., B. Williams-Ahlf, J. Naggert and S. Smith.** 1990. Molecular cloning of the mammalian fatty acid synthase gene and identification of the promoter region. Biochem. J. *271*:675-679.
4. **Asins, G., J.L. Rosa, D. Serra, G. Gil-Gomez, J. Ayté, R. Bartrons, A. Tauler and F.G. Hegardt.** 1994. Gene expression of enzymes regulating ketogenesis and fatty acid metabolism in regenerating rat liver. Biochem. J. *299*:65-69.
5. **Auffrey, C. and F. Rougeon.** 1980. Purification of mouse immunoglobulin heavy-chain messenger RNAs from total myeloma tumor RNA. Eur. J. Biochem. *107*:303-314.
6. **Barlow, J.J., A.P. Mathias and R. Williamson.** 1963. A simple method for the quantitative isolation of undegraded high molecular weight ribonucleic acid. Biochem. Biophys. Res. Commun. *13*:61-66.
7. **Barnard, E.A.** 1964. The unfolding and refolding of ribonuclease in urea solutions. J. Mol. Biol. *10*:235-262.
8. **Bauchart, D.** 1993. Lipid absorption and transport in ruminants. J. Dairy Sci. *76*:3864-3881.
9. **Bertaux, F., M. Noel, Y. Malthiery and P. Fragu.** 1991. Demonstration of a heterogeneous transcription pattern of thyroglobulin mRNA in human thyroid tissues. Biochem. Biophys. Res. Commun. *178*:586-592.
10. **Birnbaum, R.S., R.R. Bowsher and K.M. Wiren.** 1995. Changes in IGF-I and -II expression and secretion during the proliferation and differentiation of normal rat osteoblasts. J. Endocrinol. *144*:251-259.
11. **Bone, E., L.D. Kohn and P. Chomczynski.** 1986. Thyroglobulin gene activation by thyrotropin and cAMP in hormonally depleted FRTL-5 thyroid cells. Biochem. Biophys. Res. Commun. *141*:1261-1266.
12. **Cathala, G., J.F. Savouret, B. Mendez, B.L. West, M. Karin, J.A. Martial and J.D. Baxter.** 1983. A method for isolation of intact, translationally active ribonucleic acid. DNA *2*:329-335.
13. **Chebath, J., O. Chabaud, J.L. Bergé-Lefranc, G. Cartouzou and S. Lissitzky.** 1977. Molecular weight of the thyroglobulin messenger RNA of sheep thyroid gland. Biochem. Biophys. Res. Commun. *79*:267-273.
14. **Chirgwin, J.M., A.E. Przybyla, R.J. MacDonald and W.J. Rutter.** 1979. Isolation of biologically active ribonucleic acid from sources enriched in ribonuclease. Biochemistry *18*:5294-5299.
15. **Chomczynski, P. and N. Sacchi.** 1987. Single-step method of RNA isolation by acid guanidinium thiocyanate-phenol-chloroform extraction. Anal. Biochem. *162*:156-159.
16. **Cox, R.A.** 1968. The use of guanidinium chloride in the isolation of nucleic acids. Methods Enzymol. *12*:120-129.
17. **Edwall, D., M. Schalling, E. Jennische and G. Norstedt.** 1989. Induction of insulin-like growth factor I messenger ribonucleic acid during regeneration of rat skeletal muscle. Endocrinology *124*:820-825.
18. **Elisei, R., A. Pinchera, C. Romei, M. Gryczynska, V. Pohl, C. Maenhaut, L. Fugazzola and F. Pacini.** 1994. Expression of thyroglobulin receptor (TSH-R), thyroglobulin, thyroperoxidase, and calcitonin messenger ribonucleic acids in thyroid carcinomas: evidence of TSH-R gene transcript in medullary histotype. J. Clin. Endocrinol. Metab. *78*:867-871.
19. **Feliciello, A., P. Giuliano, A. Porcellini, C. Garbi, S. Obici, E. Mele, E. Angotti, D. Grieco et al.** 1996. The v-ki-ras oncogene alters cAMP nuclear signaling by regulating the location and the expression of cAMP-dependent protein kinase IIb. J. Biol. Chem. *271*:25350-25359.
20. **Foufelle, F., B. Gouhot, D. Perdereau, J. Girard and P. Ferre.** 1994. Regulation of lipogenic enzyme and phosphoenolpyruvate carboxykinase gene expression in cultured white adipose tissue. Eur. J. Biochem. *223*:893-900.
21. **Gibbons, G.F.** 1990. Assembly and secretion of hepatic very-low-density lipoprotein. Biochem. J. *268*:1-13.
22. **Glisin, V., R. Crkvenjakov and C. Byus.** 1974. Ribonucleic acid isolated by cesium chloride centrifugation. Biochemistry *13*:2633-2637.
23. **Gruffat, D., C. Piot, D. Durand and D. Bauchart.** 1996. Comparison of four methods for isolating large mRNA: apolipoprotein B mRNA in bovine and rat livers. Anal. Biochem. *242*:77-83.
24. **Haines, M.E., N.H. Carey and R.D. Palmiter.** 1974. Purification and properties of ovalbumin messenger RNA. Eur. J. Biochem. *43*:549-560.

25. **Han, J.H., C. Stratowa and W.J. Rutter.** 1987. Isolation of full-length putative rat lysophospholipase cDNA using improved methods for mRNA isolation and cDNA cloning. Biochemistry *26*:1617-1625.
26. **Hepler, J.E., J.J. Van Wyk and P.K. Lund.** 1990. Different half-lives of insulin-like growth factor I mRNAs that differ in length of 3′ untranslated sequence. Endocrinology *127*:1550-1552.
27. **Hilz, H., U. Wiegers and P. Adamietz.** 1975. Stimulation of proteinase K action by denaturing agents: application to the isolation of nucleic acids and the degradation of "masked" proteins. Eur. J. Biochem. *56*:103-108.
28. **Höppener, J.W.M., S. Mosselman, P.J.M. Roholl, C. Lambrechts, R.J.C. Slebos, P. de Pagter-Holthuizen, C.J.M Lips, H.S. Jansz and J.S. Sussenbach.** 1988. Expression of insulin-like growth factor-I and -II genes in human smooth muscle tumours. EMBO J. *7*:1379-1385.
29. **Isozaki, O., T. Tsushima, N. Emoto, M. Saji, Y. Tsuchiya, H. Demura, Y. Sato, K. Shizume, S. Kimura and L.D. Kohn.** 1991. Methimazole regulation of thyroglobulin biosynthesis and gene transcription in rat FRTL-5 thyroid cells. Endocrinology *128*:3113-3121.
30. **Kirby, K.S.** 1965. Isolation and characterization of ribosomal ribonucleic acid. Biochem. J. *96*:266-269.
31. **Kirby, K.S.** 1968. Isolation of nucleic acids with phenolic solvents. Methods Enzymol. *12*:87-99.
32. **Kruh, J.** 1967. Preparation of RNA from rabbit reticulocytes and liver. Methods Enzymol. *12*:609-613.
33. **Lai, C.-C., T.H. Chiu, H.C. Rosenberg and W.-h. Huang.** 1993. Improved proteinase K digestion for the rapid isolation of mRNA from mammalian tissues. BioTechniques *15*:620-626.
34. **Lund, P.K., B.M. Moats-Staats, M.A. Hynes, J.G. Simmons, M. Jansen, A.J. D'Ercole and J.J. Van Wyk.** 1986. Somatomedin-C/insulin-like growth factor-I and insulin-like growth factor-II mRNAs in rat fetal and adult tissues. J. Biol. Chem. *261*:14539-14544.
35. **MacCarthy, T.L., M. Centrella and E. Canalis.** 1989. Parathyroid hormone enhances the transcript and polypeptide levels of insulin-like growth factor I in osteoblast-enriched cultures from fetal rat bone. Endocrinology *124*:1247-1253.
36. **MacDonald, R.J., G.H. Swift, A.E. Przybyla and J.M. Chirgwin.** 1987. Isolation of RNA using guanidinium salts. Methods Enzymol. *152*:219-227.
37. **Miller, J.F. and D.W. Bolen.** 1978. A guanidine hydrochloride induced change in ribonuclease without gross unfolding. Biochem. Biophys. Res. Commun. *81*:610-615.
38. **Nozaki, Y. and C. Tanford.** 1970. The solubility of amino acids, diglycine, and triglycine in aqueous guanidine hydrochloride solutions. J. Biol. Chem. *245*:1648-1652.
39. **Prip-Buus, C., D. Perdereau, F. Foufelle, J. Maury, P. Ferre and J. Girard.** 1995. Induction of fatty-acid-synthase gene expression by glucose in primary culture of rat hepatocytes. Eur. J. Biochem. *230*:309-315.
40. **Rami, J., W. Stenzel, S.M. Sasic, C. Puel-M'Rini, J.P. Besombes, J.A. Elias and S.A. Rooney.** 1994. Fatty-acid synthase activity and mRNA level in hypertrophic type II cells from silica-treated rats. Am. J. Physiol. *267*:L128-L136.
41. **Raymond, Y. and G.C. Shore.** 1979. The precursor for carbamyl phosphate synthase is transported to mitochondria via a cytosolic route. J. Biol. Chem. *254*:9335-9338.
42. **Rhoads, R.E.** 1975. Ovalbumin messenger ribonucleic acid. J. Biol. Chem. *250*:8088-8097.
43. **Scharla, S.H., D.D. Strong, S. Mohan, D.J. Baylink and T.A. Linkhart.** 1991. 1,25-dihydroxyvitamin D3 differentially regulates the production of insulin-like growth factor I and IGF-binding protein-4 in mouse osteoblasts. Endocrinology *129*:3139-3146.
44. **Schimke, R.T., R. Palacios, D. Sullivan, M.L. Kiely, C. Gonzales and J.M. Taylor.** 1974. Immunoadsorption of ovalbumin synthesizing polysomes and partial purification of ovalbumin messenger RNA. Methods Enzymol. *30*:631-648.
45. **Shore, G.C. and J.R. Tata.** 1977. Two fractions of rough endoplasmic reticulum from rat liver. J. Cell Biol. *72*:726-743.
46. **Soncini, M., S.F. Yet, Y. Moon, J.Y. Chun and H.S. Sul.** 1995. Hormonal and nutritional control of fatty acid synthase promoter in transgenic mouse. J. Biol. Chem. *270*:30339-30343.
47. **Steenberg, P.H., A.M.C.B. Koonen-Reemst, C.B.J.M. Cleutjens and J.S. Sussenbach.** 1991. Complete nucleotide sequence of the high molecular weight human IGF-I mRNA. Biochem. Biophys. Res. Commun. *175*:507-514.
48. **Tang, K.T., L.E. Braverman and W.J. DeVito.** 1995. Tumor necrosis factor-α and interferon-γ modulate gene expression of type I 5′-deiodinase, thyroid peroxidase, and thyroglobulin in FRTL-5 rat thyroid cells. Endocrinology *136*:881-888.
49. **Taylor, J.M.** 1979. The isolation of eukaryotic messenger RNA. Annu. Rev. Biochem. *48*:681-717.
50. **Ullrich, A., J. Shine, J. Chirgwin, R. Pictet, E. Tischer, W.J. Rutter and H.M. Goodman.** 1977. Rat insulin genes: construction of plasmids containing the coding sequences. Science *196*:1313-1319.
51. **Vassart, G., L. Verstreken and C. Dinsart.** 1977. Molecular weight of thyroglobulin 33S messenger RNA as determined by polyacrylamide gel electrophoresis in the presence of formamide. FEBS Lett. *79*:15-18.
52. **Xu, Z.X., W. Stenzel, S.M. Sasic, D.A. Smart and S.A. Rooney.** 1993. Glucocorticoid regulation of fatty acid synthase gene expression in fetal lung. Am. J. Physiol. *265*:L140-L147.

5 | RT-PCR without RNA Isolation

Robert J. Klebe, George M. Grant, Anne M. Grant, Melissa A. Garcia and Troy A. Giambernardi
Department of Cellular and Structural Biology, University of Texas Health Science Center, San Antonio, TX, USA

OVERVIEW

Reverse transcription polymerase chain reaction (RT-PCR) has traditionally required time-consuming RNA extraction and purification. In this article, we demonstrate that one can completely avoid the RNA extraction step in RT-PCR by basing the comparison of samples on cell number rather than micrograms of total RNA.

A new method for lysing cells while preserving RNA is described. RT-PCR is carried out by: *(i)* rapidly freezing cells in the presence of RNase inhibitor and *(ii)* using extracts of 1000 or fewer cells directly in the RT-PCR assay. Aldolase mRNA extracted by freezing cells in the presence of RNase inhibitor was found to be stable at 42°C for over 3 h. Because the RT step can be completed within 1 h, it can be carried out with minimal degradation of mRNA. This simple procedure avoids the use of harsh reagents that might inhibit enzymes involved in RT-PCR and produces results virtually identical to methods that use guanidinium thiocyanate and phenol for RNA extraction. Optimized conditions for each parameter of the procedure are described, which permit one to amplify mRNA from as few as four cells. We have shown that the procedure described can be used with minor modifications to perform RT-PCR with RNA isolated from frozen sections.

Because RNase inhibitor is the most expensive reagent in the RT-PCR protocol, we have developed a simple procedure for isolating RNase inhibitor from bovine liver. The preparative method uses a streamlined procedure for clarification of the liver extract and affinity chromatography on RNase-A-Sepharose (Sepharose® from Amersham Pharmacia Biotech, Piscataway, NJ, USA). The method presented here has been described in the primary literature (21).

BACKGROUND

The time required for extraction of total RNA often exceeds the time involved in the RT-PCR procedure itself. We report an exceptionally simple method for performing RT-PCR that involves: *(i)* lysis of cells by a rapid freeze-thaw cycle in the presence of RNase inhibitor and 5 mM dithiothreitol (DTT) and *(ii)* the use of extracts of 1000 or fewer cells directly in the RT-PCR assay. The method described entirely avoids RNA extraction and thereby eliminates the most time-consuming and error-prone step in RT-PCR.

PROTOCOLS

Cell Culture

Cells were maintained in 50% Dulbecco's modified Eagle medium (DMEM), 50% F-12 medium containing 10% newborn calf serum plus 100 U/mL penicillin, 100 µg/mL streptomycin and 50 µg/mL gentamicin sulfate. MG-63 human osteosarcoma cells were obtained from ATCC (Rockville, MD, USA). Following trypsinization, note that cells were resuspended in an isotonic saline solution just before freezing (step 3 in the Procedure section).

RT-PCR

The aldolase (7) and fibronectin (26) primer sets and the methods used have been described in detail (8). In brief, the RT step is carried out with avian myeloblastosis virus (AMV) reverse transcriptase (Life Sciences, St. Petersburg, FL, USA), and the PCR step was performed with *Taq* DNA polymerase (Promega, Madison, WI, USA). The entire RT-PCR is carried out with six stock solutions as previously described (8). AMV reverse transcriptase and *Taq* DNA polymerase are contained in two stock solutions that are frozen in unit-of-use aliquots (8). In the study presented here, human aldolase primers were used (7), except where otherwise noted. The sense primers were end-labeled with T4 kinase as described (6), and primers were separated from unreacted [γ-^{32}P]ATP with a QIAquick® Nucleotide Removal Kit (Qiagen, Chatsworth, CA, USA).

As described in the Procedure section, rapid freezing of cells was accomplished by inserting a thin-walled 0.2-mL PCR tube (USA Plastics, Ocala, FL, USA) containing a cell suspension into a pipet tip box filled with 95% ethanol that had been prechilled to -70°C. Using a pipet tip box intended for 10-µL tips, 0.2-mL tubes were approximately half submerged. Depending on the intended experiment, different volumes of cell suspension can be used as long as an equal volume of 2× RNase inhibitor solution is used. Note that cells remain viable until they are frozen. Once the cells are thawed, the RNase inhibitor, which is frozen with the cells, protects RNA from degradation.

The optimized procedure described in the text is presented in the Procedure section. The PCR step was carried out with a Model TCX20A thermal cycler (Ericomp, San Diego, CA, USA) equipped with a 96-well plate block that accepts thin-walled 0.2-mL PCR tubes. RT-PCR products of aldolase were electrophoresed on 5% polyacrylamide gels, and the single band that resulted was excised and counted with a scintillation counter. Results are reported as the average of three determinations ± standard deviation of the counts per minute (cpm) in RT-PCR products minus background.

Reagents

All chemicals were of reagent grade. RNase inhibitor (Animal Injectable Grade RNasin®, supplied as a 100 U/μL solution; Promega) was obtained as a glycerol-free solution because glycerol was shown to reduce the amount of RT-PCR product obtained (Figure 1F). Because RNase inhibitor is the most expensive reagent in this procedure, we developed a simple procedure for purifying RNase inhibitor from bovine liver (see Results and Discussion sections for a brief description of this method). Torula yeast RNA (Catalog No. R3629; Sigma Chemical, St. Louis, MO, USA) was subjected to two ethanol precipitations to remove low-molecular-weight RNA prior to its use in the RNase inhibitor assay of Blackburn et al. (3).

RT-PCR Using Frozen-Thawed Cells

Materials and Reagents

- Freezing solution: 0.15 M NaCl, 10 mM Tris, pH 8.0
- 2× RNase inhibitor stock: 0.15 M NaCl, 10 mM Tris, pH 8.0, 1 μL RNase inhibitor (100 U/μL)/20 μL, 5 mM DTT. Add 2 μL of RNase inhibitor to 18 μL of 0.15 M NaCl, 10 mM Tris, pH 8.0, 5 mM DTT. Store at -20°C in 2–10-μL aliquots. This solution should not be repetitively frozen and thawed (Table 1).
- Mixture A: This solution contains all reagents for the RT step except primers (8) and is previously mixed, separated into aliquots and frozen. Mixture A consists of 0.5 μL of 1 M Tris, pH 8.3, 0.4 μL of 1 M KCl, 0.8 μL of 0.1 M $MgCl_2$, 0.5 μL of 5 mg/mL bovine serum albumin (BSA), 0.25 μL of 0.5 M DTT, 0.25 μL of 40 U/L RNasin, 0.2 μL of 25 mM dNTP, 0.025 μL of 20 U/μL AMV reverse transcriptase and 0.275 μL of diethyl pyrocarbonate (DEPC)-treated H_2O. Note that RNase inhibitor was often deleted from the mixture A formulation in this study.
- Mixture T: This solution contains all reagents for the PCR step except primers (8) and is previously mixed, separated into aliquots and frozen. Mixture T consists of 3.2 μL of 500 mM KCl, 0.8 μL of 100 mM Tris, pH 9.0, 2.4 μL of 25 mM $MgCl_2$, 0.4 μL of 25 mM dNTP, 0.25 μL of 5 U/μL *Taq* DNA polymerase and 31.95 μL of DNase-free H_2O.
- Primers: 100 ng of sense and 200 ng antisense primer/50-μL reaction

Procedure

This procedure is for a 50-μL RT-PCR mixture.

1. Trypsinize and count cells.
2. Add 10^5 cells to a culture tube and spin at approximately 400× *g* for 4 min to gently pellet cells.
3. Resuspend cells in 0.5 mL freezing solution to adjust cell concentration to 2×10^5 cells/mL.
4. Place 10 μL of resuspended cells in a 0.2-mL microcentrifuge tube and add 10 μL of 2× RNase inhibitor stock.
   Note: Final cell concentration is now 10^5 cells/mL.
5. Freeze immediately at -70°C in a pipet tip box containing 95% ethanol

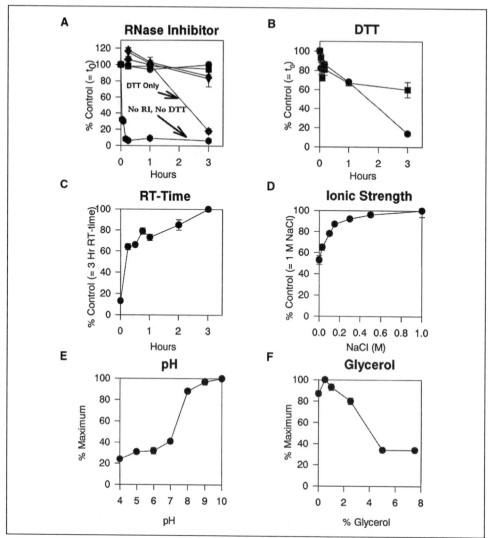

Figure 1. Optimization of variables of the RNase inhibitor method. In all cases, cells were prepared as described in the Procedure section, except that individual components in the freezing solution or 2× RNase inhibitor solution were varied as noted. In the studies presented below, the mRNA released from 250 cells was subjected to RT-PCR. The number of counts in RT-PCR products was normalized to the control indicated on the y axis of each graph. The results presented represent the mean of three trials ± standard error. (A) Concentration of RNase inhibitor in the 2× RNase inhibitor solution. In the presence of 5 mM DTT, the concentration of RNase inhibitor was varied. Extracts from 250 cells were maintained at 42°C for the times indicated prior to a 1-h RT. The 2× RNase inhibitor solutions used to prepare extracts contained the following amounts of RNase inhibitor per 20 µL of solution: 0 µL (tilted squares), 0.25 µL (inverted triangles), 0.5 µL (triangles), 1 µL (squares) and 2 µL (circles). (B) DTT effect. Cell extracts were incubated at 42°C with (squares) or without (circles) 5 mM DTT in the presence of a suboptimal amount of RNase inhibitor (0.2 µL RNase inhibitor/20 µL 2× RNase inhibitor stock). Each time point represents the amount of time that the cell lysate was incubated at 42°C prior to a 1-h RT. DTT inhibition of RNA degradation was evident only after a prolonged incubation period. (C) RT time. The effect of RT time at 42°C was studied. It was found that 64% of aldolase mRNA could be converted to cDNA within 15 min. A 1-h RT time was chosen for subsequent studies. (D) Ionic strength of the freezing solution. The effect of ionic strength on the RT-PCR was investigated using NaCl solutions of the indicated concentrations, each containing 10 mM Tris, pH 8.0. (E) pH of the freezing solution. The influence of pH of the freezing solution was examined. The pH of 0.15 M NaCl was adjusted with a buffer with broad pH range and constant ionic strength (14). The buffer contained 3.3 mM succinic acid, 4.4 mM imidazole and 4.4 mM DEA at the pHs indicated. (F) Effect of cryoprotective agents. Glycerol was found to reduce the yield of RT-PCR products (closed circles) at concentrations at which it is an active cryoprotective agent. The concentration of glycerol presented is the final glycerol concentration.

RT-PCR without RNA Isolation

(maintained in a -70°C freezer). An ethanol-filled 10-µL pipet tip box will allow a 0.2-mL PCR tube to be about 50% submerged in ethanol.

6. To a new 0.2-mL PCR tube, add 3.2 µL mixture A and 0.5 µL antisense primer (200 ng/µL); maintain at 4°C until cells are added.
7. Thaw cells rapidly in a room-temperature water bath, vortex mix for 10 s, spin tubes for a few seconds to collect all liquid in the bottom of tubes.
8. Add desired number of cells to tubes (note that 2.5 µL of 10^5 cells/mL = 250 cells etc.). Bring volume to 10 µL with DEPC-treated water.
9. Place tubes in a PCR thermal cycler preheated to 42°C; incubate for 1 h to complete the RT step.
10. Add 39 µL of mixture T and 1 µL of sense primer (100 ng/µL) to perform the PCR step at a 50-µL final volume.
11. Analyze PCR products on a 5% polyacrylamide gel.

Precautions

1. Separate the 2× RNase inhibitor stock, mixture A and mixture T into unit-of-use aliquots. Do not refreeze any unused solution. See Table 1, which describes the adverse effects of repetitive freezing and thawing on the activity of 2× RNase inhibitor stock.
2. Use of premixed stock solutions expedites analysis and is essential for ensuring reproducibility (8).
3. The following modified hot-start protocol can eliminate artifacts. False priming can be decreased and stringency increased by: *(i)* decreasing the time interval between the addition of PCR reagents (mixture T and sense primer) and placement of the tubes into the PCR thermal cycler and *(ii)* preheating the PCR cycler block to 80°C before loading the sample into the PCR tubes to reduce the amount of time that a PCR is maintained at lower temperatures at which false priming can occur.

RESULTS

That mRNA, rather than genomic DNA, is amplified in the system studied below is indicated by the fact that: *(i)* no RT-PCR product is observed when AMV reverse transcriptase is not used (data not shown), *(ii)* RNase inhibitor protects the target mRNA from degradation (Figure 1A) and *(iii)* use of fibronectin ED-A primers (26) results in two RT-PCR products representing the presence or absence of an alternatively spliced exon of the fibronectin primary transcript (data not shown).

Optimization of the RNase Inhibitor Solution

The optimized procedure for RT-PCR without RNA extraction is presented in detail in the Procedure section. As described below, each parameter of this procedure was examined by varying one parameter of the optimized procedure at a time (Figure 1). The RNase inhibitor solution used contains RNase inhibitor and DTT plus the components of the freezing solution (see Materials and Reagents section). In all

Table 1. Stability of mRNA Following Freeze-Thaw Cycles

Number of Cycles	cpm in PCR Product	% RNase Inhibitor Remaining
1	1701 ± 50	46% ± 9.3%
2	1205 ± 38	22% ± 0.3%
3	748 ± 21	12% ± 3.2%
4	188 ± 9	5% ± 0.5%
5	144 ± 15	4% ± 1.2%

Cells were frozen as described in the Procedure section and subjected to the number of freeze-thaw cycles indicated prior to RT-PCR. Radiolabeled PCR products were counted as described in the Protocols section and are presented as the mean of three determinations ± standard error. The results indicate that one can detect aldolase mRNA even after 3 freeze-thaw cycles; however, substantial degradation of aldolase mRNA was noted beyond 3 freeze-thaw cycles. RNase inhibitor was assayed according to the method of Blackburn et al. (3), and it was found that approximately 50% of the RNase inhibitor activity was lost after each freeze-thaw cycle. Hence, mRNA degradation is due to inactivation of RNase inhibitor during each freeze-thaw cycle. The amount of RNase inhibitor used here was sufficient to protect RNA after one freeze-thaw cycle but was insufficient for additional freeze-thaw cycles.

studies reported here, cells were frozen at a final concentration of 10^5 cells/mL, and most assays used 2.5 µL of cell suspension (250 cells).

In assays involving even a few hundred cells, RNA was found to be rapidly degraded in the absence of RNase inhibitor and/or DTT (diamonds; Figure 1, A and B). While RNA is degraded within 3 h without RNase inhibitor even in the presence of 5 mM DTT, the higher concentrations of RNase inhibitor used protected RNA for at least 3 h. When cells were frozen in the presence of sufficient amounts of RNase inhibitor, aldolase mRNA remained stable at 42°C for over 3 h (Figure 1A), which is well in excess of the time required for optimal RT (Figure 1C). In the absence of RNase inhibitor and DTT, note that over 60% of the aldolase mRNA was destroyed in only 2.5 min (Figure 1A). RNase inhibitor (0.25 µL/20 µL of cell extract) was found to protect aldolase mRNA for 3 h (Figure 1B); however, some degradation occurred unless a higher amount of RNase inhibitor was used (Figure 1A). Thus, in subsequent studies, 1 µL RNase inhibitor/20 µL of cells was used. It has been established that RNase inhibitor requires a high concentration of reducing agent to avoid its irreversible inactivation (3). As expected, the protective effect of DTT on RNase inhibitor was observed when incubation of RNA was carried out for long periods of time (Figure 1B). It should be noted that the effects of DTT in the absence of RNase inhibitor were subject to some variability. While 15 min of RT yield more than 64% of the aldolase cDNA generated by 3 h of RT (Figure 1C), we used 1 h of RT to expedite the procedure in subsequent studies.

Optimization of Freezing Solution

The freezing solution (0.15 M NaCl, 10 mM Tris, pH 8.0) used to wash and resuspend cells was optimized as follows. While a hypertonic sodium chloride concentration improved cDNA yield slightly, hypotonic sodium chloride resulted in a sharp decline in RT-PCR product obtained (Figure 1D). Hypotonic solutions could prematurely lyse cells or disrupt lysosomes and thereby degrade RNA prior to the addition of DTT and RNase inhibitor. To maintain high viability of cells during the washing step with freezing solution, we chose to use isotonic saline in the final version of the procedure. The results indicate that isotonic and hypertonic conditions yield better results, and therefore 0.15 M NaCl was used in subsequent studies.

Alkaline pH was found to be much more effective than acidic or neutral conditions for preservation of RNA (Figure 1E). Cleland demonstrated that DTT is 10-fold more active as a reducing agent at pH 8.0 than at pH 7.0 (11). Thus, the decreased yield of RT-PCR product from cells lysed under acidic or neutral solutions is probably explained by the enhanced activity of DTT at alkaline pH rather than any direct effect on RNA stability. Because of the short exposure of cells to the freezing solution, one would not expect the mildly acidic to neutral versions of the freezing solution to result in RNA hydrolysis. The broad-range buffer permitted the use of a buffer system that differed only in pH while buffer ions and ionic strength remained constant. The buffer produced an ionic strength of 10 mM. The results indicate that the yield of RT-PCR products is markedly reduced below pH 7.0, and thus 10 mM Tris (pH 8.0) was used in subsequent experiments. The decrease in the efficiency of RT-PCR when samples were prepared at pH 4.0 might be due to inhibition of the RT-PCR enzymes. The reduction in yield of RT-PCR products under mildly acidic to neutral conditions might be due to a decreased reductive efficiency of DTT (11).

Commercial RNase inhibitors are usually supplied in 50% glycerol. Because glycerol is a well-known cryoprotective agent for mammalian cells (23), its effects were examined. Marked inhibition of RT-PCR product formation was observed when freezing solutions contained greater than 5% glycerol (Figure 1F) or dimethyl sulfoxide (DMSO) (data not shown). Thus, the use of RNase inhibitor solutions or other enzymes that contain high concentrations of glycerol is not recommended during the lysis of cells by freezing. A commercial preparation of RNase inhibitor is available that does not contain glycerol (animal-injectable grade RNasin). Glycerol-free RNasin was used in all studies reported here. Following the rapid freeze-rapid thaw cycle used, cells at all concentrations of glycerol studied were permeabilized as judged by erythrosin-B vital staining (27).

Using the optimized cell lysis procedure, the linearity of the assay was assessed. When assays are carried out in the linear range of amplification, it has previously been shown that the amount of RT-PCR product generated is halved as the amount of total RNA is serially diluted by half (6). We examined the amount of RT-PCR product observed with serial twofold dilutions of cells (Figure 2). When amplification was carried out within the linear range of amplification (24 PCR cycles), we found that the number of counts in the RT-PCR product was halved as the number of cells was halved. RT-PCR product from approximately 62 cells could be observed above background. When amplification was carried out for 30 PCR cycles, the linear relationship between cell number and RT-PCR product generated deteriorated because the assay had entered the saturation range. However, at 30 cycles, the RT-PCR product resulting from fewer than four cells could be detected (Figure 2).

Stability of mRNA in Frozen Cells

Aldolase mRNA levels declined following repeated freeze-thaw cycles (Table 1). Thus, one can use an aliquot of cells several times in qualitative, but not quantitative, analyses. Although RNA is stable for as long as 3 h following the first freeze-thaw cycle (Figure 1A), repetitive freeze-thaw cycles obviously result in the inactivation of either RNA or RNase inhibitor. We determined that RNase inhibitor is the material inactivated by repetitive freezing and thawing (Table 1). It should be noted that the commercial glycerol-free RNase inhibitor (animal-injectible grade RNasin) is provided in a solution lacking both glycerol and DTT; hence, repetitive freezing and thawing is not recommended for this product. In contrast, the glycerol-containing version of RNasin appears to be quite stable to repetitive freeze-thaw cycles; however, glycerol lowers the yield of RT-PCR product (Figure 1F).

Table 2. Comparison of Traditional RNA Isolation Method with RNase Inhibitor Method

Method	cpm/250 cells
Guanidine thiocyanate–phenol–chloroform	5130 ± 501
RNase inhibitor	5275 ± 283

Total RNA was isolated from 10^6 MG-63 cells by guanidinium thiocyanate lysis followed by phenol extraction and ethanol precipitation (9,10). A volume of total RNA derived from 250 cells was compared to 250 cells prepared by the RNase inhibitor method. The results indicate that the methods yield comparable results. Results presented represent the mean ± standard error of three determinations.

Comparison of the RNase Inhibitor Method with Established Methods for RNA Isolation

Total RNA was extracted from 10^6 MG-63 cells by the guanidinium thiocyanate–phenol method (9,10) and resuspended in 20 µL of DEPC-treated water. A volume of the total RNA solution representing 250 cells was compared to 250 cells prepared according to the RNase inhibitor method described above. It was found that the two methods yielded comparable results (Table 2).

Using the RNase inhibitor method, we have examined several mRNAs that were previously studied with the guanidinium thiocyanate–phenol RNA extraction method. The human genes studied, in addition to aldolase, were the fibronectin alternatively spliced exons ED-A and ED-B (26) as well as several matrix metalloproteinase genes and their inhibitors (19), which are of moderate abundance (i.e., MMP-1, MMP-2, MMP-3, MMP-7, MMP-10, TIMP-1 and TIMP-2). In all cases, the RNase inhibitor method succeeded. Because one normally uses 1 µg of total RNA (equivalent to about 2×10^4 cells) pre-

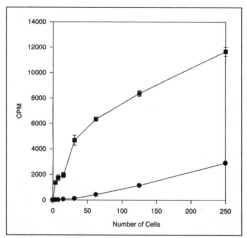

Figure 2. Sensitivity of RT-PCR using RNA prepared with the RNase inhibitor method. Cells were serially diluted twofold following a freeze-thaw cycle as described in the text. RT-PCR products were prepared using ^{32}P end-labeled primers and collected during the exponential range of PCR amplification (cycle 24) (circles) or as the PCR assay entered saturation phase (cycle 30) (squares). At cycle 24, the amount of RT-PCR product decreased 50% as the number of cells amplified decreased 50% between 250 and 62 cells. PCR product was not observed below about 15 cells at cycle 24. At cycle 30, while the linear relationship between cells amplified and RT-PCR product recovered deteriorated, RT-PCR product could be detected from as few as 4 cells.

pared by the guanidinium thiocyanate–phenol method and fewer than 10^3 cells when using the RNase inhibitor method, several additional PCR cycles are required when one converts an established RT-PCR procedure to an assay using the RNase inhibitor method described here.

Use of the RNase Inhibitor Method for Analysis of RNA from Frozen Sections

We have used the procedure described above to analyze RNA in frozen sections (32). RNA was found to be stable in sections maintained in a dry state for at least 2 weeks. In brief, RNA was extracted from a 1 × 0.5-cm 10-µm frozen section by placing the frozen section in 50 µL 1× RNase inhibitor stock (see Materials and Reagents) under humidified conditions. After 10 min at room temperature and an additional 3-min incubation on a platform shaker operated at 60 rpm, the RNA-containing solution was withdrawn and analyzed by RT-PCR as described above. RNase inhibitor was again shown to be essential for recovery of RNA (32).

Preparation of RNase Inhibitor from Bovine Liver by Affinity Chromatography

Because RNase inhibitor is the most costly reagent involved in this RT-PCR procedure, we developed an improved method for the affinity chromatography procedure described by Blackburn (2) for the isolation of RNase inhibitor from bovine liver. Using inexpensive reagents and a simple protocol (17), 40 000 U of RNase inhibitor can be purified to homogeneity from 400 g of bovine liver within two days. The RNase inhibitor purification procedure involves: *(i)* a facile procedure for the initial clarification of the liver extract, *(ii)* affinity chromatography on RNase-A-Sepharose and *(iii)* concentration of RNase inhibitor (17).

DISCUSSION

In the past, RT-PCR has been performed by: *(i)* extracting total RNA from cells, *(ii)* spectrophotometrically determining the amount of total RNA recovered, *(iii)* performing RT of mRNA to yield a cDNA and *(iv)* amplifying the cDNA by PCR (4,5,8). The extraction of total RNA and its quantitation are the most labor-intensive aspects of the RT-PCR procedure. In addition, errors are often encountered in the spectrophotometric determination of RNA if trace amounts of phenol (and other agents that absorb at 260 nm) are not completely removed during the extraction procedure. The procedure described eliminates many of these sources of error and also streamlines the RT-PCR assay.

RT-PCR results are usually reported as the amount of PCR product generated from 1 µg of total RNA (6). In contrast, the method described here bases comparisons on cell number, which can be determined quite accurately with an electronic cell counter. The use of cell number rather than mass of total RNA to standardize results should result in more accurate determinations.

Several methods for preparation of RNA for RT-PCR have been described. RNA extraction is most often carried out by lysis of cells in guanidinium thiocyanate in the presence of detergent followed by phenol–chloroform extraction and ethanol precipitation (9,10). This now traditional method is quite time-consuming and necessitates the removal of several reagents that could interfere with the RT-PCR procedure. Alternatively, RNA can be released from cells by heating cells to 90°C followed by treatment with proteinase K (29). Unfortunately, heating RNA in the presence of divalent cations can lead to RNA degradation (28). The freeze-thaw procedure described here is simple and permits the examination of RNA species in extremely small numbers of cells (Figure 2). A rapid freeze-thaw cycle has been known for many years to be an effective means to lyse both mammalian and bacterial cells (12,25,31). Recently, it has been shown that RT-PCR can be performed with 20 frozen-thawed cells using *Tth* as a polymerase but without RNase inhibitor (33). Because RNA is degraded within a few minutes when RNase inhibitor is absent (Figure 1A), the use of RNase inhibitor is important no matter which polymerase is used and would be critical if more than 20 cells were used.

In studies involving RNA, the problem of both endogenous and exogenous RNase contamination has been well-recognized. Because of extensive structural analyses of RNase, a detailed knowledge of methods to inhibit RNase has been gained. Although RNase can easily be irreversibly inhibited by reductive alkylation (16), this procedure will contaminate a sample with compounds that can inhibit many other enzymes. More than 5 min at 95°C are required to irreversibly inactivate RNase (35); however, such conditions would be quite deleterious for RNA (28). Placental RNase inhibitor is a well-studied protein that binds avidly to many members of the RNase superfamily but not to several bacterially derived RNases (1,24). A binding constant of 3×10^{-10} M for RNase A has been reported (3). Use of placental RNase inhibitor to form a stable enzyme–inhibitor complex is probably the gentlest method that can be used to inhibit RNase.

Although RNase inhibitor from human placenta can inhibit insect RNases (18), it should be noted that not all mammalian RNases are inhibited by mammalian RNase inhibitor. For example, RNases specific for certain mRNAs (mRNases) have been described (20). It is quite possible that numerous highly specific RNases are present in mammalian cells because it has been established that *E. coli* possesses at least 18 RNases, of which some display specificity for certain RNA classes (13). Thus, one should use the RNase inhibitor method described here only after establishing that this method yields results similar to the guanidinium thiocyanate methods (9,10) (Table 2).

Although it has been known for many years that glycerol and DMSO cryoprotect cells from freezing damage (15,22,30,34), the mechanism by which glycerol and other structurally diverse cryoprotective agents (23) act is still unknown. That freezing-induced lysis is the basis of the method described is demonstrated by the fact that at concentrations that cryoprotect cells, both glycerol and DMSO markedly reduce the generation of RT-PCR products (Figure 1F). Because cryoprotective agents such as glycerol decrease freezing-induced cell lysis, glycerol-free RNase inhibitor preparations should be used. The freeze-thaw RNA extraction method described here should greatly expedite RT-PCR studies by eliminating the time-consuming procedures involved in RNA isolation.

ACKNOWLEDGMENTS

This study was supported in part by Grants Nos. DE08144 and DE00152 from the National Institutes of Health.

REFERENCES

1. Beintema, J.J., C. Schuller, M. Irie and A. Carsana. 1988. Molecular evolution of the ribonuclease superfamily. Prog. Biophys. Mol. Biol. *51*:165-192.
2. Blackburn, P. 1979. Ribonuclease inhibitor from human placenta: rapid purification and assay. J. Biol. Chem. *254*:12484-12487.
3. Blackburn, P., G. Wilson and S. Moore. 1977. Ribonuclease inhibitor from human placenta. J. Biol. Chem. *252*:5904-5910.
4. Bouaboula, M., P. Legoux, B. Pessegue, B. Delpech, X. Dumont, M. Piechaczyk, P. Casellas and D. Shire. 1992. Standardization of mRNA titration using a polymerase chain reaction method involving co-amplification with a multispecific internal control. J. Biol. Chem. *267*:21830-21838.
5. Chelly, J., D. Montarras, C. Pinset, Y. Berwald-Netter, J.C. Kaplan and J.A. Kahn. 1990. Quantitative estimation of minor mRNAs by cDNA-polymerase chain reaction. Application to dystrophin mRNA in cultured myogenic and brain cells. Eur. J. Biochem. *187*:691-698.
6. Chen, D. and R.J. Klebe. 1993. Controls for validation of relative reverse transcription-polymerase chain reaction (RT-PCR) assays. PCR Methods Appl. *3*:127-129.
7. Chen, D., V.L. Magnuson, S. Hill, C. Arnaud, B. Steffensen and R.J. Klebe. 1992. Regulation of integrin gene expression by substrate adherence. J. Biol. Chem. *267*:23502-23506.
8. Chen, D., V.L. Magnuson, B. Steffensen and R.J. Klebe. 1993. Use of stock solutions to simplify mRNA quantitation by reverse transcription-polymerase chain reaction (RT-PCR) assays. PCR Methods Appl. *2*:351-353.
9. Chirgwin, J.M., A.E. Przybyla, R.J. MacDonald and W.J. Rutter. 1979. Isolation of biologically active ribonucleic acid from sources enriched in ribonuclease. Biochemistry *18*:5294-5299.
10. Chomczynski, P. and N. Sacchi. 1987. Single-step method of RNA isolation by acid guanidinium thio-cyanate-phenol-chloroform extraction. Anal. Biochem. *162*:156-159.
11. Cleland, W.W. 1964. Dithiothreitol, a new protective reagent for SH groups. Biochemistry *3*:480-482.
12. Craine, B.L. and C.S. Rupert. 1978. Identification of a biochemically unique DNA-membrane interaction involving the *Escherichia coli* origin of replication. J. Bacteriol. *134*:193-199.
13. Deutscher, M.P. 1993. Ribonuclease multiplicity, diversity, and complexity. J. Biol. Chem. *268*:13011-13014.
14. Ellis, K.J. and J.F. Morrison. 1982. Buffers of constant ionic strength for studying pH-dependent processes. Methods Enzymol. *87*:405-426.
15. Fallani, A., A. Arcangeli and S. Ruggieri. 1987. Lipid changes associated with erythroid differentiation of Friend erythroleukemia cells. Toxicol. Pathol. *15*:170-177.
16. Fujita, Y. and Y. Noda. 1991. Effect of reductive alkylation on termal stability of ribonuclease A and chymotrypsinogen A. Int. J. Pept. Protein Res. *38*:445-452.
17. Garcia, M.A. and R.J. Klebe. 1997. Affinity chromatography of RNase inhibitor. Mol. Biol. Rep. *24*:231-233.
18. Garcia-Segura, J.M., J.M. Fominaya, M.M. Orozco and J.G. Gavilanes. 1985. Alkaline ribonuclease from the insect *Ceratitis capitata*. Biochim. Biophys. Acta *826*:129-136.
19. Grant, G.M., J.K. Cobb, B. Castillo and R.J. Klebe. 1996. Regulation of matrix metalloproteinases following cellular transformation. J. Cell. Physiol. *167*:177-183.
20. Hua, J., R. Garner and V. Paetkau. 1993. An RNasin-resistant ribonuclease selective for interleukin 2 mRNA. Nucleic Acids Res. *21*:155-162.
21. Klebe, R.J., G.M. Grant, A.M. Grant, M.A. Garcia, T.A. Giambernardi and G.P. Taylor. 1996. RT-PCR without RNA isolation. BioTechniques *21*:1094-1100.
22. Klebe, R.J., D.P. Hanson, J.V. Harriss and K.L. Bentley. 1986. Uptake by cells of nucleic acids promoted by compounds sharing the pleiotropic effects of poly(ethylene glycol). Teratog. Carcinog. Mutagen. *6*:245-250.
23. Klebe, R.J. and M.G. Mancuso. 1983. Identification of new cryoprotective agents for cultured mammalian cells. In Vitro *19*:167-170.
24. Lee, F.S. and B.L. Vallee. 1993. Structure and action of mammalian ribonuclease (angiogenin) inhibitor. Prog. Nucleic Acid Res. Mol. Biol. *44*:1-30.
25. Lovelock, J.E. 1953. HET mechanism of the protective action of glycerol against haemolysis by freezing and thawing. Biochim. Biophys. Acta *11*:28-36.
26. Magnuson, V.L., M. Young, D.G. Schattenberg, M.A. Mancini, D. Chen, B. Steffensen and R.J. Klebe. 1991. The alternative splicing of fibronectin pre-mRNA is altered during aging and in response to growth factors. J. Biol. Chem. *266*:14654-14662.
27. Merchant, D.J., R.H. Kahn and W.H. Murphy. 1964. Techniques and procedures, p. 157-158. *In* Handbook of Cell and Organ Culture. Burgess Publishing, Minneapolis.
28. Myers, T.W. and D.H. Gelfand. 1991. Reverse transcription and DNA amplification by a *Thermus thermophilus* DNA polymerase. Biochemistry *30*:7661-7666.

29. **O'Brien, D.P., D. Billadeau and B. Van Ness.** 1994. RT-PCR assay for detection of transcripts from very few cells using whole cell lysates. BioTechniques *16*:586-590.
30. **Rubin, R.A. and H.S. Earl.** 1983. Dimethyl sulfoxide stimulates tyrosine residue phosphorylation of rat liver epidermal growth factor receptor. Science *219*:60-63.
31. **Staudenbauer, W.I.** 1976. Replication of small plasmids in extracts of *Escherichia coli*. Mol. Gen. Genet. *145*:273-280.
32. **Taylor, G.P., D.A. Troyer, T.A. Giambernardi and R.J. Klebe.** 1998. A direct method for extraction of RNA from frozen sections. J. Pathol. (In press).
33. **von Eggeling, F. and W. Ballhausen.** 1995. Freezing of isolated cells provides free mRNA for RT-PCR amplification. BioTechniques *18*:408-411.
34. **Yamamoto, N.** 1989. Effect of dimethyl sulfoxide on cytosolic ionized calcium concentration and cytoskeletal organization of hepatocytes in a primary culture. Cell Struct. Funct. *14*:75-85.
35. **Zale, S.E. and A.M. Klibanov.** 1986. Why does ribonuclease irreversibly inactivate at high temperatures? Biochemistry *25*:5432-5444.

Section II

Quantitative RT-PCR

6. Absolute Quantification of Messenger RNA Using Multiplex RT-PCR

David E. Dostal, Anna M. Kempinski, Thomas J. Motel and Kenneth M. Baker
Weis Center for Research, Henry Hood M.D. Research Program, Pennsylvania State University College of Medicine, Danville, PA, USA

OVERVIEW

Over the last decade, several molecular approaches involving detection and quantification of individual RNA molecules have been introduced. The most sensitive of these techniques has been reverse transcription polymerase chain reaction (RT-PCR), which has the advantage of detecting low-copy mRNAs in small numbers of cells. However, the lack of methods to correct for differences in RNA loading, RT and amplification efficiency has seriously limited the use of RT-PCR for mRNA quantification. Because the amplification process is exponential, small differences in any of the variables that control amplification can greatly affect the amount of DNA produced. Two general methods have been reported for quantification of mRNA. One approach compares the RT-PCR signal from a target mRNA to that of a housekeeping mRNA (5). This method controls for RNA loading, degradation and efficiencies of RT. Although control and target sequences do not compete for the same primers, a shortcoming of this approach is that the two primer sets might not amplify with the same efficiency. The second approach uses an exogenous RNA or DNA that has the same primer requirements as the cDNA of the endogenous experimental mRNA but differs in product size or by the presence of restriction sites (3). The amount of standard mRNA is titrated against a constant amount of experimental mRNA. The concentration at which the amounts of amplification products from both targets are equivalent is taken to be the starting concentration of the experimental mRNA. This approach provides an absolute quantification of the experimental mRNA but does not control for RNA loading or degradation. The multiplex competitive RT-PCR titration assay (MPTA) combines the strengths of these two approaches to produce a very sensitive and accurate method to measure expression levels of target mRNAs (6). Using MPTA, a 20%–50% change in expression levels of mRNAs can be readily determined in a small amount of sample RNA (1–2 μg). As an example, we describe the construction, testing and use of MPTA for quantification of angiotensin II type 1 (AT_{1A} and AT_{1B}) receptor mRNAs expressed by primary cultures of neonatal rat cardiac fibroblasts and myocytes.

BACKGROUND

Many studies of gene expression require quantitative analysis of target mRNA to better understand cellular function. Under circumstances in which the amount of sample RNA is limiting and/or the target mRNA has low abundance, RT-PCR is the only method that can be used to measure changes in gene expression. Quantitation using RT-PCR, however, poses several problems. In the RT step, the major sources of error result from tube-to-tube variability with respect to efficiency, RNA loading and degradation. Because amplification is exponential, small changes in temperature and availability of substrate and polymerase can markedly affect synthesis of PCR product. Several approaches to quantitative RT-PCR have been developed based on coamplification of a standard template (3,5,6,8,12,16,19–21). A common method uses an internal standard RNA transcript added to the reaction in known amounts to determine a standard curve. In this procedure, cDNA from the RNA template has the same primer binding sites as cDNA from the target mRNA but can be distinguished by length or by restriction site analysis from the target PCR product. If a housekeeping transcript is co-quantified with target mRNAs, then discrepancies among samples with respect to RNA loading, RNA degradation and efficiency of RT can be taken into account. If the levels of the housekeeper RNA are known, then results can be expressed in absolute units and compared with assays performed by other laboratories. Knowledge of the absolute levels of the housekeeper in a tissue also makes it possible to quantify target transcripts in RNA samples that are too small to quantify using UV spectrophotometry prior to RT-PCR. Below, methods will be described for developing an MPTA in which absolute amounts of three target mRNAs are simultaneously quantified. In this example, the housekeeping transcript is elongation factor 1α (EF-1α), which is highly abundant, and the two target transcripts of interest are highly homologous forms of the AT_{1A} and AT_{1B} receptor, which are derived from separate genes. Length will be used to distinguish between the competitor and target PCR products, whereas restriction site analysis will be used to distinguish between AT_{1A} and AT_{1B} receptor products.

PROTOCOLS

Materials and Reagents

- 10× PCR buffer: 500 mM KCl, 100 mM Tris-HCl (pH 8.4 at room temperature)
- dNTPs and rNTPs: neutralized, 100 mM solutions (Amersham Pharmacia Biotech, Piscataway, NJ, USA). Components of dNTPs or rNTPs are combined to make a 25 mM stock solution.
- RNase inhibitor: PRIME RNase INHIBITOR™ (1.0 U/μL; 5 Prime→3 Prime, Boulder, CO, USA)
- Random hexamers (Amersham Pharmacia Biotech): diluted to 50 μM stock solutions with H_2O

- PCR primers: usually 18–22 bases, dissolved in water at 400–800 pmol/μL
- Reverse transcriptase: Maloney murine leukemia virus (MMLV) (200 U/μL; Promega, Madison, WI, USA)
- *Taq* DNA polymerase: 5 U/μL, 25 mM MgCl$_2$ in separate vial (Fisher Scientific, Pittsburgh, PA, USA)
- *Taq* Extender™: (5 U/μL; Stratagene, La Jolla, CA, USA)
- RQ1 DNase: (1 U/μL; Promega)
- 5× transcription buffer: 200 mM Tris-HCl, pH 7.5, 30 mM MgCl$_2$, 10 mM spermidine, 50 mM NaCl (Promega)
- 20× TBE buffer: combine 216 g Tris base, 16.4 g disodium EDTA and 110 g boric acid in filtered water to make 2 L of 10× solution (pH 8.3). Prior to use, pass the buffer through a sterile filter.
- Light mineral oil (Fisher Scientific)
- Phenol (solid form; Fisher Scientific). Solution is prepared by saturation with water.
- Chloroform (optima grade; Fisher Scientific)
- Microcentrifuge tubes: use RNase- and DNase-free tubes for RNA storage and RT-PCR. Sample and competitor RNA can be stored in 0.5-mL microcentrifuge tubes (Marsh Biomedical Products, Rochester, NY, USA). For RT-PCR, use tubes specified for the particular thermal cycler. Strip tubes for 96-well thermal cyclers can be obtained from Robbins Scientific (Sunnyvale, CA, USA).
- Pipet tips: should be high-quality aerosol barrier tips that are DNase- and RNase-free (e.g., βt™; Continental Laboratory Products, Burlington, MA, USA)
- 3.5% polyacrylamide: 30% acrylamide–bisacrylamide solution (29:1; BioRad, Hercules, CA, USA)
- Gel-loading dye: 0.125% bromophenol blue, 0.125% xylene cyanol, 15% glycerol in water
- DNA ladder: pGEM® DNA markers (Promega)
- DNA stain: Vistra™ Green (10 000× concentration; Amersham Pharmacia Biotech)
- RNA stain: Sybr® Green II RNA Gel Stain (10 000× concentration; Molecular Probes, Eugene, OR, USA)
- Isotope: [α-^{32}P]dCTP (NEN Life Science Products, Boston, MA, USA), for isotopic detection of PCR products
- X-ray film: Fuji™ RX Medical X-Ray Film (Fisher Scientific)
- Distilled, deionized water: Molecular Biology Grade Ultrapure Water (5 Prime→3 Prime) and Cellgro™ Water (Mediatech, Herndon, VA, USA) are suitable for resuspending RNA and DNA and preparation of all enzyme buffers. Autoclaved water pretreated with diethyl pyrocarbonate (DEPC) is not recommended because autoclaving can contaminate water with metals that inhibit enzyme activity, and DEPC alters UV absorption (lowers absorbance $[A]_{260}/A_{280}$ ratio) and inhibits DNA amplification during PCR.
- Tissue culture 0.2-μm filter unit (Nalge Nunc International, Rochester, NY, USA)
- Liqui-Nox® (Fisher Scientific)
- Gloves: powder-free vinyl gloves (Oak Technical, Ravenna, OH, USA)
- Vertical gel electrophoresis system: Model V16 or V16-2 apparatus with 1.5-mm comb thickness (Life Technologies, Gaithersburg, MD, USA). Samples can be loaded with a standard multichannel pipettor if a custom comb (6-mm well

width, 3 mm between wells, 17 total wells) is used.
- 2× RT reaction mixture: 40 µL 25 mM MgCl$_2$, 20 µL 10× PCR buffer, 8 µL dNTPs, 10 µL random hexamers, 10 µL (10 U) RNasin® (Promega), 2 µL H$_2$O and 10 µL (10 U) reverse transcriptase
- Primer buffer: 8 µL 25 mM MgCl$_2$, 25 µL 10× PCR buffer, 12.5 µCi [α-^{32}P]dCTP (for isotopic detection only) and water to 200 µL

Choosing and Testing Primers for Multiplex RT-PCR

The choice of PCR primers is essential to the success of MPTA. Therefore, critical analysis and testing of candidate primers is required before proceeding to subsequent steps of constructing MPTA. Like conventional PCR, primer pairs should be well-balanced, with similar optimal annealing temperatures, lack homologous and/or heterologous interactions and produce a single product. In designing PCR primers for MPTA, we have followed the recommendations of Saiki et al. (18), in which there are 20-mers with base compositions in the range of 50%–55%, while trying to match the GC content of the primers within 5%. It might be possible to compensate for greater differences in the GC content by making minor adjustments in the lengths of the primers, thus allowing greater flexibility in their placement. Also, if possible, primer sets should be selected so that they reside in separate exons to inhibit amplification of any contaminating DNA in the RNA preparation. Even if amplification of genomic sequence occurs, when primers are in different exons, the size difference between the mRNA and genomic product should be easy to distinguish.

Choosing primers for MPTA is more complicated than for conventional PCR because a greater number of interactions among primers and reaction products must be considered. Prior to testing, computer analysis should be used to stringently select and screen primers. In addition to selection of compatible primers, lengths of PCR products generated by multiplex RT-PCR is an important consideration. Successful quantification requires that all PCR products of interest (targets and competitors) be distinguished based on size. Because there is an inverse relationship between template size and amplification efficiency (12), it is important to keep the size difference between wild-type and competitor as small as possible. Larger differences in size are required for longer PCR products because separation on a gel is related to the logarithm of the size, making it more difficult to resolve longer products. For optimal results, PCR products in an MPTA should be between 150 and 500 bp, in which there is a 25–50-bp separation between competitor and target species. The methods used to resolve, detect and quantify PCR products will also partially dictate minimal size differences among PCR products. Bands must be further apart if agarose is used to resolve products instead of polyacrylamide gel and if ^{32}P incorporation and autoradiography are used to detect PCR bands instead of immunofluorescence scanning. Finally, to aid in determining specificity, primer sequences should be compared with known nucleic acid database sequences at the National Center for Biotechnology Information (NCBI). To test for possible repetitive sequences, primers can be aligned with the target sequence using the Basic Local Alignment Search Tool (BLAST) (http://www.ncbi.nlm.nih.gov/BLAST/).

Once candidate primers have been identified, pilot RT-PCRs should be performed to determine whether primers produce a single product and can be successfully multiplexed. The procedure for determining whether primers are suitable for MPTA follows.

Protocol 1: Testing Candidate Primers for Multiplex RT-PCR

Powder-free gloves should be worn for all steps of this procedure. Water used for dilution of RNA and RT-PCR should be DNase- and RNase-free, as described above in the Materials and Reagents section. Alternative methods for staggered primer set addition and isotopic detection are included.

Procedure

1. Label a duplicate set of tubes to be used for RT-PCR. Enough reactions should be performed to test all combinations of primer sets. For example, if two primer sets are to be tested (i.e., housekeeper and target mRNA), then a minimum of three reactions (i.e., housekeeper primer set, target primer set, housekeeper plus target primer set) would be required.
2. To the first set of tubes add 250 ng (10 µL suspended in water) of sample RNA.
3. Prepare the 2× RT reaction mixture (100 µL, 10 reactions) in a 0.5-mL microcentrifuge tube.
4. Mix by pipetting up and down 3 times and add 10 µL of the RT reaction mixture to each tube in step 2.
5. Mix components in tubes and transfer 10 µL of the RT reaction mixture (combined with sample) RNA into the duplicate set of tubes (labeled in step 1). Overlay each sample with 50 µL of light mineral oil.
6. Reverse transcribe sample RNA by incubating for 40 min at 40°C. Inactivate reverse transcriptase by incubating for 8 min at 96°C. Place the reaction mixture on ice (4°C).
7. Prepare primer buffer (200 µL, 10 reactions).
8. Add 20 µL of primer buffer and 5 µL of primer working stock (200 pmol of each primer diluted in water) to appropriate tubes containing reverse-transcribed RNA.
9. Perform hot start by incubating at 96°C for 8 min and cool to the optimal annealing temperature.
10. During the hot start, prepare *Taq* reaction mixture (150 µL, 10 reactions). Prepare Option 1 mixture if all primers will be added at the beginning of the PCR. Prepare Option 2 mixture if additional primers will be added part way through the PCR.

	Option 1	Option 2
25 mM MgCl$_2$	12.0 µL	4.0 µL
10× PCR buffer	15.0 µL	5.0 µL
Taq DNA polymerase	2.5 µL (12.5 U)	2.5 µL
Taq Extender	1.5 µL (12.5 U)	2.5 µL
Water	118.0 µL	36.0 µL

11. If Option 1 is followed, add 15 µL of *Taq* reaction mixture to each tube (50 µL final volume) and proceed to step 12; otherwise add 5 µL of the Option 2 *Taq* reaction mixture (35 µL final volume) and proceed to step 13.
12. Option 1: Perform the PCR for the required number of cycles using optimal

amplification conditions. If the PCR has not been optimized, a PCR profile of annealing for 1.0 min, extension for 1.5 min and denaturation for 1.0 min for 30–35 cycles should provide an initial starting point. After the last cycle, use a 7-min extension to complete primer extension and cool reaction to 4°C.

13. Option 2: Perform the PCR for the initial number of cycles and maintain the PCR at the optimal annealing temperature until 5 µL of additional primer buffer (described below) and 5 µL of primer working stock (50–200 pmol of each primer diluted in water) have been added to the reaction. Continue the PCR using the guidelines given in step 12.

    Prepare buffer for the second set of primers by combining the following components:

    8 µL 25 mM $MgCl_2$
    10 µL PCR buffer
    32 µL water

14. Add 15 µL of loading dye to each sample and store at -20°C. If samples will be subjected to restriction digestion, they can be stored for an indefinite period in the absence of dye.

Protocol 2: Separation of PCR Products Using Polyacrylamide Gel Electrophoresis

Once MPTA has been performed, the products can be resolved by gel electrophoresis. It is recommended that a 4%–8% polyacrylamide gel (1.5 mm thick) be used to separate the PCR products. The procedure for separating PCR products on a 6% polyacrylamide gel follows.

Procedure

1. Prepare a 6% polyacrylamide solution (100 mL, 2 gels) by combining the following in a 150-mL beaker: 20 mL 30% acrylamide–bisacrylamide (29:1), 75 mL deionized water, 5 mL 5× TBE.
2. Use a sterile filter (yellow-top filter unit; Nalge Nunc International) to degas and remove fluorescent debris (e.g., dust and lint).
3. During filtration of polyacrylamide solution, wash two sets of plates with detergent (Liqui-Nox), dry with a towel, remove final residue with 100% ethanol and assemble.
4. Add 600 µL of ammonium persulfate (1:10 dilution in water) and 134 µL of TEMED to the polyacrylamide solution and mix for 5 s.
5. Completely fill the space between the two glass plates of the electrophoresis apparatus with polyacrylamide solution and insert the comb.
6. Once the polyacrylamide gel has polymerized (10–20 min), complete assembly of the electrophoresis apparatus, fill buffer tanks with 1× TBE buffer and gently remove combs from the gel. Check circuitry of apparatus by applying 200 V, which should give approximately 30 mA for a single gel.
7. Flush wells and load 13 µL of PCR sample (includes loading dye) in each well. To a remaining well, add 0.25 µL of pGEM marker (in loading buffer). Sample loading can be expedited by using a multichannel pipettor. Apply current to gel

Absolute Quantification of Messenger RNA

apparatus (200 constant volts) and run for 3–6 h, depending on the size of the PCR products.

8. Remove gel from apparatus, incubate for 45 min with Vistra Green (dilute 1:10 000 in filtered 1× TBE) and scan. For isotopic detection, vacuum dry gel, expose to X-ray film and perform densitometric analysis.

Figure 1 shows an example in which EF-1α and AT_1 receptor sets have been tested using RNA isolated from primary cultures of neonatal rat cardiac fibroblasts. Primers for the AT_1 receptor produced a predicted PCR product of 224 bp, and EF-1α primers produced the expected product of 347 bp. When combined, EF-1α and AT_1 primers produced only their respective products. Based on published nucleotide sequences (10,14), the AT_1 primer set was expected to coamplify both AT_{1A} (GenBank® Accession No. X62295) and AT_{1B} (GenBank Accession No. X64052) cDNAs to produce identically sized products. Analysis of the sequences of PCR products indicates that XcmI is expected to cleave only the AT_{1A} PCR product to produce two fragments with sizes of 139 and 85 bp. As shown in Figure 1, XcmI had no effect on

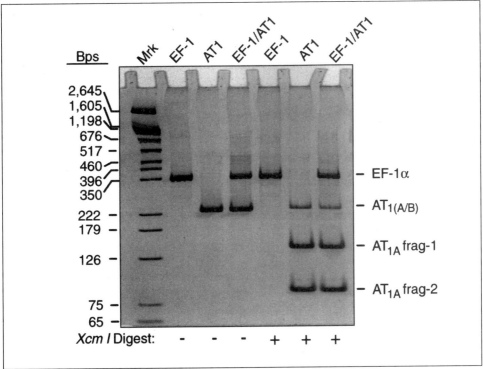

Figure 1. Testing candidate MPTA primer sets. Candidate PCR primers for EF-1α and AT_1 receptor were tested individually and combined. Sample RNA was reverse-transcribed and subsequently amplified by PCR, as described in the procedures. The sense primer 5′-GGA ATG GTG ACA ACA TGC TG-3′ (bases 635–654) and the antisense primer 5′-CGT TGA AGC CTA CAT TGT CC-3′ (base pairs 982–963) were used to generate a 347-bp product of rat EF-1α. The sense primer for the detection of rat $AT_{1(A/B)}$ cDNA spanned oligonucleotide bases 786–805 (5′-TCG AGA ACA CCA ATA TCA CA-3′), and the antisense primer spanned bases 984–1003 (5′-GCA CAA TCG CCA TAA TTA TC-3′) to produce a 224-bp product. Aliquots from each PCR were digested with XcmI for 2 h at 37°C. Samples (13 µL) and pGEM markers (0.25 µL) were separated on a 6% nondenaturing polyacrylamide gel in 1× TBE running buffer for 4 h at 200 V (constant). PCR products were of expected size from primers for EF-1α (347 bp) and AT_1 (224 bp). Upon restriction digestion, XcmI had no effect on the PCR product from EF-1α primers. Treatment with the restriction enzyme resulted in two fragments (AT_{1A} frag-1, 139 bp; AT_{1A} frag-2, 85 bp) that were of the expected sizes. The AT_1 band unaffected by XcmI restriction digestion was AT_{1B} (verified by sequencing).

EF-1α PCR bands but cleaved AT_1 PCR products to produce two AT_{1A} fragments, leaving the AT_{1B} PCR product. To verify that PCR products were authentic, bands were removed from the gel and sequenced using dideoxy chain termination and *Taq* DNA polymerase as described (11).

Generation of Competitor RNA

Once suitable primers have been identified, the next step is to construct RNA competitors. All of the above considerations are important for construction and analysis of internal standards for MPTA. Although some have recommended the use of competitor DNA templates rather than competitor RNA (2), this approach does not account for variability of the RT step, and such a procedure is not suitable when absolute quantification is required. The cDNA generated from the competitor RNA should contain the same priming sites as cDNA from the target mRNA. It is not essential that nucleotide sequence between primer pairs of the competitor RNA be similar to the endogenous mRNA. However, if the competitor RNA has characteristics similar to the target mRNA, then it is more likely to reverse transcribe and amplify with a similar efficiency. Because it is necessary to distinguish the target and competitor PCR products based on length, previously discussed size constraints need to be considered.

Several methods have been used to synthesize competitor RNAs for RT-PCR. A common approach is to clone cDNA of the target mRNA into a plasmid vector under the control of an RNA polymerase promoter (e.g., SP6, T3, T4, T7). The cDNA insert is mutated by adding or deleting an appropriate length of nucleotide sequence between the PCR priming sites. In some cases, the deletion can be performed using restriction enzymes. However, if appropriate restriction sites are not available, then PCR can be used to make a deletion in the insert. In this method, PCR primers are oriented so that extension proceeds outward from the deletion region of cDNA. This method of permutation has been dubbed "inverse" or "inside-out" PCR (15). However, an important caveat is that circular double-stranded DNA can be difficult to amplify by PCR. A possible explanation is that covalently closed circular DNA is a poor template for the first round of primer extension because of "snap-back" reannealing of the DNA strands. If the template for inverse PCR is present in limited amounts and in the form of closed double-stranded DNA, then linearization or nicking might be required.

The use of inverse PCR to produce a deletion mutation has been previously documented (6,15). In the example used for this chapter, deletion mutations were made in EF-1α (GenBank Accession No. X63561) and AT_{1A} cDNA inserts, which were cloned in pCR®II (Invitrogen, Carlsbad, CA, USA) and Bluescript® II KS(-) (Stratagene) vectors, respectively. A net deletion of 47 bases was made in the EF-1α cDNA insert using the sense primer 5'-[AGC GTC GT]-ATG CAC CAT GAA GCT TTG-3' (bases 928–945) and antisense primer 5'-[AGC GTC GT]-GAG AAC ACC AGT CTC CAC-3' (bases 864–847), which contained a random eight-base sequence at the 5' ends. A 44-base deletion was made in the AT_{1A} cDNA insert using the sense primer 5'-GGG CTT CTT GTT CCC TTT CC-3' (bases 877–896) and the antisense primer 5'-CCG AGA CTC ATA ATG AAA CGC GC-3' (bases 810–832). Prior to inverse PCR, plasmids containing EF-1α and AT_{1A} cDNA inserts were linearized in the regions to be deleted, using restriction enzymes *Bst*EII and *Xcm*I, respectively. Note that the lack of an *Xcm*I restriction site in this region of AT_{1B} can be used to distinguish between MPTA products generated from AT_{1A} and AT_{1B} mRNA. This restriction enzyme, however, will have no effect on PCR products from the AT_1 competitor because this region is deleted by inverse PCR. Once the insert has been mutated, competitor RNA is generated as follows.

Protocol 3: Synthesis of Competitor RNA

Synthesis of internal standard RNA uses the complementary strand of mutated target cDNA as a template. Prior to synthesis, the plasmid must be linearized at the 3′ end of the insert.

Procedure

1. Digest the plasmid with the appropriate restriction enzyme to linearize the plasmid near the 3′ end of the insert to be transcribed.
2. Treat the digest with 100–200 µg/mL proteinase K and 0.5% sodium dodecyl sulfate (SDS) at 50°C for 30–60 min.
3. Remove degraded protein by extracting with an equal volume of phenol–chloroform.
4. Precipitate plasmid DNA with 1/10 vol of 5 M ammonium acetate and 2.5 vol of ethanol. Incubate for 2 h at -20°C and centrifuge at 12 000× g.
5. Wash pellet with 70% ethanol to remove residual salt from the nucleic acid pellet. The volume should be sufficient to cover the pellet and wet sides of the tube. Vortex mix the sample to break up the pellet. Centrifuge for 10 min to re-collect the pellet. Remove ethanol, air dry and resuspend DNA to 1–5 mg/mL in water.
6. Set up the transcription reaction in a final volume of 100 µL in 0.5 mL DNase- and RNase-free microcentrifuge tubes. Combine the following components:
   20 µL 5× transcription buffer (pH 7.9)
   10 µg linearized plasmid
   1.0 µL 25 mM rNTPs
   1.0 µL (1 U) RNasin
   2.0 µL (40 U) RNA polymerase (e.g., T7, T4)
   Water to 100 µL
7. Incubate transcription reaction mixture at 39°C for 2 h.
8. Add 1 U of RNase-free RQ1 DNase I per µg of template DNA and incubate for 15 min at 37°C.
9. Repeat steps 3–5.
10. Check integrity of synthetic RNA by separation on a formaldehyde gel with subsequent staining with Sybr Green II diluted 1:10 000 in 1× TBE.
11. The amount of RNA transcript is calculated using this formula: C (µg/mL) = $A_{260}/0.025$, where A_{260} is the absorbance at 260 nm with a light path of 1 cm (1). High-quality RNA should have an A_{260}/A_{280} between 1.9 and 2.0.

Testing Primers Using Competitor RNA

After synthesis of competitor RNA, it should be verified that no significant artifacts have been created, and all PCR products of interest can be completely resolved by gel electrophoresis. Figure 2 shows a multiplex RT-PCR containing EF-1α and AT_1 deletion mutant competitor RNA. When competitor and endogenous targets are coamplified, artifacts that are 3 to 5 times longer than the expected PCR product can result (Figure 2). The appearance of these artifacts does not affect quantitation because signal ratios used for quantification are not significantly altered (unpublished data). However, these complexes will interfere with analysis if they cannot be resolved from competitor and target PCR products.

Protocol 4: Performing MPTA

Once suitable primers have been obtained and high-quality competitor RNA has been synthesized, MPTA can be performed. The concentration range of competitors in the assay must be determined empirically and can be refined to meet the needs of the experimental system studied. It is useful to perform an initial pilot run over a wide range of competitor to approximate the amount of transcript in a particular tissue. Using this information, a smaller range of competitor can be used to flank the point at which target and competitor RNAs are present in a 1:1 molar ratio. In the MPTA procedure described below (Figure 3), serial dilutions (1- to 64-fold) of competitor RNA are used to determine target levels of transcript in RNA samples. When optimized, this range of competitor is sufficient to accurately determine regulatory changes of most target mRNAs. The following procedures describe implementation of MPTA using a 96-well thermal cycler.

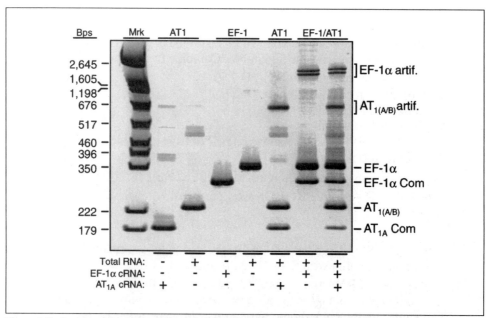

Figure 2. **Testing competitor RNA with multiplex PCR primer sets.** Competitor RNAs (EF-1α cRNA, AT_1 cRNA) were co-reverse-transcribed with and without sample RNA and amplified with EF-1α and AT_1 primers individually and combined. Samples (13 µL/well) and pGEM markers (0.25 µL/well) were separated on a 6% nondenaturing polyacrylamide gel in 1× TBE running buffer for 4 h at 200 V (constant). The gel was stained with Vistra Green (1:10 000 dilution) in 1× TBE and detected by fluorescence scanning. Competitor and sample RNA gave PCR products of expected sizes for AT_1 (178 and 224 bp, respectively) and EF-1α (300 and 347 bp, respectively).

Absolute Quantification of Messenger RNA

Procedure

1. Thaw sample RNA and working stocks of competitor RNA, 25 mM $MgCl_2$, 10× PCR buffer, dNTPs and random hexamers. Keep thawed stocks on ice (4°C).
2. Label the tubes 1 through 8 in three sets of strip tubes (e.g., A, B and C).
3. Transfer 100 μL of the highest concentration of competitor RNA to tubes 1 and 2 of strip A. To tubes 3–8, add 50 μL of water.
4. Serially dilute competitor RNA in tubes 2–8 in strip A in a 2:1 manner. This is performed by transferring 50 μL from tube 2 to tube 3 followed by mixing. This procedure is repeated with tubes 3 and 4 and remaining tubes. After serial dilution, tubes 1 and 8 will contain 100 μL of competitor RNA, and 50 μL will be contained in the remaining tubes.
5. Add 5 μL of competitor RNA from each tube in strip A to the corresponding tube in strip B.

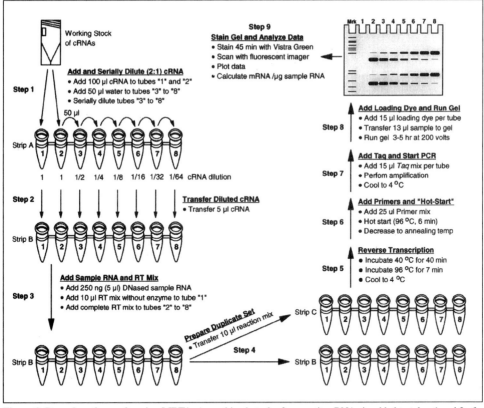

Figure 3. **Procedure for performing MPTA.** A combined stock of competitor RNAs is added to tubes 1 and 2 of a strip of tubes (strip A; step 1), after which 2:1 serial dilutions are performed. Once serial stock dilutions have been performed, 5 μL of competitor RNA (step 2) are added to a new set of strip tubes (strip B), after which 5 μL (250 ng) of DNase-free sample RNA and RT mixture are added (step 3). After splitting the reaction mixture into a duplicate set (step 4), RT is performed by incubating for 40 min at 40°C, after which the enzyme is denatured by incubating for 7 min at 96°C, and the reaction is cooled to 4°C (step 5). For amplification, primers are added, and the reaction is hot-started and cooled to the annealing temperature (step 6). Following addition of *Taq* DNA polymerase, amplification is conducted using optimal extension, melting and annealing conditions (step 7). After PCR, products are separated by PAGE (step 8) and stained, and amounts of target mRNA are determined (step 9).

6. Add 250 ng (5 µL) of DNase-treated sample RNA to each tube in strip B.
7. Prepare RT mixture (90 µL, 10 reactions) by combining the following components:

   40 µL 25 mM MgCl$_2$
   20 µL 10× PCR buffer
   8 µL dNTPs
   10 µL random hexamers
   10 µL (10 U) RNasin
   2 µL water

8. Mix by pipetting up and down 3 times.
9. Remove 9.0 µL of the RT mixture and add to tube 1 in strip B. Also, add 1.0 µL of water to the same tube. This will be the "-RT" control for MPTA (Figures 4 and 5).

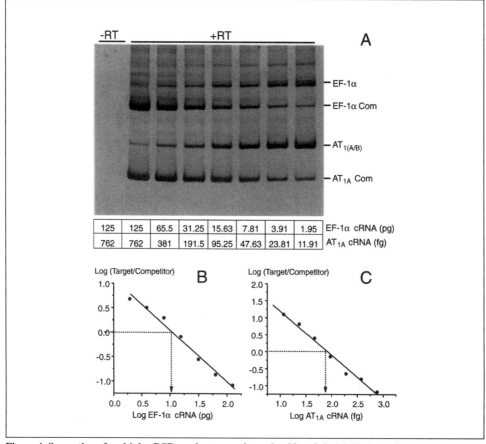

Figure 4. Separation of multiplex PCR products on polyacrylamide gel. Serial dilutions of competitor RNA for EF-1α (EF-1α cRNA, 125–1.95 pg/tube) and AT$_1$ receptor (AT$_{1A}$ cRNA, 762–11.91 fg/tube) were reverse-transcribed with 250-ng aliquots of total RNA from neonatal rat cardiac fibroblasts. Sense and antisense primers (200 pmol/reaction) were added for AT$_1$, and first-strand cDNAs were amplified for 10 cycles, after which primers for EF-1α (100 pmol/reaction) were added. The PCR was stopped at 35 cycles, and amplification products (13 µL) were separated on a 6% nondenaturing polyacrylamide gel in 1× TBE running buffer for 5 h at 200 V (constant). The gel was stained for 45 min with Vistra Green (1:10 000 dilution in 1× TBE), and PCR products were detected and quantified using fluorescence scanning.

10. Add 9.0 µL (90 U) of reverse transcriptase to the remaining portion of the premixture prepared in step 7 and mix.
11. Add 10 µL of reverse transcriptase mixture to tubes 2–8 in strip B and mix.
12. Make a duplicate set of reactions by transferring 10 µL from tubes 1–8 into corresponding tubes in strip C. After RT, this set can be used either as a duplicate, a spare or a parallel multiplex RT-PCR to quantify a second set of target mRNAs. If strip C is to be used for quantification of a different set of mRNAs, the additional RNA competitors can be added with original working stock of cRNAs (Figure 3, step 1). The first-strand cDNA reaction mixture can be stored for up to 3 weeks at -70°C prior to PCR.
13. Overlay tubes in strips B and C with 50 µL of light mineral oil and cap.
14. Perform RT by incubating for 40 min at 40°C. Inactivate the reverse transcriptase by incubating for 7 min at 96°C.
15. During the RT reaction, prepare the primer mixture (250 µL, 10 reactions) by combining the following components:
    8 µL 25 mM MgCl$_2$
    25 µL 10× PCR buffer
    50–200 pmol primers per reaction
    10 µCi [α-^{32}P]dCTP (for isotopic detection only)
    Water to 250 µL
16. After mixing, add 25 µL of primer mixture to each tube in strips B and C.
17. Hot start by incubating at 96°C for 8 min followed by cooling to the optimal annealing temperature.
18. During the hot-start procedure, prepare *Taq* reaction mixture (150 µL, 10 reactions). If all PCR primers will be added simultaneously, prepare the mixture listed under Option 1. Prepare the mixture listed under Option 2 if additional primers will be added after a given number of PCR cycles.

	Option 1	**Option 2**
25 mM MgCl$_2$	12.0 µL	4.0 µL
10× PCR buffer	15.0 µL	5.0 µL
Taq DNA polymerase	2.5 µL (12.5 U)	2.5 µL
Taq Extender	2.5 µL (12.5 U)	2.5 µL
Water	118.0 µL	36.0 µL

19. If Option 1 is followed, add 15 µL of *Taq* mixture to each tube (50 µL total volume) in strips B and C and proceed to step 20. If Option 2 is followed, add 5 µL of *Taq* mixture to each tube in strips B and C (40 µL total volume) and proceed to step 21.
20. Perform the PCR for the required number of cycles using optimal amplification conditions. If the PCR has not been optimized, a program of annealing for 1.0 min, extension for 1.5 min and denaturation for 1.0 min for 30–35 cycles should give satisfactory results. At the last cycle, use a 7-min extension to complete primer extension and cool the reaction mixture to 4°C.
21. Perform the PCR for the initial number of cycles and maintain the reaction at the optimal annealing temperature until the second primer mixture (10 µL/tube) has been added. Continue the PCR using the guidelines given in step 20. The second primer mixture (100 µL, 10 reactions) is prepared by combin-

ing the following components:
 8 µL 25 mM $MgCl_2$
 10 µL 10× PCR buffer
 50–200 pmol primers per reaction
 Water to 100 µL

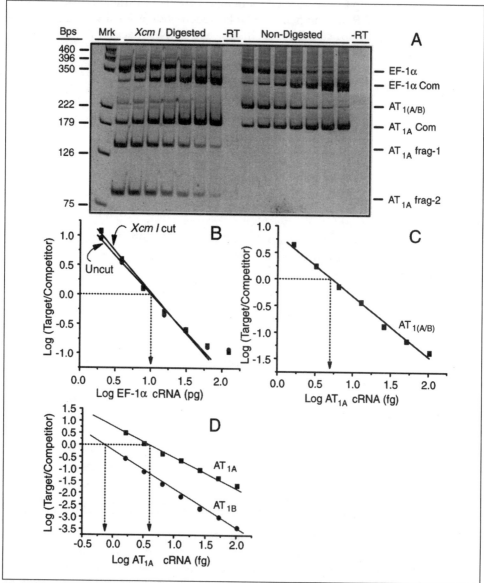

Figure 5. The use of restriction digestion to distinguish between MPTA products of identical size. MPTA was performed using cultured neonatal rat cardiac myocyte RNA (250 ng/reaction) as described in the legend of Figure 4. Following the reaction, an aliquot from each reaction was subjected to restriction digestion using $XcmI$ (2 h at 37°C), after which digested and nondigested PCR products were separated on a 6% polyacrylamide gel, stained with Vistra Green and quantified using fluorescence scanning. In panels B and C, linear regression analysis of signals from nondigested PCR product was used to determine amounts of EF-1α and $AT_{1(A/B)}$ mRNA, respectively, in the RNA sample. In panel D, analysis of signals from restriction digestion products (i.e., AT_{1A} frag-1 and AT_{1A} frag-2) and the remaining AT_1 PCR band (i.e., AT_{1B}) were used to determine amounts of AT_{1A} and AT_{1B} mRNA in the RNA sample.

22. If samples are to be stored prior to gel separation, place at -20°C. Fifteen microliters of loading dye can be added to each tube if PCR products will not be subjected to restriction digestion.
23. Separate PCR products using polyacrylamide gel electrophoresis (PAGE) and quantify using fluorescent DNA stain or autoradiography as described above in Protocol 2.

An example MPTA is shown in Figure 4A, in which EF-1α and AT$_1$ receptor transcripts have been assayed in cardiac fibroblasts. As shown, the "-RT" lane should have no bands, and the remaining bands should be clearly separated so that accurate quantification can be performed.

Mathematical Analysis

Following quantification of PCR bands, amounts of target mRNA in the sample are determined by plotting the logarithm of the target-to-competitor species ratios as a function of the logarithm of the initial amounts of competitor RNA. The linear analysis can be performed using spreadsheet or scientific plotting software. However, prior to plotting, the competitor signal must be multiplied by a correction factor to adjust for the difference in size compared to the endogenous target. If fluorescence staining is used to quantify PCR products, then the correction factor would depend on the relative lengths of the competitor and PCR product (i.e., correction factor = target length/competitor length). However, if incorporation of a radioactive nucleotide (e.g., [α-^{32}P]dCTP) is used, the correction factor is the relative ratio of that particular nucleotide in the two PCR products. Once data have been plotted and regression analysis performed, the results should be examined to determine whether proper amplification has occurred within each reaction tube. An experimental error made in one of the reactions will be manifested as a log ratio that substantially deviates from the linear relationship calculated from all other data points. If the plot is sigmoidal, end points in the plateau region should also be eliminated. Once regression analysis has been performed using appropriate data points, the initial amount of target mRNA is determined by extrapolating from the point on each curve at which the amount of amplified target is equal to the amount of amplified competitor (i.e., log 1:1 = 0). For example, in Figure 4, the initial concentrations of EF-1α (panel B) and AT$_{1(A/B)}$ receptor (panel C) mRNA in 250 ng of total RNA were determined to be 11.2 pg (i.e., $10^{1.05}$) and 51.3 fg (i.e., $10^{1.71}$), respectively. To correct for RNA degradation and loading, the amount of target mRNA is divided by the amount of housekeeper mRNA in the sample. If multiple assays have been performed, then the average amount of housekeeper mRNA (per microgram of total RNA) can be determined for a particular tissue or cell type. In the present example, when MPTA was performed on 10 primary dispersions of cultured neonatal rat cardiac fibroblasts, the average level of EF-1α mRNA was 66.7 ± 7.8 pg/µg total RNA. Given this information, AT$_{1(A/B)}$ receptor mRNA levels in Figure 4 can be expressed as 307.7 fg/µg total RNA (i.e., [51.2 fg AT$_{1(A/B)}$/11.1 pg EF-1α] × 66.7 pg EF-1α/µg total RNA). Because expression levels of housekeeper mRNA can slightly vary with tissue type, species and pathological condition, it is preferable to present levels of the target mRNA as mass (or copy number) per microgram of total RNA. Normalization of target mRNA levels in this manner is required for statistical analysis and provides absolute numbers that can be compared to those reported in the literature.

Use of Restriction Digestion to Quantify Overlapping PCR Products

In some cases, it might not be possible to separate MPTA products according to size. In Figure 5, we demonstrate the use of restriction digestion to aid in distinguishing between PCR products that cannot be separated by conventional gel electrophoresis. In this example, we illustrate quantification of AT_{1A} and AT_{1B} mRNA levels in total RNA isolated from cultured neonatal rat cardiac myocytes. The reactions were performed using the same amounts of RNA sample (250 ng) and competitor RNA as described for cardiac fibroblasts in Figure 4. As shown in Figure 5A, there is complete overlap of AT_{1A} and AT_{1B} PCR products. However, treatment with the restriction enzyme *Xcm*I diminished the $AT_{1(A/B)}$ band and formed two restriction products with sizes corresponding to AT_{1A} "frag-1" and AT_{1A} "frag-2". When data were analyzed using regression analysis, amounts of EF-1α in the RNA sample (Figure 5B) were the same before (10.1 pg/250 ng total RNA) and after restriction digestion (10.5 pg/250 ng total RNA). The level of EF-1α (ca. 40 pg/μg total RNA) is similar to the average level of this transcript (45.9 ± 9.4 pg/μg total RNA) obtained from 8 primary dispersions of cultured neonatal rat cardiac myocytes (data not shown). The cumulative amounts of mRNA estimated for AT_{1A} (4.7 fg/250 ng total RNA) and AT_{1B} (0.75 fg/250 ng total RNA; Figure 5D) were essentially the same as that

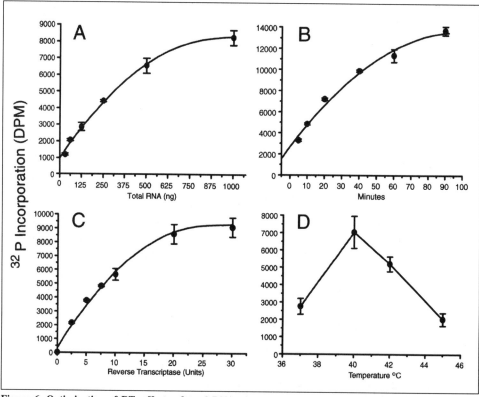

Figure 6. Optimization of RT: effects of total RNA, time, enzyme and temperature. Incorporation of [α-^{32}P]dCTP into cDNA was used as a measure of reverse transcriptase activity. The effects of sample RNA (A), incubation time (B), enzyme concentration (C) and incubation temperature (D) are shown. Reactions, unless conditions varied as indicated, were performed at 40°C for 40 min using 250 ng total RNA and 10 U of reverse transcriptase. Each point is the mean ± standard error of 6 data points.

calculated for $AT_{1(A/B)}$ (5.3 fg/250 ng total RNA; Figure 5C). As described above, the data can be further normalized by dividing these amounts of $AT_{1(A/B)}$, AT_{1A} and AT_{1B} mRNA by that of EF-1α in the sample. The amount of AT_{1A}, AT_{1B} and $AT_{1(A/B)}$, when expressed as amounts per microgram of RNA (based on average EF-1α levels in cardiac myocytes) would therefore be 20.5, 3.28 and 23.17 fg/μg of total RNA, respectively. These results indicate that restriction digestion, when combined with MPTA, can be used to quantify PCR products that have identical sizes.

RESULTS AND DISCUSSION

MPTA has application to a variety of clinical and basic research studies requiring accurate measures of particular transcripts. The procedure allows for detailed analysis of the regulation of transcripts of interest in microlocalization studies and in tissues or cells with low copy numbers. In a clinical setting, this aspect of the procedure is particularly attractive because it is difficult or impossible to examine regulation of gene expression using more conventional methods when there are limiting amounts of tissue, such as from needle aspirations, biopsies and small isolated vessels. This technology also has the advantage of simultaneous measurements of high copy numbers of transcript in small amounts of RNA (1–2 μg). Also, because results from MPTA are absolute rather than relative, samples to be compared can be assayed at different times without a loss of experimental accuracy. Although relative quantification can be used to assess the efficacy of clinical treatment, absolute quantification is a better choice.

Optimization of MPTA

Prior to quantification of RNA, it is important to optimize RNA extraction procedures as well as RT and amplification steps of MPTA. Genomic DNA should be eliminated by treating total RNA samples with DNase I, as described (6). After selection of the most appropriate method of RNA preparation, care should be taken to optimize the RT of RNA. MMLV and avian myeloblastosis virus (AMV) are most commonly used for transcription. In Figure 6, the relationships between total input RNA (panel A), incubation time (panel B), amount of reverse transcriptase (panel C) and temperature (panel D) are given when MMLV and random hexamers are used to synthesize first-strand cDNA. All of these parameters can be adjusted to maximize the amount of product obtained by RT. However, there is a marginal increase in product when greater than 500 ng of total RNA or 20 U of reverse transcriptase are added to the reaction. The benefits of using higher amounts of reverse transcriptase, however, are offset by the increased cost of performing the assay. It is recommended that temperature be initially optimized because changing this parameter does not require additional resources (i.e., enzyme, RNA etc.) and can markedly affect the yield from RT (Figure 6D).

Although MPTA can accommodate several-fold changes in experimental mRNA levels, it is optimal to flank the initial concentrations of mRNA species with competitor RNA (6,16). It is usually assumed that amplification efficiencies of experimental and competitor cDNAs are independent of relative concentrations (8,20). However, differences between the initial concentrations of competitor and target have been shown to adversely affect accuracy and precision in nontitration competitive

PCR (6). This error occurs because the more abundant component has a higher efficiency, thus favoring production of one species over the other. When concentrations of competitor RNA evenly flank initial amounts of experimental mRNA, this bias is minimized. Thus, ranges of competitor concentrations should be optimized for each tissue before performing regulatory studies.

The amplification step of MPTA also should be tailored to the specific targets of interest. The concentration of enzyme, primers and nucleotides, the number of cycles, and denaturation, annealing and extension times have all been shown to affect the sensitivity, specificity and yield of the PCR (7). The number of cycles will depend on expression levels of the transcript. Reliable data can be obtained from MPTA when it is terminated in either the plateau or linear phases of amplification (6).

A common problem associated with multiplex PCR is that of uneven amplification among endogenous targets. The more efficiently amplified (or more abundant) species inhibit amplification of less efficient (or less abundant) species. This occurs because PCR has a limited supply of enzyme and nucleotides, and all products compete for the same pool of substrate. Two approaches have been used to overcome this problem (6,9). One approach, presented in this chapter, is to add primers for highly abundant species at a later cycle in the PCR (6). A second approach is to change proportions of the various primers in the reaction. In the latter method, amounts of primer for the weaker signal are increased, whereas amounts of primer for stronger signals are decreased (9). Both methods are effective, and their use will partially depend on the species coamplified. However, regardless of which approach is used, preliminary experiments will be required to determine optimal cycle numbers and/or primer concentrations. These conditions are likely to be different for each brand of thermal cycler. For example, when Henegariu et al. (9) used the same PCR program to perform multiplex PCR, very reproducible results were obtained from the same machine or model, whereas thermal cyclers from different manufacturers gave markedly different results. However, upon optimization of amplification parameters for multiplex PCR, each brand of thermal cycler gave similar results.

Special care should also be taken to optimize the annealing temperature. This is one of the most important parameters of amplification because slight variations in the annealing temperature can greatly influence both yield and specificity of the reaction (17). The theoretical optimal annealing temperature, based on nearest neighbor analysis, can be determined using commercial software or calculated as described (4,17). However, we and others (9) have found that the optimal annealing temperature for multiplex PCR is slightly lower (2°–6°C) than for a single species. Because lowering the annealing temperature can also increase unwanted hybridization events (13), it is recommended that the PCR be started at a high temperature (i.e., hot start) to prevent formation of artifacts.

Limits and Accuracy

PCR is becoming a well-established method that has much greater sensitivity than Northern blot analysis and ribonuclease protection assay. Optimized multiplex RT-PCR using competitor RNA can significantly (at the level of 95% confidence) distinguish expression levels that differ by as little as 10%–20% (6,19,21). However, it is important not to overestimate the accuracy of competitive PCR, because standard deviations between 10% and 25% in these studies (6,19,21) represent analysis of replicate portions of the same sample on different occasions. Therefore, for reliable discrimination of changes among samples, it is preferable to calculate a mean of replicate determinations from the same RNA sample.

In summary, MPTA has application to a variety of clinical and experimental studies requiring accurate measurements of target transcripts. This procedure is particularly useful for determining mRNA levels in microlocalization studies and in tissues or cells with low copy numbers of these transcripts. This method can also be tailored to simultaneously measure high and low copy numbers of transcript in small amounts (ca. 2 µg) of total RNA. Because absolute results are obtained using MPTA, samples to be compared can be assayed at different times without a loss in experimental accuracy. These aspects make MPTA a rapid and powerful method for examining gene expression.

ACKNOWLEDGMENTS

This work was supported by grants from the National Institutes of Health (K.M.B.; No. HL44883), the American Heart Association (K.M.B.; No. 91003020), the Pennsylvania Affiliate of the American Heart Association (D.E.D.) and the Pennsylvania State University College of Medicine. Dr. Kenneth M. Baker is an Established Investigator of the American Heart Association.

REFERENCES

1. **Ausubel, F.M., R. Brent, R.E. Kingston, D.D. Moore, J.G. Seidman, J.A. Smith and K. Struhl.** 1994. Quantification of DNA and RNA with absorption and fluorescence spectroscopy. Appendix A, p. D1-D3. *In* Current Protocols in Molecular Biology (Suppl 28), John Wiley & Sons, New York.
2. **Babu, J.S., S. Kanangat and B.T. Rouse.** 1993. Limitations and modifications of quantitative polymerase chain reaction. Application to measurement of multiple mRNAs present in small amounts of sample RNA. J. Immunol. Methods *165*:207-216.
3. **Becker-Andre, M. and K. Hahlbrock.** 1989. Absolute mRNA quantification using the polymerase chain reaction (PCR). A novel approach by a PCR aided transcript titration assay (PATTY). Nucleic Acids Res. *17*:9437-9446.
4. **Breslauer, K.J., R. Frank, H. Blocker and L.A. Marky.** 1986. Predicting DNA duplex stability from the base sequence. Proc. Natl. Acad. Sci. USA *83*:3746-3750.
5. **Chelly, J., D. Montarras, C. Pinset, Y. Berwald-Netter, J.C. Kaplan and A. Kahn.** 1990. Quantitative estimation of minor mRNAs by cDNA-polymerase chain reaction. Application to dystrophin mRNA in cultured myogenic and brain cells. Eur. J. Biochem. *187*:691-698.
6. **Dostal, D.E., K.N. Rothblum and K.M. Baker.** 1994. An improved method for absolute quantification of mRNA using multiplex polymerase chain reaction: determination of renin and angiotensinogen mRNA levels in various tissues. Anal. Biochem. *223*:239-250.
7. **Erlich, H.A., D. Gelfand and J.J. Sninsky.** 1991. Recent advances in the polymerase chain reaction. Science *252*:1643-1651.
8. **Gilliland, G., S. Perrin, K. Blanchard and H.F. Bunn.** 1990. Analysis of cytokine mRNA and DNA: detection and quantitation by competitive polymerase chain reaction. Proc. Natl. Acad. Sci. USA *87*:2725-2729.
9. **Henegariu, O., N.A. Heerema, S.R. Dlouhy, G.H. Vance and P.H. Vogt.** 1997. Multiplex PCR: critical parameters and step-by-step protocol. BioTechniques *23*:504-511.
10. **Iwai, N. and T. Inagami.** 1992. Identification of two subtypes in the rat type I angiotensin receptor. FEBS Lett. *298*:257-260.
11. **Lee, J.S.** 1991. Alternative dideoxy sequencing of double-stranded DNA by cyclic reactions using *Taq* polymerase. DNA Cell Biol. *10*:67-73.
12. **McCulloch, R.K., C.S. Choong and D.M. Hurley.** 1995. An evaluation of competitor type and size for use in the determination of mRNA by competitive PCR. PCR Methods Appl. *4*:219-226.
13. **Mullis, K.B.** 1991. The polymerase chain reaction in an anemic mode: how to avoid cold oligodeoxyribonuclear fusion. PCR Methods Appl. *1*:1-4.

14. **Murphy, T.J., R.W. Alexander, K.K. Griendling, M.S. Runge and K.E. Bernstein.** 1991. Isolation of a cDNA encoding the vascular type-1 angiotensin II receptor. Nature *351*:233-236.
15. **Ochman, H., A.S. Gerber and D.L. Hartl.** 1988. Genetic applications of an inverse polymerase chain reaction. Genetics *120*:621-623.
16. **Raeymaekers, L.** 1993. Quantitative PCR: theoretical considerations with practical implications. Anal. Biochem. *214*:582-585.
17. **Rychlik, W., W.J. Spencer and R.E. Rhoads.** 1990. Optimization of the annealing temperature for DNA amplification in vitro. Nucleic Acids Res. *18*:6409-6412.
18. **Saiki, R.** A practical approach. *In* K.E. Davies (Ed.), Genome Analysis. IRL Press at Oxford University Press, Oxford.
19. **Souazé, F., A. Ntodou-Thomé, C.Y. Tran, W. Rostène and P. Forgez.** 1996. Quantitative RT-PCR: limits and accuracy. BioTechniques *21*:280-285.
20. **Wang, A.M., M.V. Doyle and D.F. Mark.** 1989. Quantitation of mRNA by the polymerase chain reaction. Proc. Natl. Acad. Sci. USA *86*:9717-9721.
21. **Xia, H.-Z., C.L. Kepley, K. Sakai, K. Chelliah, A.-M.A. Irani and L.B. Schwartz.** 1995. Quantitation of tryptase, chymase, FcεRIα and FcεRIγ mRNAs in human mast cells and basophils by competitive reverse transcriptase chain reaction. J. Immunol. *154*:5472-5480.

7 | Standard Curve Quantitative Competitive RT-PCR (SC-QC-RT-PCR): A Simple Method to Quantify Absolute Concentration of mRNA from Limited Amounts of Sample

Shaw-Jenq Tsai[1] and Milo C. Wiltbank[2]
[1]*Department of Physiology, College of Medicine, National Cheng Kung University, Tainan, Taiwan and* [2]*Endocrinology–Reproductive Physiology Program and Department of Dairy Science, University of Wisconsin–Madison, Madison, WI, USA*

OVERVIEW

The use of reverse transcription polymerase chain reaction (RT-PCR) with internal RNA competitive standards (competitors) provides a means for measuring absolute amounts of mRNA transcripts in small numbers of cells. Most quantitative RT-PCR methods require analysis of multiple reactions to determine the equimolar point of the products from mRNA vs. competitor RNA. We present a method to produce one standard curve for each assay with all unknown samples compared directly to this standard curve. The standard curve is produced with differing amounts of standard RNA (native) amplified with one constant amount of competitor RNA. The number of transcripts in an unknown mRNA sample can be directly determined by RT-PCR of the sample with the same amount of competitor RNA and comparison of the ratio of products to the standard curve. This method has been used to quantify expression of multiple gene products from cultured cells or limited amounts of tissues and was found to be straightforward, sensitive, repeatable and quantitative. A complete protocol for producing standard and competitor RNA, subsequent quantitative competitive (QC)-RT-PCR steps and an evaluation of accuracy, sensitivity and precision for this assay are described using bovine prostaglandin $F_{2\alpha}$ receptor mRNA as an example.

BACKGROUND

Many different techniques have been used to evaluate concentrations of specific mRNA; however, most of these procedures yield results that are either nonquantitative or semiquantitative. Recently, QC-RT-PCR has been used to quantify absolute amounts of mRNA (4,11,17). Most QC-RT-PCR applications use serial dilutions of competitor with a constant amount of unknown mRNA in the RT-PCR (11). This procedure requires 3–5 reactions for each sample to plot the ratio of products against the amount of added competitor to determine the equimolar amount for each unknown sample. Obviously the procedure can be fairly labor-intensive, particularly if multiple samples and gene products are being analyzed. Recently, a procedure has been introduced that allows a quantitative, single-tube RT-PCR procedure (5); however, this procedure still relies on comparison of results to a quantified competitor. Comparison to a competitor that differs in any way from the native mRNA provides the possibility that any amplification differences between native and competitor mRNA will produce nonquantitative results (9). We have developed a procedure that uses a competitor RNA to account for variation between PCR assay tubes but with final quantitative results obtained by comparison to a standard curve that uses a native RNA with sequence identity to the mRNA of interest. The procedure minimizes the amounts of mRNA needed because multiple reactions at different concentrations of competitor RNA are not needed for each unknown sample. We have used this technique to quantify 3 different mRNA transcripts (prostaglandin G/H synthase-2 [PGHS-2], prostaglandin $F_{2\alpha}$ receptor [FP receptor] and prostaglandin E_2 receptor EP_3 subtype [EP_3 receptor]) in mRNA samples isolated from fewer than 25 000 bovine luteal cells or cultured granulosa cells (15). We report the standard curve (SC)-QC-RT-PCR method using bovine FP receptor as an example.

PROTOCOLS

The whole procedure for SC-QC-RT-PCR is schematically drawn in Figure 1. A step-by-step protocol is discussed in detail in the text.

Materials and Reagents

- 5× PCR buffer: 250 mM Tris-HCl, pH 8.3, 375 mM KCl, 15 mM $MgCl_2$
- Oligo(dT)$_{12-18}$ primer (Promega, Madison, WI, USA)
- dNTP stock, 10 mM each
- SUPERSCRIPT™ RNase H⁻ Reverse Transcriptase (Life Technologies, Gaithersburg, MD, USA)
- PCR cloning vector (e.g., pCR®2.1 [Invitrogen, Carlsbad, CA, USA] or pGEM®-T [Promega])
- Proper restriction enzymes
- T7, T3 or SP6 RNA polymerase
- rNTP stock, 5 mM each
- RNase-free DNase I
- RNase inhibitor
- Phenol saturated with TE buffer (10 mM Tris-HCl, 1 mM EDTA, pH 8.0)

Standard Curve Quantitative Competitive RT-PCR

- Chloroform
- 100% and 70% ethanol
- 2 M sodium acetate, pH 4.0
- *Taq* DNA polymerase and 10× PCR buffer (10 mM Tris-HCl, 50 mM KCl, 1.5 mM $MgCl_2$, 0.1% [wt/vol] Triton® X-100) (Promega)
- 5% nonreducing polyacrylamide gel
- 1× TBE buffer: 0.09 M Tris, 0.09 M boric acid, 0.001 M EDTA, pH 8.0
- Thermal cycler
- Microcentrifuge
- Electrophoresis equipment (Mini-PROTEAN® II Electrophoresis System; Bio-Rad, Hercules, CA, USA)
- Densitometer or image analysis equipment (Collage™ software; Fotodyne, Hartland, WI, USA)

Choice of Primers

Primers should be chosen according to the general guidelines used for other applications. We usually use 18–22-mers with 45%–65% GC content and minimal overlap with known sequences by GenBank® analysis. Primers are chosen to give fragments of 300–600 bp so we can easily design competitors. We usually design primers to flank at least one intron so that the amplified products can be readily distinguished from genomic DNA contamination.

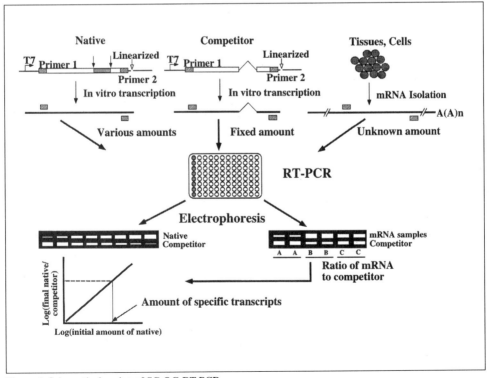

Figure 1. Schematic drawing of SC-QC-RT-PCR.

Preparation of Native and Competitor Templates

We usually use RT-PCR to amplify cDNA products and clone them into a PCR cloning vector (pCR2.1 or pGEM-T) to generate plasmids containing native sequence. Competitor plasmids have been generated using either restriction enzymes to delete an internal fragment or PCR approaches using internal primers (3). We sequence the native and competitor plasmid to confirm the sequence and orientation of the insert. Plasmids are linearized by the proper restriction enzyme and transcribed in vitro by T7 RNA polymerase according to the manufacturer's protocol. To produce quantitative results, it is critical that the plasmid be completely linearized. Any nonlinearized plasmid will generate a large RNA transcript after T7 RNA polymerase transcription, and an erroneously high value for the number of specific RNA transcripts will be calculated from the RNA concentration. After in vitro transcription, plasmids are removed by treatment with DNase I (RNase-free) at 37°C for 15 min. RNA that has been transcribed in vitro is precipitated twice with 0.3 M sodium acetate (pH 4.0) and 2 vol of 100% ethanol after phenol–chloroform extraction. The RNA concentration is determined by optical density $(OD)_{260}$ absorption. RNA is then separated into aliquots and stored at -80°C until used. For accuracy, each RNA aliquot is thawed and used for only one assay.

Protocol 1: Reverse Transcription

To minimize tube-to-tube variation due to pipetting error, we recommend adding competitor RNA to the reaction master mixture. Also, prepare the master mixture for planned reactions and allow some extra for pipetting loss. For example, if we are performing 20 reactions, we usually prepare for 21 reactions to be sure that there is enough for all of the 20 reactions. All procedures should be performed on ice.

Procedure

	Master mixture	Vol/reaction
	5× PCR buffer	4 µL
	DTT (100 mM)	2 µL
	random primer (100 µM)	1 µL
or	specific downstream primer (10 µM)	1 µL
or	oligo(dT) primer (100 ng/µL)	1 µL
	dNTPs (10 mM each)	0.4 µL
	SUPERSCRIPT (200 U/µL)	0.2 µL
	DEPC-treated ddH$_2$O	8.4 µL
	competitor RNA (fixed concentration)	2 µL

1. Dispense 18 µL of the RT master mixture to individual thin-walled tubes.
2. To tubes 1–9, add 2 µL of native RNA (serially diluted). These serve as the standard curve.
3. Add 2 µL of sample RNA to each of the rest of the tubes.
4. Overlay one drop of mineral oil to each tube using a 1-mL Pipetman® (Gilson, Villiers-Bel, France).
5. Perform RT reaction at 42°C for 1 h, then heat to 95°C for 10 min and immediately ramp to 4°C.

6. The cDNA is ready for PCR amplification; otherwise, it should be stored at -80°C until used.

Protocol 2: Polymerase Chain Reaction

Procedure

Master mixture	Vol/reaction	Final concentration
10× reaction buffer	2 µL	
dNTPs (10 mM each)	0.4 µL	0.2 mM each
upstream primer (20 µM)	0.4 µL	0.4 µM
downstream primer (20 µM)	0.4 µL	0.4 µM
Taq DNA polymerase (5 U/µL)	0.1 µL	0.5 U/reaction
ddH$_2$O	11.7 µL	

1. Dispense 15 µL of PCR master mixture to each 0.2-mL thin-walled tube.
2. Add 5 µL of RT cDNA to the PCR master mixture (final volume 20 µL/tube).
3. Add one drop of mineral oil to each tube using a 1-mL Pipetman.
4. Set up PCR using the following program:

Step 1	95°C	30 s
Step 2	57°C[a]	30 s
Step 3	72°C	30 s
Step 4	Repeat Step 1 29 times[b]	
Step 5	72°C	5 min
Step 6	4°C	0 s
Step 7	End	

5. Hold the thermal cycler at 95°C, place tubes in the thermal cycler and wait for 30 s.
6. Start the PCR.

[a]Annealing temperature should be adjusted according to the primer's melting temperature.
[b]Repeating 29 times gives a total of 30 cycles. The PCR cycle number should also be adjusted according to the amount of template added in the reaction.

Protocol 3: Electrophoresis and Image Analysis

Procedure

1. PCR products (10 µL) are directly separated on a 5% polyacrylamide gel with 1× TBE buffer (0.09 M Tris, 0.09 M boric acid, 0.001 M EDTA, pH 8.0) at 110 V for 40 min using the Mini-PROTEAN II electrophoresis system.
2. The gel is then stained with ethidium bromide and placed on a UV illuminator equipped with a camera connected to a Macintosh® computer (Apple Computer, Cupertino, CA, USA).
3. The gel image is analyzed using Collage software or another image analysis system.

4. A ratio is calculated for the intensity of native vs. competitor bands on each lane of the gels.
5. For constructing a standard curve, logarithmically transformed ratios of native-to-competitor band intensity are plotted against logarithmically transformed amounts of native RNA initially added in the RT reaction (Figures 1 and 2).
6. Logarithmically transformed ratio of sample to competitor is compared to standard curve to calculate amounts of transcripts in each sample (see schematic in Figure 1).

RESULTS AND DISCUSSION

Bovine FP receptor cDNA (native) partial sequence and internally deleted DNA (competitor) are schematically shown in Figure 2A. The sequence of competitor (12), bPGF1Δ55, is identical to the sequence of native, bPGF1, except for the removal of a 55-bp *Hin*fI fragment. Both native and competitor had consensus primer sequences (indicated as hatched boxes) and can be amplified by a single primer pair (Figure 2B). In this study, we chose to produce our competitor by restriction enzyme digestion and re-ligation of the products. Other PCR-based methods for the production of

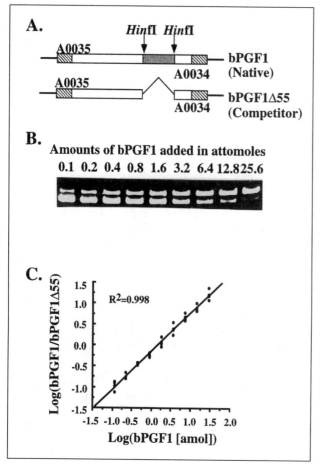

Figure 2. Schematic drawing of bovine FP receptor cDNA fragment (bPGF1) and internally deleted DNA (bPGF1Δ55) (A), ethidium bromide-stained PCR products (B) and the plot of standard curves (C). Panel A shows that the native and competitor have exactly the same sequence, except for removal of a 55-bp *Hin*fI fragment and hence can be amplified by the same set of primers (A0034 and A0035, indicated as hatched boxes). Panel B shows the bands for 0.1–25.6 amol of native (bPGF1) co-amplified with 1 amol of competitor (bPGF1Δ55) for 30 cycles and separated on a 5% polyacrylamide gel. Panel C shows the composite of 4 standard curves obtained from 4 independent experiments. The log ratio of native to competitor product was plotted against the log amount of initial native added to the RT-PCR.

competitors have been described (3,4), and the procedure of Celi et al. (3) has been particularly efficient for producing competitors to many gene products that we analyze in our laboratory. One advantage of this procedure is that appropriate restriction enzyme sites do not need to be identified for deletion of the internal fragment.

For this assay, the band intensities of the PCR products are directly quantified after gel electrophoresis. Figure 2 shows a typical gel image from a standard curve that contains 9 different amounts of native bPGF1 standard (0.1–25.6 amol) with one constant amount of competitor bPGF1Δ55 RNA (1 amol). There is both an increase in amount of native product (upper band) and a decrease in amount of competitor product (lower band) with increasing amounts of initial standard RNA used (Figure 2B). Use of a constant amount of competitor in every amplification tube accounts for possible between-tube differences in amplification efficiency with the same set of primers used for amplifying native RNA. Figure 2C shows a combination of standard curves compiled from 4 different days. It is clear that the standard curve is highly repeatable and linear over more than 2 orders of magnitude. We used RNA instead of cDNA templates for both native and competitor to prevent underestimation of mRNA transcripts due to potentially low RT efficiency (18). To prevent the potential degradation of RNA, the native and competitor RNAs were separated into aliquots, stored at -80°C and used on only one occasion. We found that the single use of RNA produced very consistent results over several months.

This assay uses a standard curve methodology similar to that used for enzyme-linked immunosorbent assays (ELISAs), radioimmunoassays and other quantitative assays, and similar procedures can be used to assess validity of results. For example, we have prepared standard preparations of mRNA containing low (0.1 ng), medium (1 ng) or high (10 ng) concentrations of FP mRNA for use as quality control samples in all assays. The coefficient of variation (CV) can be calculated within an assay and between assays as a measure of assay precision. In our assay with FP receptor, the intra- and inter-assay CVs were 8.7% and 9.9% (mean of high, medium and low mRNA samples evaluated in triplicate on 4 occasions). As expected, the lowest intra- and inter-assay CVs were found near the point of equimolar ratio of mRNA to competitor with larger CVs found when

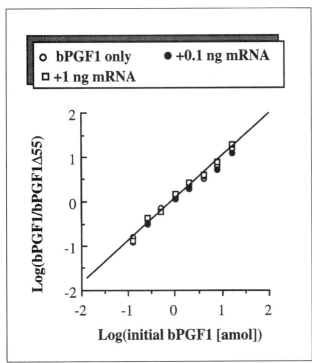

Figure 3. Effects of nonspecific mRNA on the slope and intercept of standard curve. RT master mixtures were prepared according to the standard procedure with the addition of 0, 0.1 or 1 ng of mRNA isolated from bovine liver and subjected to RT-PCR amplification.

either mRNA or competitor was in excess. Thus, precision of SC-QC-RT-PCR assays can and should be evaluated by providing quality control CVs for each assay.

A prerequisite for equimolar QC-RT-PCR is similar RT and PCR efficiencies between samples and competitor RNA. Amplification efficiency differences of greater than 10-fold have been found between competitors of different lengths (9). Use of competitors with different amplification efficiencies would obviously result in substantial variation in the calculated equivalence points unless corrections are made (9). The present assay using standard curve methodology did not require that the amplification efficiency of native and competitor RNA be equivalent. This was because the absolute amount of competitor RNA was not used in any of the calculated values. The slope of the standard curve will be related to the amplification efficiency of the competitor vs. native (15). Obviously, the amount of competitor must be constant among all wells, and this is assured by adding the competitor to the RT master mixture. In addition, accurate quantification requires that the amplification efficiency of the native RNA used in the standard curve and the unknown mRNA be similar. In this study, no direct comparison of amplification efficiency was made; however, both amplified products are of identical size and nucleotide sequence, making it likely that amplification efficiency is similar (9). Addition of known amounts of standard RNA to a background of mRNA (0.1 and 1 ng liver mRNA) did not change the slope or intercept of the standard curve (Figure 3), providing evidence that nonspecific mRNA does not alter amplification efficiency. It cannot be ruled out that differences in size of native mRNA vs. standard RNA could produce differences in RT efficiency and thus alter the absolute, but not relative, accuracy of final results. As with an ELISA or radioimmunoassay, the similarity of standard and unknown samples can be evaluated by comparing the parallelism of standard curves as a measure of assay accuracy. Figure 4 shows that parallel lines are produced using either the standard preparation of RNA from the bPGF1 plasmid or using mRNA isolated from bovine corpus luteum (2-fold serial dilutions of mRNA or bPGF1 RNA). It is apparent that a standard, quantified pool of mRNA could be used for the standard curve in place of the bPGF1 RNA.

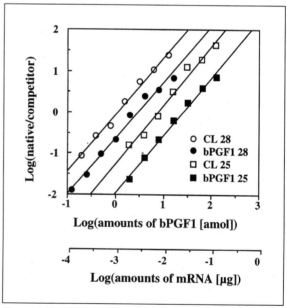

Figure 4. Parallelism of QC-RT-PCR between standard RNA (bPGF1) and mRNA. Serial dilutions of mRNA from bovine corpus luteum (open circle: 0.2–12.5 ng; open square: 1.56–100 ng) or bPGF1 (filled circle: 0.125–16 amol; filled square: 2–128 amol) were used in RT-PCR in the presence of 2 amol (28 cycles, circles) or 16 amol (25 cycles, squares) of bPGF1Δ55.

Figures 4 and 5 also show that increased sensitivity can be obtained by increasing the number of PCR cycles. Amounts of competitor and standard RNA are decreased to maintain quantifiable band intensities. Acceptable standard curves were obtained between 25 and 34 cycles of PCR (Figure 5). It is important that amounts of standard RNA, competitor RNA and unknown mRNA and PCR cycle number be carefully selected in a particular assay to give bands with an intensity that can be accurately quantified by the imaging equipment. It is difficult to provide a precise measure of sensitivity for these assays; however, if the ratio at -1 log unit is arbitrarily selected as a measure of sensitivity, then the sensitivity increases from about 1 million molecules at 25 cycles to 3360 molecules at 34 cycles. This sensitivity is similar to what has been reported with NASBA (16). The present study was not designed to determine the limit of sensitivity for this assay, but theoretically this limit could be only a few molecules with sufficient cycles of PCR amplification.

Heteroduplex formation can occur at higher numbers of PCR cycles (13), and this can seriously confound results with the equimolar QC-RT-PCR method (9). In our assays, we have attempted to eliminate overamplification by minimizing PCR cycle number or amount of starting RNA, and this practice virtually eliminates heteroduplex formation. We have found that greater numbers of PCR cycles produced larger variation in final results, regardless of heteroduplex formation. Our sensitive imaging equipment allows accurate assessment of very low-intensity bands, reducing required amplifications. In addition, possible interference of ethidium bromide-stained bands should be minimized by having sufficiently separated standard and competitor products. Obviously, other methods can be used to quantify standard and competitive PCR products (1,6,8,11), such as: *(i)* fluorescently labeled primers quantified with an automated DNA sequencer (10), *(ii)* electrochemiluminescence detection (16), *(iii)* UV absorbance detection after HPLC (5) or *(iv)* DNA stains with greater sensitivity (13).

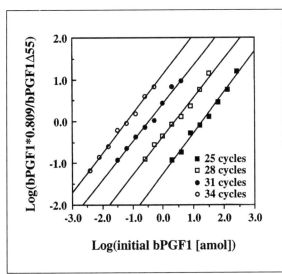

Figure 5. **Sensitivity of QC-RT-PCR can be increased by using more PCR cycles.** More than 4 orders of magnitude of bPGF1 (256–0.004 amol) in 2-fold serial dilutions were quantified using 25 amol (25 cycles, open circle), 4 amol (28 cycles, filled circle), 0.5 amol (31 cycles, open square) and 0.05 amol (34 cycles, filled square) of bPGF1Δ55. Note that the *y*-axis values were justified with a constant, 0.809 (233/288), to reflect the actual band intensity because longer DNA tends to cause more intercalation of ethidium bromide. This is required in calculating results from the equimolar QC-RT-PCR procedure but is not necessary in SC-QC-RT-PCR because the constant is cancelled out when comparing the unknown to the standard curve.

The SC-QC-RT-PCR protocol can easily be adapted to quantify mRNA concentration from normal cell cultures. In our previous experiments (15), we have directly lysed cells in a homogenization buffer, and mRNA was isolated using magnetic beads attached to oligo(dT)$_{25}$. We have quantified genomic DNA in the supernatant using Hoechst 33258 dye with calf thymus DNA as a standard (7) after isolation of mRNA. The DNA values can be converted to cell number using a constant that we had previously determined in our laboratory (2). Therefore, final results can be accurately expressed as number of specific mRNA transcripts per cell. The sensitivity of this procedure allows many different specific mRNAs to be quantified from the same mRNA preparation (14,15).

We have calculated the amounts of FP receptor mRNA

in 5 different cell preparations using SC-QC-RT-PCR and equimolar QC-RT-PCR. The calculated values for these samples were similar (statistically) with a correlation of 94% between the 2 methods. Equimolar QC-RT-PCR required five times more mRNA sample, more lanes of gel space if more than 2 samples were being analyzed and greater analysis time because the equivalence point was determined for each sample. In addition, the discrepancy of amplification efficiencies of native vs. competitor could influence the outcome in equimolar QC-RT-PCR (9). Thus, SC-QC-RT-PCR should yield results that are similar to or more accurate than equimolar QC-RT-PCR with need for less mRNA sample, fewer gels and less analysis time.

In conclusion, QC-RT-PCR using a standard curve methodology is a simplified method that allows the quantification of absolute amount of mRNA transcripts in an unknown sample. The whole procedure including RT, PCR and image analysis is straightforward and can be done in a normal working day. Multiple samples can be analyzed by this procedure without a great deal of extra effort. The procedure is extremely sensitive and can be used to evaluate gene expression from limited amounts of experimental material. The sensitivity can be increased by simply increasing PCR cycles, thus eliminating the need for radioisotopes. Moreover, the results can be expressed in a biologically meaningful way as specific transcripts per cell. The procedure was found to be highly repeatable, and validity of results could be tested by procedures that are used with ELISAs or radioimmunoassays. Most standard laboratories should be able to adopt this technology because the only specialized equipment required is a densitometer equipped with image analysis software.

REFERENCES

1. **Alard, P., O. Lantz, M. Sebagh, C.F. Calvo, D. Weill, G. Chavanel, A. Senik and B. Charpentier.** 1993. A versatile ELISA-PCR assay for mRNA quantitation from a few cells. BioTechniques *15*:730-737.
2. **Bodensteiner, K.J., M.C. Wiltbank, D.R. Bergfelt and O.J. Ginther.** 1996. Alteration in follicular estradiol and gonadotropin receptors during development of bovine antral follicles. Theriogenology *45*:499-507.
3. **Celi, F.S., M.E. Zenilman and A.R. Shuldiner.** 1993. A rapid and versatile method to synthesize internal standard for competitive PCR. Nucleic Acids Res. *21*:1047.
4. **Gilliland, G., S. Perrin, K. Blanchard and F. Bunn.** 1990. Analysis of cytokine mRNA and DNA: detection and quantitation by competitive polymerase chain reaction absolute quantitative measurement of gene expression by single reaction. Proc. Natl. Acad. Sci. USA *87*:2725-2729.
5. **Hayward-Lester, A., P.J. Oefner, S. Sabatini and P.A. Doris.** 1995. Accurate and absolute quantitative measurement of gene expression by single-tube RT-PCR and HPLC. Genome Res. *5*:494-499.
6. **Kohsaka, H., A. Taniguchi, D.D. Richman and D.A. Carson.** 1993. Microtiter format gene quantification by covalent capture of competitive PCR products: application to HIV-1 detection. Nucleic Acids Res. *21*:3469-3472.
7. **Labarca, C. and K. Paigen.** 1980. A simple, rapid, and sensitive DNA assay procedure. Anal. Biochem. *102*:344-352.
8. **Lear, W., M. McDonnell, S. Kashyap and P.H. Boer.** 1995. Random primer $p(dN)_6$-digoxigenin labeling for quantitation of mRNA by Q-RT-PCR and ELISA. BioTechniques *18*:78-83.
9. **McCulloch, R.K., C.S. Choong and D.M. Hurley.** 1995. An evaluation of competitor type and size for use in the determination of mRNA by competitive PCR. PCR Methods Appl. *4*:219-226.
10. **Porcher, C., M.-C. Malinge, C. Picat and B. Grandchamp.** 1992. A simplified method for determination of specific DNA or RNA copy number using quantitative PCR and an automatic DNA sequencer. BioTechniques *13*:106-114.
11. **Reischl, U. and B. Kochanowski.** 1995. Quantitative PCR: a survey of the present technology. Mol. Biotechnol. *3*:55-71.
12. **Sakamoto, K., T. Ezashi, K. Miwa, E. Okuda-Ashitaka, T. Houtani, T. Sugimoto, S. Ito and O. Hayaishi.** 1994. Molecular cloning and expression of a cDNA of the bovine prostaglandin F_2 alpha receptor. J. Biol. Chem. *269*:3881-3886.
13. **Schneeberger, C., P. Speiser, F. Kury and R. Zeillinger.** 1995. Quantitation detection of reverse transcriptase-PCR products by means of a novel and sensitive DNA stain. PCR Methods Appl. *4*:234-238.

14. **Tsai, S.-J. and M.C. Wiltbank.** 1997. Prostaglandin $F_{2\alpha}$ induces expression of prostaglandin G/H synthase-2 in the ovine corpus luteum: a positive feedback loop during luteolysis. Biol. Reprod. *57*:1016-1022.
15. **Tsai, S.-J., M.C. Wiltbank and K.J. Bodensteiner.** 1996. Distinct mechanisms regulate induction of messenger ribonucleic acid for prostaglandin (PG) G/H synthase-2, PGE (EP_3) receptor, and $PGF_{2\alpha}$ receptor in bovine preovulatory follicles. Endocrinology *137*:3348-3355.
16. **Van Gemen, B., P.V.D. Wiel, R. Van Beuningen, P. Sillekens, S. Jurriaans, C. Dries, R. Schoones and T. Kievits.** 1995. The one-tube quantitative HIV-1 RNA NASBA: precision, accuracy, and application. PCR Methods Appl. *4*:S177-S184.
17. **Wang, A.M., M.V. Doyle and D.F. Mark.** 1989. Quantification of mRNA by the polymerase chain reaction. Proc. Natl. Acad. Sci. USA *86*:9717-9721.
18. **Zenilman, M.E., W. Graham, K. Tanner and A.R. Shuldiner.** 1995. Competitive reverse-transcriptase polymerase chain reaction without an artificial internal standard. Anal. Biochem. *224*:339-346.

8. Design and Use of an Exogenously Added Competitive RNA Multiplex Template for Measurement of mRNA Levels by RT-PCR

Roy W. Tarnuzzer
Division of Endocrinology and Metabolism, Department of Medicine, University of Florida, Gainesville, FL, USA

OVERVIEW

Detection of low-abundance mRNAs by reverse transcription polymerase chain reaction (RT-PCR) has become a standard technique to determine gene expression by tissues and cells in culture. The ability to determine relative or absolute copy number of specific mRNAs has been difficult because of inadequate internal standards to control for sample-to-sample variation. The use of a synthetic RNA standard with identical sequences to the PCR primers allows reproducible quantitation between samples and assays. By designing multiplex templates, several specific mRNAs can be quantified using a single template. Addition of multiple templates to a single RT reaction allows the quantitation of a large number of targets from as few as 4 µg of total RNA. We present a method describing the design of competitive RNA templates to detect and quantitate specific messages in total RNA samples.

BACKGROUND

Standard RT-PCR is most useful to determine if a specific mRNA is present in sufficient numbers to be amplified to a detectable level and is a more sensitive alternative to Northern blot analysis (2,3). Relative or semiquantitative RT-PCR uses an endogenous standard (such as β-actin or glyceraldehyde-3-phosphate dehydrogenase [*GAPD*]) and can determine relative message levels provided the endogenous standard remains unchanged. Actual numbers of mRNA molecules in a sample cannot be determined, and data from two different experiments cannot be compared accurately (6,7). Competitive RT-PCR is more versatile because it permits the determination of the number of mRNA molecules in a sample. Data obtained by this technique can be compared between different experiments (1,5).

Competitive RT-PCR is based on the principal that upstream and downstream oligonucleotide primers will compete equally for authentic cellular cDNA molecules and synthetic template cDNA during the amplification reaction. Competition will be established provided the exogenous synthetic template contains identical complementary sequences found in the authentic mRNA. As the concentration of the synthetic template RNA changes in relation to the authentic mRNA, the number of amplified PCR products of the two species will also change proportionally (10). By designing the synthetic template so that it generates a PCR product of a different size than those of the true mRNA, the products can be easily separated and identified on agarose gels.

The quantitative RT-PCR procedure we describe allows the measurement of up to twenty individual mRNAs from as few as 4 µg of total RNA when multiple competitive RNA templates are included in the reaction. The requirement for small amounts of total RNA, the ability to measure large numbers of distinct mRNA targets in a single sample and the fidelity to amplify sequences of different origin and length with equal efficiency make these templates and this competitive RT-PCR technique cost-effective and applicable to many research models. In our experience, this technique allows measurement of mRNA levels in a nonradioactive protocol using agarose gel electrophoresis. Measurements from ethidium bromide-stained agarose gels have proven to be reproducible within the same RNA sample and between samples from similar tissues or cells (9,12). The high sensitivity of this assay system and the ability to measure numerous mRNA targets from a single, small total RNA sample make this approach to mRNA quantitation easily applicable to small tissue biopsies or small numbers of cultured cells.

PROTOCOLS

Materials and Reagents

- dNTPs (Promega, Madison, WI, USA) (stock made up to 10 mM, final working concentration 200 µM)
- Oligo(dT) (Life Technologies, Gaithersburg, MD, USA) (0.5 µg/mL, working concentration 0.125 µg/µg RNA)
- Moloney murine leukemia virus (MMLV) Reverse Transcriptase (Life Technologies) (200 U/µL, working concentration 200 U/µg RNA)
- Placental RNase Inhibitor (Life Technologies) (10 U/µL, working concentration 2.5 U/µg RNA)
- 10× PCR buffer: 100 mM Tris-HCl, pH 8.3, 500 mM KCl, 15 mM $MgCl_2$ (use at 1×)
- AmpliTaq® DNA Polymerase (Perkin-Elmer, Norwalk, CT, USA) (5 U/µL, working concentration 1.25 U/reaction)

Note: These manufacturers' reagents are what we typically use for our assays. Other manufacturers' products should work equally as well with optimized protocols.

PCR Primer and Synthetic Template Design

The first step in the design of a synthetic multiplex template is to determine the mRNAs of interest and how many targets are going to be incorporated into the

Table 1. Insert Size, PCR Product Size and Recommended Target Size for Various Numbers of Target mRNAs

Number of Targets	Insert Size (bp)	PCR Product Size (bp)	Recommended Target Size (bp)
10	400	220	320
11	440	240	340
12	480	260	360
13	520	280	380
14	560	300	400
15	600	320	420
16	640	340	440

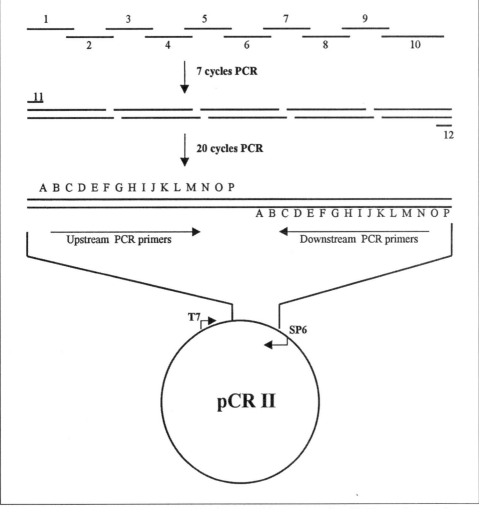

Figure 1. Map and scheme for the construction of a competitive template plasmid. The template contains complementary sequences homologous to the PCR primers for up to 16 different genes. The construct is generated by amplification of up to 8 overlapping oligonucleotides by PCR and subsequent cloning into a TA Cloning vector.

design. Some of our earlier designs incorporated internal detection probe oligonucleotide sequences for detection using radioisotopes and hybridization. Ethidium bromide staining of agarose gels proved sensitive and reproducible enough to forego detection sequences. Thus, current template design allows a larger number, up to 16, of target primer pairs to be incorporated into a single multiplex template. We routinely use PCR primers that are 20 bases long for our amplification reactions. These primers are chosen so that their PCR products are approximately 100 bp larger than the template product. Therefore, an insert containing 5′ and 3′ primers of 20 bases for 16 total target sequences would generate a total insert size of 640 bp (Table 1). Because most oligonucleotide synthesis facilities recommend oligonucleotide synthesis of fewer than 100 bases for reasonable efficiency and yield of full-length product, we design overlapping oligonucleotides of fewer than 90 bases. The strategy for template construction requires several 80–90-base oligonucleotides to be synthesized that share 25-base terminal complementary sequences. Therefore, to span 640 bases would require five 86-mers and five 87-mers (Figure 1). When the primer sequences are incorporated into the template sequence, the order of the 5′ and 3′ primers is kept the same so that a PCR product of similar size is produced regardless of which primer pair is used in the PCR. Always test the PCR primers before incorporating their sequences into the template design.

Synthetic Supertemplate Plasmid Construction

Overlapping oligonucleotides (five 86-mers and five 87-mers as in the above example) containing the complementary sequences to the sense and antisense PCR primers (Figure 1) and 2 flanking 21-base oligonucleotides were synthesized by a DNA synthesis facility. This is a modification of a procedure described by Dillon and Rosen (4). The template synthesis reaction is set up as follows.

Procedure

Step one

1. Add 0.5 µg of each of the template olignucleotides to a 500-µL microcentrifuge tube.
2. Add 10 µL of 10× PCR buffer.
3. Add 2 µL of 10 mM dNTP mixture.
4. Add 0.5 µL AmpliTaq DNA polymerase.
5. Adjust volume to 100 µL with sterile, distilled H_2O.
6. Amplify by PCR for 7 cycles (94°C for 1 min, 55°C for 2 min and 72°C for 3 min).

Step two

1. Add 1 µL of the reaction mixture from Step one to a new 500-µL microcentrifuge tube.
2. Add 1 µg of the flanking 20-base oligonucleotides.
3. Add 10 µL of 10× PCR buffer.
4. Add 2 µL of 10 mM dNTP mixture.

Design and Use of Competitive RNA Multiplex Template

Table 2. Master Reagent Mixture for RT Reactions

No. of Reactions	10	15	20	25	30	35	40	45	50
10× PCR buffer	50	75	100	125	150	175	200	225	250
10 mM dNTP	10	15	20	25	30	35	40	45	50
Oligo(dT)	2.5	3.8	5	6.3	7.5	8.8	10	11.3	12.5
RNase inhibitor	2.5	3.8	5	6.3	7.5	8.8	10	11.3	12.5
MMLV reverse transcriptase	10	15	20	25	30	35	40	45	50
DEPC-treated, distilled H$_2$O	375	562.4	750	977.4	1125	1312.4	1540	1687.4	1875
Total volume	450	675	900	1125	1350	1575	1800	2025	2250

All numbers represent volumes in µL.

5. Add 0.5 µL AmpliTaq DNA polymerase.
6. Adjust volume to 100 µL with sterile, distilled H$_2$O.
7. Amplify by PCR for 25 cycles (94°C for 1 min, 55°C for 2 min and 72°C for 3 min).

Twenty microliters of this reaction mixture are electrophoresed on a 2% low-melting-point agarose gel, and the amplified band is excised and heated to 65°C for 10 min. From this point, any standard PCR fragment cloning protocol and fragment isolation technique can be used. We routinely use the pCR®II TA Cloning® Kit (Invitrogen, Carlsbad, CA, USA) to clone the template inserts. Once a positive clone is isolated, a second cloning step is carried out to insert a poly(A) tail downstream of the template. Depending on the vector used to clone the template, an oligonucleotide cassette is designed for insertion into a unique restriction site 3′ of the template sequence (Figure 2). Again, positive clones should be identified and a large-scale plasmid isolation performed.

Preparation of Competitive RNA Template

Competitive template RNA is generated from the template plasmid by transcription with T7 or SP6 RNA polymerase after linearization with a restriction enzyme that cuts immediately downstream of the template insert's poly(A) tail. Several kits are available, and we routinely use the Promega T7 or SP6 transcription kits and follow the manufacturer's protocol. Following transcription, the reaction mixture is digested with proteinase K at 37°C for 1 h, extracted 2 times with an equal volume of phenol–

Figure 2. Oligonucleotide cassette for cloning of poly(A) tail into the template. Restriction endonuclease sites were engineered at the 5′ and 3′ ends to facilitate cloning into the vector of choice, in this case *Bam*HI and *Hin*dIII (underlined). An *Xba*I site was incorporated for screening for the insertion of the cassette (italics).

Table 3. Master Reagent Mixture for PCR Amplifications

No. of Reactions	10	15	20	25	30	35	40	45	50
10× PCR Buffer	50	75	100	125	150	175	200	225	250
10 mM dNTP	10	15	20	25	30	35	40	45	50
AmpliTaq DNA Polymerase	2.5	3.8	5	6.3	7.5	8.8	10	11.3	12.5
PCR primers	10	15	20	25	30	35	40	45	50
DEPC-treated, distilled H_2O	377.5	566.2	755	983.7	1132.5	1321.2	1550	1698.7	1887.5
Total volume	450	675	900	1125	1350	1575	1800	2025	2250

All numbers represent volumes in microliters.

Table 4. Calculations for Normalized Band Ratios

Lane	Template Copies	Peak Area Template	Peak Area Target	Normalized Values Template	Normalized Values Target	Ratio
1	6.1×10^6	2168	184	6.45	0.34	18.97
2	6.1×10^5	1366	502	4.07	0.93	4.38
3	6.1×10^4	1423	449	4.23	0.83	5.09
4	6.1×10^3	668	390	1.99	0.72	2.76
5	6.1×10^2	343	538	1.02	0.99	1.03

chloroform–isoamyl alcohol (1:1:0.2) and the poly(A^+) template purified by oligo(dT) chromatography. RNA is quantified by absorbance at 260 nm, the number of molecules per microliter is calculated, and the RNA is stored in small volumes at -80°C.

Competitive RT-PCR

Procedure

Reverse transcription

1. Label a series of 500-µL microcentrifuge tubes: 10^{-1}, 10^{-2}, 10^{-3}, 10^{-4}, 10^{-5}, 10^{-6}, 10^{-7}, 10^{-8}.
2. Make 10-fold serial dilutions of the RNA template with diethyl pyrocarbonate (DEPC)-treated, distilled H_2O.
3. Add 1 µL of each dilution to a separate reaction tube (7 tubes).
4. Add 2 µg of total sample RNA.
5. Bring volume in tube to 5 µL with DEPC-treated, distilled H_2O.
6. Make a master reaction mixture containing all RT reagents as outlined in Table 2.
7. Add 45 µL of master reagent mixture to each tube.
8. Incubate reaction tubes at 25°C for 10 min, 37°C for 60 min and 94°C for 5 min.

Polymerase chain reaction

1. Add 5 μL from each RT reaction mixture to a new microcentrifuge tube.
2. Make a master reaction mixture containing all RT reagents as outlined in Table 3.
3. Add 45 μL of master reagent mixture to each tube.
4. Amplify by PCR for 1 cycle at 94°C for 5 min, 58°C for 2 min and 72°C for 3 min, followed by 40 cycles at 94°C for 1.5 min, 58°C for 2 min and 72°C for 3 min, and 1 cycle at 72°C for 10 min.
5. Electrophorese 15 μL of each reaction mixture on a 2% agarose gel and photograph.

Quantitation of PCR Products

Procedure

The first step in the analysis is to generate a digitized image of the stained agarose gel (Figure 3). This can be done directly with a digital camera or by scanning a photograph of the gel. The image can then be imported into NIH Image (**http://rsb.info.nih.gov/nih-image**) or another suitable gel analysis software package. The objective is to generate a value proportional to the intensity of the ethidium bromide-stained band.

Upon opening the image file in a gel analysis program, box each lane separately. Again, each lane should contain two bands—one for the target and one for the template. Plot the histogram of the lanes and draw a baseline through the peaks similar to that shown in Figure 4. Then integrate the area under the peaks. This is the value used

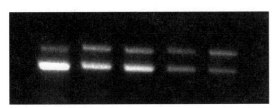

Figure 3. Ethidium bromide-stained gel showing competitive RT-PCR results. The upper band is the native cDNA product, and the lower band is the template product.

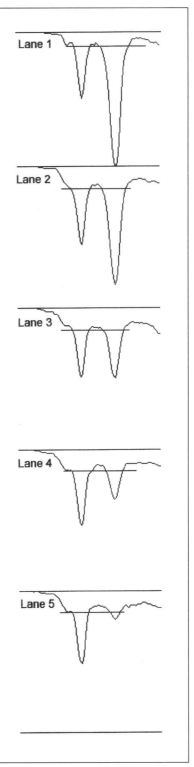

Figure 4. Histogram of band areas from a scan of an agarose gel.

to normalize for the molecular weights of the PCR products and calculate a ratio of band intensities. For example, from the values generated from the histogram in Figure 4, the following calculations can be made (Table 4):

- Lane 1 has: a 541-bp band with an area of 184 pixels^2 and
  a 336-bp band with an area of 2168 pixels^2.
- Lane 2 has: a 541-bp band with an area of 502 pixels^2 and
  a 336-bp band with an area of 1366 pixels^2.
- To calculate the normalized band intensities:
  lane 1: 2168/336 = 6.45 and 184/541 = 0.34;
  lane 2: 1366/336 = 4.07 and 502/541 = 0.93.
- The ratio of band intensities will equal:
  lane 1: 6.45/0.34 = 18.97 and lane 2: 4.07/0.93 = 4.38.
- Repeat the analysis for all measured lanes.

Finally, plot these calculated values against the starting copy number of the template added to each reaction as shown in Figure 5. This is usually expressed as copy number/micrograms total RNA and calculated as follows.

Starting as in the above example with 1.22×10^{10} copies/µL of RNA template, if 1 µL of a 100-fold dilution were added to the equivalent of 2 µg of sample RNA, it would generate 6.1×10^7 copies/µg total RNA.

It is recommended that each point be repeated in triplicate and the average of the values with standard deviations be plotted as in Figure 5.

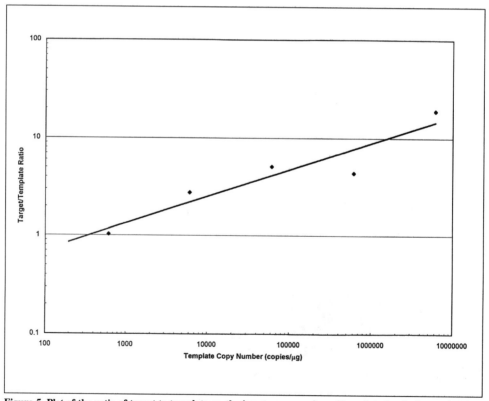

Figure 5. Plot of the ratio of target to template vs. the input copy number of template (copies/µg total RNA).
The copy number of the message can be determined where the plot of the line passes through 1.0.

Interpretation of Data

From the plot in Figure 5, where the line of the input copy number versus template-to-target ratio intersects 1, the copy number of the mRNA can be determined by extrapolating to the x axis. In this example, there would be 3.5×10^2 copies/µg of RNA.

RESULTS AND DISCUSSION

Regardless of whether the target mRNAs are amplified in a separate RT-PCR from total RNA, the results should show that the template and its respective mRNA targets for the selected primer pairs amplify at the same rate. This would be demonstrated by the fact that the slopes of the lines are parallel regardless of the internal sequence differences between the target and template RNAs, message size or the presence of multiple templates in the same reaction tube (12). The ability to use this type of competition reaction to quantify the levels of mRNAs in cellular RNA requires that certain criteria be met. The efficiency of extension in the amplification reaction of the target and the competitive template must be the same regardless of internal sequence differences and overall length of the products. If the efficiencies of annealing and polymerization are equal, then the slope of the line determined by plotting the log of the ratio of template-to-target band intensities versus the log of the input template copy number must equal one (8,11). Thus, the efficiencies of annealing and polymerization for the targets on the templates must be equal regardless of sequence, size and abundance of the message. In some instances, the slopes might deviate from a value of 1.0. If the slopes of the reactions for the samples one is comparing are equal, then the analysis should remain robust. Again, by repeating the analysis in triplicate, many of these anomalies can be corrected.

In the construction of new templates, a problem arose with one design attempted involving closely related mRNA sequences. When the sequences were aligned, apparently the random sequence generated was prone to form a hairpin loop, and repeated deletion of a portion of the insert was encountered upon cloning of the template. This might be avoided by examining the aligned sequence of the template with a nucleic acid folding/secondary structure analysis program. If such a problem arises, try shuffling the sequences and rerunning the analysis until the problem is solved.

In our hands, this approach to quantitation of mRNA levels has proven to be reliable and reproducible. These templates represent a self-renewing and powerful tool applicable to any system that might express the target mRNAs of interest. Once constructed, the templates represent a rapid and reliable method for the measurement of mRNA levels from small RNA samples.

ACKNOWLEDGMENTS

We would like to thank the University of Florida Interdisciplinary Center for Biotechnology Research's DNA synthesis core for the rapid and reliable production of our oligonucleotides. This project was supported in part by funds from NIH Grant No. EY05587, the US Army Medical Research and Development Command Contract No. DAMD17-91-C-1095 and the Swedish Medical Research Council Project No. 9881.

REFERENCES

1. **Bouaboula, M., P. Legoux, B. Pessegue, B. Delpech, X. Dumont, M. Piechaczyk, P. Casella and D. Shire.** 1992. Standardization of mRNA titration using a polymerase chain reaction method involving co-amplification with a multispecific internal control. J. Biol. Chem. *267*:21830-21838.
2. **Chelly, J., J.-C. Kaplan, P. Maire, S. Gautron and A. Kahn.** 1988. Transcription of the dystrophin in human muscle and non-muscle tissue. Nature *338*:858-860.
3. **Delidow, B.C., J.J. Peluso and B.A. White.** 1989. Quantitative measurement of mRNA by polymerase chain reaction. Gene Anal. Tech. *6*:120-124.
4. **Dillon, P.J. and C.A. Rosen.** 1990. A rapid method for the construction of synthetic genes using the polymerase chain reaction. BioTechniques *9*:298-300.
5. **Funk, C.D. and G.A. FitzGerald.** 1991. Eicosanoid forming enzyme mRNA in human tissues. Analysis by quantitative polymerase chain reaction. J. Biol. Chem. *266*:12508-12513.
6. **Hoof, T., J.R. Riordan and B. Tummler.** 1991. Quantitation of mRNA by the kinetic polymerase chain reaction assay: A tool for monitoring P-glycoprotein gene expression. Anal. Biochem. *196*:161-169.
7. **Murphy, L.D., C.E. Herzog, J.B. Rudick, A.T. Fojo and S.E. Bates.** 1990. Use of polymerase chain reaction in the quantitation of *mdr-1* gene expression. Biochemistry *29*:10351-10356.
8. **Raeymaekers, L.** 1994. Comments on quantitative PCR. Eur. Cytokine Netw. *5*:57.
9. **Riedy, M.C., E.A. Timm, Jr. and C.C. Stewart.** 1995. Quantitative RT-PCR for measuring gene expression. BioTechniques *18*:70-76.
10. **Scadden, D.T., Z. Wang and J.E. Groopman.** 1992. Quantitation of plasma immunodeficiency virus type 1 by competitive polymerase chain reaction. J. Infect. Dis. *165*:1119-1123.
11. **Shire, D.** 1994. Quantitative PCR: reply to L. Raeymaeker's comments. Eur. Cytokine Netw. *5*:59-60.
12. **Tarnuzzer, R.W., S.P. Macauley, W.G. Farmerie, S. Caballero, M.R. Ghassemifar, J.T. Anderson, C.P. Robinson, M.B. Grant et al.** 1996. Competitive RNA templates for the detection and quantitation of growth factors, cytokines, extracellular matrix components and matrix metalloproteinases by RT-PCR. BioTechniques *20*:670-674.

9 Alternative Quantitative PCR Method

Antonino Nicoletti and Caroline Sassy-Prigent
INSERM U430, Hôpital Broussais, Paris, France

OVERVIEW

The exponential nature of polymerase chain reaction (PCR) makes quantitation of amplified products possible only if the efficiency of the enzymatic steps is estimated or cast off. In this chapter, we present a technique that permits a relative quantitation of reverse transcription (RT)-PCR products using a series of progressive dilutions achieved by mixing RNA solutions of the samples to be compared. After a standard RT-PCR, this range of concentrations permits elimination of the tube-to-tube efficiency variations. To illustrate this method, the renal erythropoietin (EPO) expression level was compared in anemic and control rats. RT was performed using specific primers for *EPO* and glyceraldehyde-3-phosphate dehydrogenase (*GAPD*) genes. The EPO mRNA expression was also checked by Northern blotting. Quantitative PCR indicated that anemic rats produced 23 times more EPO mRNA than control rats. The results from Northern blotting matched those of PCR. This simple new method does not provide absolute amounts of nucleic acid but rather relative ones, and it works with any set of primers. It is a valid alternative to methods like competitive RT-PCR.

BACKGROUND

It is difficult to quantify products amplified by RT-PCR because of: (*i*) the two sequential enzymatic steps involved (the synthesis of cDNA from the RNA template [RT] and the amplification of cDNA [PCR]) and (*ii*) the exponential nature of PCR. The reaction can be described by the following equation:

$$N = N_0(1 + \text{eff})^n$$

where N is the final amount of the amplified PCR product, N_0 the initial amount of target cDNA, eff the amplification efficiency and n the number of amplification cycles. Thus, a small difference in efficiency can lead to a large change in the amount of final PCR products, even if the initial amount of cDNA template in all the tubes is the same. When the efficiency remains constant, the amount of amplified product is proportional to the initial amount of cDNA, hence it should be feasible to compare samples.

Many studies have described several methods of obtaining quantitative results. Some provide relative data on the starting amounts of target cDNA in several samples, and others estimate the absolute initial amount of target cDNA. Relative comparisons can be performed without using internal standards when the amplification efficiencies are the same for all samples, and all data are obtained in the exponential phase of the reaction

Table 1. Simulated Data

	Example 1		Example 2	
	A	B	A	B
QX ([X])	10 (0.1)	15 (0.15)	10 (0.1)	15 (0.15)
Q HK ([HK])	20 (0.2)	20 (0.2)	15 (0.15)	20 (0.2)
X_A/X_B	0.667		0.667	
X/HK (normalization)	0.5	0.75	0.667	0.75
X_A/X_B (after normalization)	0.667		0.889	

Example 1: solutions A and B have the same amount of HK RNAs; example 2: the amounts of HK RNAs in each solution are different. The amounts of X RNA in solutions A and B can be compared after correction to the HK in solutions A and B (X, HK: amount of X and HK RNAs; [X], [HK]: concentration of X and HK RNAs fixed arbitrarily).

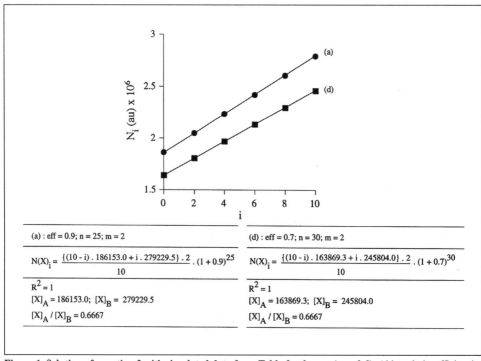

(a) : eff = 0.9; n = 25; m = 2

$$N(X)_i = \frac{\{(10-i) \cdot 186153.0 + i \cdot 279229.5\} \cdot 2}{10} \cdot (1+0.9)^{25}$$

$R^2 = 1$
$[X]_A = 186153.0; [X]_B = 279229.5$
$[X]_A / [X]_B = 0.6667$

(d) : eff = 0.7; n = 30; m = 2

$$N(X)_i = \frac{\{(10-i) \cdot 163869.3 + i \cdot 245804.0\} \cdot 2}{10} \cdot (1+0.7)^{30}$$

$R^2 = 1$
$[X]_A = 163869.3; [X]_B = 245804.0$
$[X]_A / [X]_B = 0.6667$

Figure 1. Solution of equation 3 with simulated data from Table 2 columns (a and d). Although the efficiencies and the cycle number are different in columns a and d, the ratio $[X]_A/[X]_B$ is 0.6667 in every case.

Table 2. Simulated Amplification

Volume of Solution A	Volume of Solution B	Initial Amounts of X (Equation 1)	Initial Amounts of HK (Equation 1) Example 1	Initial Amounts of HK (Equation 1) Example 2	Amount of Amplified PCR Products (Equation 3) n = 30; eff = 0.7; m = 2 X (a)	HK Example 1 (b)	HK Example 2 (c)	n = 25; eff = 0.9; m = 2 X (d)	HK Example 1 (e)	HK Example 2 (f)	Ratio (a)/(f)
10	0	1.0	2	1.5	1638693.15	3277386.29	2458039.72	1861529.91	3723059.83	2792294.87	0.59
8	2	1.1	2	1.6	1802562.46	3277386.29	2621909.03	2047682.91	3723059.83	2978447.86	0.61
6	4	1.2	2	1.7	1966431.77	3277386.29	2785778.35	2233835.90	3723059.83	3164600.85	0.62
4	6	1.3	2	1.8	2130301.09	3277386.29	2949647.66	2419988.89	3723059.83	3350753.84	0.64
2	8	1.4	2	1.9	2294170.40	3277386.29	3113516.98	2606141.88	3723059.83	3536906.84	0.65
0	10	1.5	2	2.0	2458039.72	3277386.29	3277386.29	2792294.87	3723059.83	3723059.83	0.66

The concentrations of X and HK RNAs in Table 1 (simulated data) were used to calculate the initial amounts of X and HK RNAs in mixtures of solutions A and B using equation 1. Equation 3 was used to calculate the theoretical amounts of amplified products with n = 30, eff = 0.7, m = 2 [(a), (b), (c)] or n = 25, eff = 0.9, m = 2 [(d), (e), (f)]. The ratio (a)/(f) is the case in which the efficiency and the cycle number are different for the X and HK RNAs and the initial HK levels were different in solution A and B.

[titration analysis (16) and kinetic analysis (14)]. Most studies have used internal standards to correct for tube-to-tube variations in amplification efficiency. An endogenous sequence that is usually present in a constant amount in series of samples can be used [housekeeping genes such as β-actin, *GAPD*, $β_2$-microglobulin, hypoxanthine phosphoribosyl transferase (*HPRT*) (7,10,12,13) or ribosomal RNA (9)]. However, interference can occur because two sets of primers are used. Several reports have attempted to avoid this shift by constructing exogenous internal standards that are added to the target sample and amplified simultaneously with a single set of primers (17). These standards differ from the target sequences only by the presence or absence of small sequences or restriction sites (1,2,6). These differences allow discrimination between the amplified sequences. Others have used standards that possess a different sequence of target RNA but share primer binding sites and are similar in size to the template RNA (3,8). The main problem with all these techniques is that they require considerable development, and standard template, when used, is specific for only one target gene.

This chapter describes a simple method for the relative quantitation of PCR products. The method uses a range of progressive RNA dilutions obtained by mixing solutions to compare together. This range of concentrations overcomes variations in the efficiency and thus allows samples to be compared without the need for specific standards. The new method has been used to examine the expression of the *EPO* mRNA in the kidneys of anemic and normal control rats. The validity of the method has been checked using Northern blot performed on the same samples as in PCR.

PROTOCOLS

Principle

The general equation describing PCR amplification is $N = N_0(1 + eff)^n$, where eff is the efficiency, n the cycle number and N the amount of amplified product from N_0, the initial amount of template. This general equation can be used to simulate the amplification of a target template X by PCR. Thus, for two solutions A and B with different concentrations of the target mRNA X ($[X]_A$ and $[X]_B$), the final constant volume V_f of a range of mixtures of these two solutions is $V_f = V_A + V_B$ (V_A = volume of solution A; V_B = volume of solution B), and so $V_A = V_f - i$ if $V_B = i$. A mixture of these two solutions in a final constant volume V_f will contain the X mRNA amount $N_0(X)_i$:

[Eq. 1] $$N_0(X)_i = (V_f - i)[X]_A + i[X]_B$$

From this mixture, m microliters contain $[N_0(X)_i \cdot m]/V_f$ mRNA. Using the general equation of PCR amplification after n cycles as a mathematical model, the amplification of m microliters will provide the final amount of amplified product $N(X)_i$:

[Eq. 2] $$N(X)_i = \frac{N_0(X)_i \cdot m}{V_f} \cdot (1 + eff)^n$$

From equations 1 and 2, the following equation can be written:

[Eq. 3] $$N(X)_i = \frac{\{(V_f - i)[X]_A + i[X]_B\} \cdot m}{V_f} \cdot (1 + eff)^n$$

Equations 1 and 3 have been used to plot (Figure 1) the simulated data from Tables 1 and 2, where $[X]_A$ was arbitrarily fixed at 0.1 and $[X]_B$ at 0.15. In Figure 1, the *y* axis is the calculated $N(X)_i$, and the *x* axis is i, the volume of solution B. A regression line can be drawn through these points. Resolution of equation 3 gives the values $[X]_A$ and $[X]_B$. These values will vary with the efficiency and the cycle number, but the ratio $[X]_A/[X]_B$ remains constant; it is 0.667 in the present example, as predicted from Table 1, whatever the efficiency and the cycle number (equations 1 and 2), provided that amplification remains exponential. The model can be simplified to:

[Eq. 4] $$N(X)_i = \frac{\{(V_f - i)[X]_A + i[X]_B\} \cdot m}{V_f}$$

Given that m/V_f is constant in an experiment, the following equation can be used:

[Eq. 5] $N(X)_i = (V_f - i)[X]_A + i[X]_B$

The alignment of the points along a regression line indicates that the efficiency is equivalent in all tubes. If the data for one tube lie off the line, the experiment should be repeated once more. The term "efficiency" includes the efficiency of RT and the efficiency of PCR. Hence, differences in the efficiency during the entire experiment will be detected. This method cannot be quantitative without an estimation of the initial amounts of total RNA in solutions A and B. A PCR performed with primers for an invariant mRNA, such as *GAPD* [housekeeping (HK) gene] in the same dilution tubes is sufficient to estimate the initial amounts of total RNA and thus to normalize the amounts of X with HK. A PCR carried out on dilution mixtures with primers for an HK gene will follow equation 5:

[Eq. 6] $N(HK)_i = (V_f - i)[HK]_A + i[HK]_B$

As shown in Figure 2, the direct normalization of $[X]_A$ and $[X]_B$ with values of $[HK]_A$ and $[HK]_B$ calculated with equation 6 provides the expected results even when the efficiencies and cycle numbers of X and HK amplification are different. The ratios $[X]_A/[HK]_A$ and $[X]_B/[HK]_B$ are two values that can be compared to obtain the difference between the initial amounts of X RNA in solutions A and B. All regression curves were checked with a Student's t test, where t was calculated as follows (comparison of two observed means of a small number of samples in paired series):

$$t = \frac{\text{mean}(d)}{\sqrt{\frac{\left(\frac{SS_{obs} + SS_{pred}}{n_{obs} + n_{pred} - 2}\right)}{n_{points}}}}$$

with $d = |\text{predicted value} - \text{observed value}|$,
$SS_{obs} = \Sigma[\text{observed value} - \text{mean(observed values)}]^2$,
$SS_{pred} = \Sigma[\text{predicted value} - \text{mean(predicted values)}]^2$,
n_{obs} = number of observed values,
n_{pred} = number of predicted values,
and n_{points} = number of points in the regression.

This value was compared to the value of t in the Student's table:

$t^{(n-1) \text{ degree of freedom}}_{\alpha = 0.05}$

Points were considered aligned along the regression curves if the calculated t was lower than $t^{(n-1)}_{\alpha = 0.05}$.

Materials and Reagents

Animals and Tissues

Adult male Brown-Norway rats (250–300 g) were made anemic using a protocol derived from Lacombe et al. (11). They were irradiated with a dose of 7 Grays on the first day, followed by three daily subcutaneous injections of a neutralized solution of phenylhydrazine (Sigma Chemical, St. Louis, MO, USA) (total dose: 75

Table 3. Dilution Table Indicating the Volumes of RNA Solutions

Sample No.	a Control Rats Solution	b Anemic Rats Solution (i)	c Anemic Rats Solution Diluted 1:10 (i)
1	12	0	0
2	10	2	2
3	8	4	4
4	6	6	6
5	4	8	8
6	2	10	10
7	0	12	12

Dilutions were taken from control (a) and anemic (b) rats and from control and anemic rats diluted 1:10 (c) mixed in each RNA sample (final volume: 12 µL). All volumes indicated are in microliters.

mg/kg body weight). The hematocrit of the treated animal decreased to 20%. The kidneys of both control and anemic rats were rapidly removed and frozen in liquid nitrogen. One hundred milligrams of these frozen tissues were crushed to a powder and homogenized in 4 M guanidinium isothiocyanate. Total RNAs were extracted using the method of Chomczynski and Sacchi (4). The RNAs were dissolved in sterile water and their concentration determined by absorbance at 260 nm.

Reverse Transcription

RNA solutions were pooled in each group, and aliquots (1 µg/4 µL) were prepared. A range of dilutions between RNA from anemic rats (solution b) and RNA from control rats (solution a) was prepared (Table 3) in a constant final volume of 12 µL. Further, RNA from anemic rats was diluted 1:10 (solution c), and a range of dilutions between the sample solution c and the RNA solution from the control rats was prepared. This allowed comparison of the expression level of the gene of interest in solutions a and b and in solutions a and c, and thus an indirect comparison of solutions b and c. Aliquots (4 µL) of each dilution mixture were incubated with 200 U of Moloney murine leukemia virus (MMLV) Reverse Transcriptase (Life Technologies, Gaithersburg, MD, USA) in a buffer containing 50 mM Tris-HCl, pH 8.3, 75 mM KCl, 10 mM dithiothreitol (DTT), 19 U RNase inhibitor (Amersham Pharmacia Biotech, Piscataway, NJ, USA), 1 µg oligo(dT)$_{12-18}$ (Amersham Pharmacia Biotech) and 0.5 mM each dNTP in a final volume of 20 µL.

Polymerase Chain Reaction

Two microliters of each cDNA were amplified in a total volume of 25 µL containing 50 pmol oligonucleotide primers, 10 mM each dNTP, 1× PCR buffer (10 mM Tris-HCl, pH 8.3, 50 mM KCl), 40% dimethyl sulfoxide (DMSO), 1 mM MgCl$_2$ and 2.5 U AmpliTaq® DNA Polymerase (Perkin-Elmer, Norwalk, CT, USA). The primers used for the amplification of rat *EPO* were: 5′-ACC ACT CCC

AAC CCT CAT CAA-3' (sense) and 5'-CGT CCA GCA CCC CGT AAA TAG-3' (antisense). A housekeeping gene, *GAPD*, was amplified to estimate the initial amounts of total RNAs. The primers used for the amplification of *GAPD* were: 5'-GTG AAG GTC GGA GTC AAC G-3' (sense) and 5'-GGT GAA GAC GCC

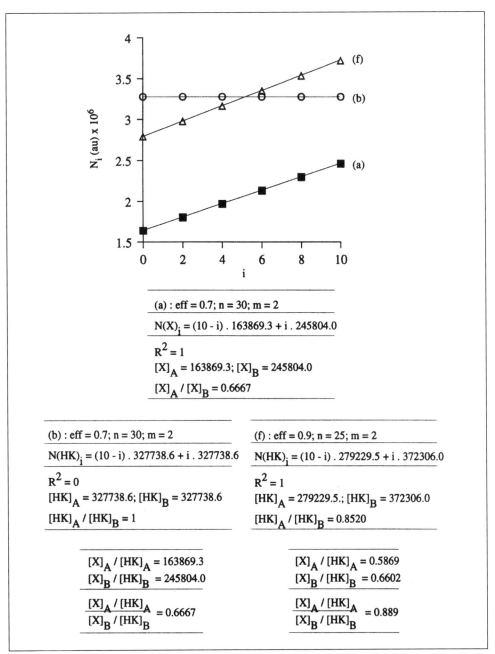

Figure 2. Solution of equation 5 with simulated data from Table 2 (columns a, b and f). In columns a and b, the efficiency and cycle number for X and HK are equal, and the normalized ratio of [X] to [HK] is 0.6667, as predicted from Table 1. In columns a and f, the efficiency and cycle number for X and HK are different, but even in this case the normalized ratio of [X] to [HK] is 0.889, as predicted from Table 1.

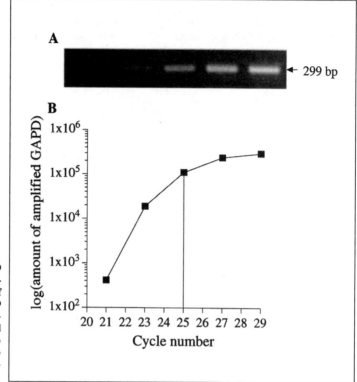

Figure 3. Kinetics of *GAPD* amplification: cDNA underwent sequential cycles of amplification with the *GAPD* primer set. Amplified products (15 µL) were visualized by electrophoresis on a 2% agarose gel containing 0.1% ethidium bromide (A) and analyzed by densitometry (B).

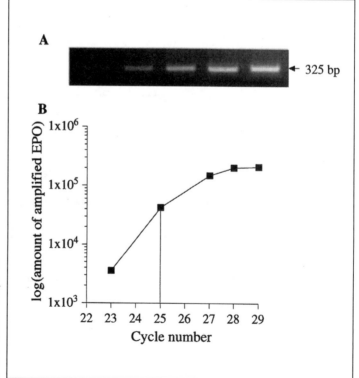

Figure 4. Kinetics of *EPO* amplification: cDNA underwent sequential cycles of amplification with the *EPO* primer set. Amplified products (15 µL) were visualized by electrophoresis on a 2% agarose gel containing 0.1% ethidium bromide (A) and analyzed by densitometry (B).

AGT GGA CTC-3′ (antisense). The expected PCR product size was 325 bp for *EPO* and 299 bp for *GAPD*. PCR was carried out in a Perkin-Elmer DNA Thermal Cycler with the following sequence: 1 min at 94°C, 2 min at 55°C and 3 min at 72°C. The appropriate number of cycles for remaining within the exponential phase for further amplifications was determined for both primers with the most concentrated sample. Fifteen microliters of each PCR product were then run on an ethidium bromide-stained agarose gel.

Northern Blot Analysis

The validity of the PCR quantitation was checked by a Northern blot analysis of samples 1, 3, 5 and 7 from the mixture between solution a and solution b (Table 3). Total RNA (15 µg per sample) was fractionated on a denaturing agarose gel and transferred by capillary blotting to a nylon filter (Hybond™ N+; Amersham Pharmacia Biotech). Equivalent loading of RNA, absence of degradation and position of the 28S and 18S ribosomal RNA were determined by ethidium bromide staining. RNAs were cross-linked to the filter by UV light. The filter was prehybridized at 42°C for 4 h in a buffer containing 50% formamide (deionized), 10× Denhardt's reagent, 0.05 M Tris-HCl, pH 7.5, 1 M NaCl, 1% sodium dodecyl sulfate (SDS), 5% dextran, 0.1% sodium pyrophosphate and 300 µg/mL denatured salmon sperm DNA. A 500-bp rat *EPO* cDNA probe (generously provided by C. Lacombe, INSERM U152) was labeled with [^{32}P]dCTP (3000 Ci/mmol; Amersham Pharmacia Biotech) by random priming. The labeled probe was then added to the buffer and the filter hybridized at 42°C for 18 h. The filter was washed twice for 15 min in 0.5× standard saline citrate (SSC)/0.1% SDS at room temperature, twice for 30 min in 0.5× SSC, 0.1% SDS at 65°C and once for 15 min in 0.1× SSC, 0.1% SDS at 65°C. The filter was exposed at -70°C to Kodak X-OMAT AR film (Scientific Imaging Systems [Eastman Kodak], New Haven, CT, USA) using intensifying screens for 48 h.

Quantitation of Detected Signals

Polaroid® photographs of ethidium bromide-stained gels and Northern blot autoradiographs were digitized into gray-scale images. The amounts of nucleic acids were determined by densitometry. The amount of nucleic acid was proportional to the log of the optical density. The sum of the logarithms of the pixel values was used to estimate the amount of nucleic acid in a band. Analysis was performed on a Macintosh® Quadra® 840AV computer (Apple Computer, Cupertino, CA, USA) using the public domain NIH Image 1.51 program (developed at the U.S. National Institutes of Health and available from the Internet by anonymous file transfer protocol [FTP] from **zippy.nimh.nih.gov**).

RESULTS

Optimization of PCR Cycle Number

The number of PCR cycles was optimized for each primer set. Aliquots (2 µL) of cDNA were subjected to varying numbers of amplification cycles (Figures 3 and 4).

Amplification was exponential up to 27 cycles for the *GAPD* and *EPO* primer sets. Further amplifications were performed at 25-round cycles.

Amplification of the Dilution Mixtures

GAPD Primers

Samples (2 µL) of dilution mixtures of nucleic acid solutions from anemic (solutions b and c) and control (solution a) rats were amplified with the *GAPD* primer set. When comparing solutions a and b (Figure 5, A and B), the amounts of *GAPD* in control rats (solution a; left point), in anemic rats (solution b; right point) and in all dilution tubes within the range were identical. Because solution c was obtained by diluting 1:10 solution b, we expected approximatively a 10-fold increase in *GAPD* level in solution a as compared to that in solution c. Figure 6 (panels A and B) shows that there was an eightfold increase in the level of *GAPD* in solution a compared to that in solution c. Curves were fitted iteratively using equation 5. All points were aligned along a straight regression line, demonstrating that the efficiencies in all tubes were similar (Figure 5: $t = 2.28 < t^6_{\alpha = 0.05} = 2.447$; Figure 6: $t = 0.49 < t^6_{\alpha = 0.05} = 2.447$.

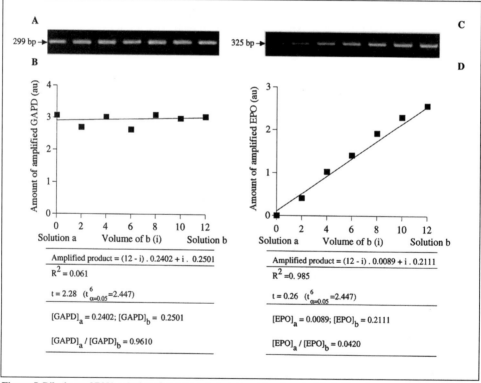

Figure 5. Dilutions of RNA solutions from anemic (solution b) and control (solution a) rats were prepared in 12 µL as described in Table 3. Aliquots (4 µL) were reverse-transcribed into cDNA in a final volume of 20 µL, and 2 µL of each cDNA solution were amplified for 25 cycles with the *GAPD* (A, B) and *EPO* (C, D) primer sets. Amplified products (15 µL) were visualized by electrophoresis on a 2% agarose gel containing 0.1% ethidium bromide (A, C) and analyzed by densitometry (B, D). The amounts of amplified products were plotted against the volumes of anemic RNA solution (solution b) in the dilution mixtures (Table 3). Equation 5 was used for curve fitting.

EPO Primers

In contrast, EPO mRNA expression in anemic rats (Figure 5, solution a; right point) was different from that in control rats (solution b; left point). EPO gene expression was at a low basal level in the control rat kidneys (solution a) but was higher in the anemic rat kidneys (solution b). The amount of amplified product was plotted against i, the volume of anemic RNA solution b in the dilution mixtures (Table 3). The amplified products in the dilution tubes of control and anemic rats lay on a regression line, indicating that the efficiencies were equal in all tubes ($t = 0.26 < t^6_{\alpha = 0.05} = 2.447$). A first estimate of the ratio [EPO]A/[EPO]B showed that the *EPO* gene was 24-fold more active in anemic rats (solution b) than in control rats (solution a). Furthermore, there was 3.3-fold more amplified EPO in solution c than in solution a (Figure 6, C and D). But these estimates did not take into account the variations in GAPD levels.

Mathematical Modeling of GAPD Normalization

The amounts of amplified *EPO* cDNA in each dilution tube were divided by the

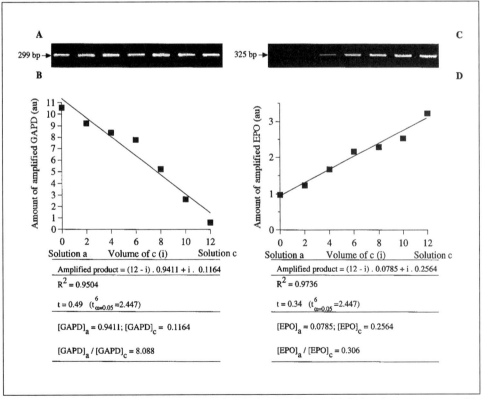

Figure 6. Dilutions of the RNA solution from anemic (solution c) and control (solution a) rats were prepared in 12 µL as described in Table 3. Aliquots (4 µL) were reverse-transcribed into cDNA in a final volume of 20 µL, and 2 µL of each cDNA solution were amplified for 25 cycles with the *GAPD* (A, B) and *EPO* (C, D) primer sets. Amplified products (15 µL) were visualized by electrophoresis on a 2% agarose gel containing 0.1% ethidium bromide (A, C) and analyzed by densitometry (B, D). The amounts of amplified products were plotted against the volumes of solution c added in the dilution mixtures (Table 3). Equation 5 was used for curve fitting.

respective amounts of amplified *GAPD* cDNA to correct for variations in the initial RNA content and thus normalize EPO:

- solution a vs. solution b

    ratio 1: $[EPO]_a/[GAPD]_a = 0.0371$
    ratio 2: $[EPO]_b/[GAPD]_b = 0.8441$
    ratio 3: $[EPO]_a/[EPO]_b = 0.0420$
    ratio 4: ratio 1/ratio 2 = 0.0439

When comparing solution a to solution b without taking into account the variations in the GAPD levels (ratio 3), we could conclude that EPO is expressed 24-fold more in anemic rats (solution b) than in control rats (solution a). However, if variations in GAPD levels are considered (ratio 4), it can be concluded that EPO is expressed 23-fold more in anemic rats than in control rats. This is not greatly different from the previous estimation because the GAPD levels were similar between solutions a and b.

- solution a vs. solution c

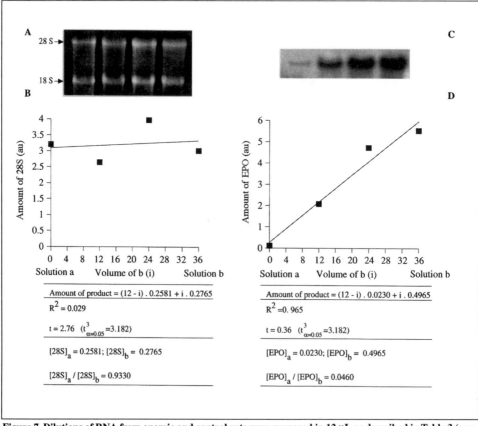

Figure 7. Dilutions of RNA from anemic and control rats were prepared in 12 µL as described in Table 3 (samples 1, 3, 5 and 7). Samples were electrophoresed in a 1% denaturing agarose gel. After blotting, a photograph of the nylon membrane (A) was digitized to quantify the 28S RNA band. The blot was hybridized with a ^{32}P-labeled *EPO* probe, the autoradiogram was exposed for 48 h (C) and digitized, and the *EPO* RNA band was quantified. The amounts of 28S RNA (B) and *EPO* RNA (D) were plotted against the volume of anemic RNA in the dilution mixtures (Table 3). Equation 5 was used for curve fitting.

ratio 5: $[EPO]_a/[GAPD]_a = 0.0834$

ratio 6: $[EPO]_c/[GAPD]_c = 2.2028$

ratio 7: $[EPO]_a/[EPO]_c = 0.3061$

ratio 8: ratio 5/ratio 6 = 0.0379

When comparing the EPO expression in solutions a and c without taking into account the variations in the GAPD levels (ratio 7), it can be concluded that EPO is expressed 3.3-fold more in solution a than in solution c. When comparing the EPO expression in solutions a and c, taking into account the variations in the GAPD levels (ratio 8), the dilution factor should be corrected. Thus, the difference in the EPO expression between solution a and solution b should be, after correction of the GAPD, equal to the difference between solution a and solution c. This is precisely what could be inferred from ratio 8, showing that EPO is expressed 26.4-fold more in solution a than in solution c and confirming the results obtained when comparing solution a to solution b. This example shows that the mathematical modeling is valid even when GAPD levels in each sample to compare are different.

Northern Blot Analysis

The intensity of the 28S ribosomal RNA band of each sample was measured, and the amount of the 28S RNA was plotted against the volume of the anemic rat RNA solution in the dilution mixture (i). Equation 5 was used for curve fitting. Equivalent 28S RNA band intensities indicated that equivalent amounts of RNA were loaded in each lane (Figure 7, A and B). The amount of *EPO* (Figure 7, C and D) increased from the tube containing only RNA from control rats (sample 1) to the tube containing only RNA from anemic rats (sample 7). Plotting the amount of RNA against i provides a first estimate of the ratio $[EPO]_a/[EPO]_b$ using equation 5. This ratio showed that there was 21.6-fold more *EPO* in anemic rats than in control rats. This estimate was corrected by variations in the 28S RNA by dividing the amount of *EPO* by the amount of 28S RNAs for each sample, and the following ratio was calculated:

$$\frac{[EPO]_a/[28S]_a}{[EPO]_b/[28S]_b} = \frac{0.0230/0.2581}{0.4965/0.2765} = 0.0496$$

This shows that the *EPO* gene was expressed 20.2-fold more in anemic rats than in control rats. Given that the difference in the EPO expression between solutions a and c was expected to be weak (ca. 2), the difference between the samples in the dilution range was too small to be detected efficiently by Northern blotting.

Comparison between RT-PCR and Northern Blot Results

Because samples 1, 3, 5 and 7 (Table 3) used in PCR and Northern blots were the same, the quantitations by the two methods should agree. Figure 8 shows that there was a significant correlation ($r^2 = 0.982$; $P < 0.01$) between estimates by PCR and Northern blotting.

DISCUSSION

The quantitative PCR method described here accurately determines the amounts of specific mRNAs in a sample relative to those in another sample. The method is easy to set up and can be used to quantify any mRNA without the need for specific standard RNA.

To demonstrate its accuracy, EPO gene expression was analyzed in a model in which it is known to be up-regulated (in the kidney of anemic rats) (11). The low basal expression of the *EPO* gene in normal rats is enhanced several fold when rats are made anemic by X-ray irradiation.

A quantitative PCR method must estimate the changes that can occur at several steps in the protocol. The first source of variation is RNA extraction, in which a difference in the efficiency of extraction can alter the total extracted RNA. These potential differences were estimated by dividing the final amounts by the amount of a stable endogenous internal standard, *GAPD* for the PCR method and 28S rRNA for Northern blotting. The PCR can be corrected even if the efficiencies and the cycle numbers for the target RNA (*EPO*) and the endogenous internal standard RNA (*GAPD*) are different. A difference during RT and/or amplification can also lead to differences in final estimated amounts. Many variables could influence the efficiency of PCR amplification. Some of the parameters that can be easily controlled are the concentrations of the template, dNTP, $MgCl_2$, primers, polymerase and the PCR cycling profile. However, differences in primer efficiency are difficult to regulate, especially when templates have different sequences and/or sizes, as is the case in methods using external standards. This method used mixtures containing dilutions of the RNAs to be compared. In theory, each assay solution within the range of dilution contains an amount of RNA that lies on a straight line, linking the amounts in the initial solutions, irrespective of the difference in the concentrations of the target RNAs in each initial solution. The ratio between the initial amounts before PCR will be equal to the ratio of their respective amplified products. The amounts of amplified products in each solution will also be on a straight line. When all points are aligned on the regression, this demonstrates that efficiencies are equal in all the tubes. One point that does not lie on the regression is sufficient to invalidate an experiment. The dilution mixtures were also made before RT. This allowed us to ascertain that the efficiencies of RT and amplification were equivalent in all the dilution tubes. As a result, samples could be compared directly if extraction efficiencies were equal in each sample. Differences in extraction efficiencies can be corrected by dividing amplified *EPO* data by amplified *GAPD* data. The agreement between the Northern blot and PCR quantitative data indi-

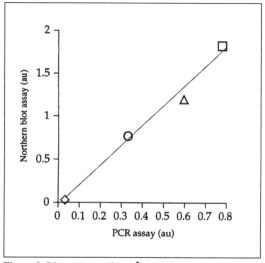

Figure 8. Linear regression ($r^2 = 0.982$, $P < 0.01$) between PCR and Northern blot data for the same dilutions (◇: sample 1; ○: sample 3; △: sample 5; □: sample 7) (Table 3).

cates that the PCR data are reliable (Figure 8). The amount of *EPO* mRNA in anemic rats was increased 23-fold as determined by PCR and 20.2-fold by Northern blot analysis. These results are quite similar, considering the great difference between the two methods. To further demonstrate the accuracy of this method, and particularly to show its validity to compare samples whose concentrations are close to each other, we have diluted 10-fold the RNA solution from the anemic rats (solution b) and compared this new solution (solution c) to the solution from control rats (solution a). In this case, the *EPO* level appears to be increased 26.4-fold in anemic rats, further showing the reliability of this method. This also demonstrates that this method can be used to compare more than two samples. For instance, to compare four samples (a, b, c and d), sample a can be compared to sample b, sample b to sample c, and sample c to sample d. From these comparisons, the relationships between the missing comparisons can be deduced. Alternatively, one sample can be selected as a reference, and all other samples can be compared to the reference sample. However, a large number of samples to be compared can be a limitation of the method (the real technical limit is the number of samples the thermal cycler is able to simultaneously amplify).

In conclusion, this new method is a simple alternative to methods like competitive RT-PCR (2,5,6,15). It can overcome variations in all steps of the RT-PCR protocol and provides accurate comparative data on the amount of an mRNA in one sample compared to that of another. This method does not provide absolute amounts of nucleic acid, only relative ones. Nevertheless, these comparative data are sufficient for most biological purposes. The method works with any PCR primer and any sequence, without the need for expensive and time-consuming construction of standard sequences. Finally, it can be used in conjunction with any other methods that do not use densitometry to quantify the amount of amplified products.

ACKNOWLEDGMENTS

We thank Dr. D. Heudes for the image analyses and advice on mathematical modeling and statistical analysis.

REFERENCES

1. **Ballagi-Pordany, A. and K. Funa.** 1991. Quantitative determination of mRNA phenotypes by the polymerase chain reaction. Anal. Biochem. *196*:89-94.
2. **Becker-André, M. and K. Hahlbrock.** 1989. Absolute mRNA quantitation using the polymerase chain reaction (PCR). A novel approach by a PCR aided transcript titration assay (PATTY). Nucleic Acids Res. *17*:9437-9446.
3. **Bouaboula, M., P. Legoux, B. Pésségué, B. Delpech, X. Dumont, M. Piechaczyk, P. Casellas and D. Shire.** 1992. Standardization of mRNA titration using a polymerase chain reaction method involving co-amplification with a multispecific internal control. J. Biol. Chem. *267*:21830-21838.
4. **Chomczynski, P. and N. Sacchi.** 1987. Single-step method of RNA isolation by acid guanidinium thiocyanate-phenol-chloroform extraction. Anal. Biochem. *162*:156-159.
5. **Diviacco, S., P. Norio, L. Zentilin, S. Menzo, M. Clementi, G. Biamonti, S. Riva, A. Falaschi and M. Giacca.** 1992. A novel procedure for quantitative polymerase chain reaction by coamplification of competitive templates. Gene *122*:313-320.
6. **Gilliland, G., S. Perrin, K. Blanchard and H.F. Bunn.** 1990. Analysis of cytokine mRNA and DNA: detection and quantitation by competitive polymerase chain reaction. Proc. Natl. Acad. Sci. USA *87*:2725-2729.
7. **Horikoshi, T., K.D. Danenberg, T.H.W. Stadlbauer, M. Volkenandt, L.C.C. Shea, K. Aigner, B. Gustavsson, L. Leichman et al.** 1992. Quantitation of thimidylate synthase, dihydrofolate reductase, and

DT-diaphoroase gene expression in human tumors using the polymerase chain reaction. Cancer Res. 52:108-116.
8. **Kanangat, S., A. Solomon and B.T. Rouse.** 1992. Use of quantitative polymerase chain reaction to quantitate cytokine messenger RNA molecules. Mol. Immunol. 29:1229-1236.
9. **Khan, I., T. Tabb, R.E. Garfield and A.K. Grover.** 1992. Polymerase chain reaction assay of mRNA using 28S rRNA as internal standard. Neurosci. Lett. 147:114-117.
10. **Kinoshita, T., J. Imamura, H. Nagai and K. Shimotohno.** 1992. Quantification of gene expression over a wide range by the polymerase chain reaction. Anal. Biochem. 206:231-235.
11. **Lacombe, C., J.-L. Da Silva, P. Bruneval, J.-G. Fournier, F. Wendling, N. Casadevall, J.-P. Camilleri, J. Bariety, B. Varet and P. Tambourin.** 1988. Peritubular cells are the site of erythropoietin synthesis in the murine hypoxic kidney. J. Clin. Invest. 81:620-623.
12. **Murphy, L.D., C.E. Herzog, J.B. Rudick, A.T. Fojo and S.E. Bates.** 1990. Use of the polymerase chain reaction in the quantitation of *mdr*-1 gene expression. Biochemistry 29:10351-10356.
13. **Noonan, K.E., C. Beck, T.A. Holzmayer, J.E. Chin, J.S. Wunder, I.L. Andrulis, A.F. Gazdar, C.L. Willman et al.** 1990. Quantitative analysis of *MDR1* (multidrug resistance) gene expression in human tumors by polymerase chain reaction. Proc. Natl. Acad. Sci. USA 87:7160-7164.
14. **Salomon, R.N., R. Underwood, M.V. Doyle, A. Wang and P. Libby.** 1992. Increased apolipoprotein E and c-*fms* gene expression without elevated interleukin 1 or 6 mRNA levels indicates selective activation of macrophage functions in advanced human atheroma. Proc. Natl. Acad. Sci. USA 89:2814-2818.
15. **Siebert, P.D. and J.W. Larrick.** 1993. PCR MIMICS: competitive DNA fragments for use as internal standards in quantitative PCR. BioTechniques 14:244-249.
16. **Singer-Sam, J., M.O. Robinson, A.R. Bellve, M.I. Simon and A.D. Riggs.** 1990. Measurement by quantitative PCR of changes in HPRT, PGK-1, PGK-2, APRT, MTase, and *Zfy* gene transcripts during mouse spermatogenesis. Nucleic Acids Res. 18:1255-1259.
17. **Wang, A.M., M.V. Doyle and D.F. Mark.** 1989. Quantification of mRNA by the polymerase chain reaction. Proc. Natl. Acad. Sci. USA 86:9717-9721.

10 Quantitation of PCR Products Using DNA Hybridization Assays in Microplates

Christoph Berndt[1], Johann Gross[1] and Wolfgang Henke[2]
[1]*Institute of Laboratory Medicine and Pathobiochemistry and*
[2]*Institute of Biochemistry, Medical Faculty Charité, Humboldt University, Berlin, Germany*

OVERVIEW

In the following, general aspects of hybridization assays are briefly summarized, focusing on the four main steps: *(i)* DNA binding to solid phases, *(ii)* denaturation of double-stranded amplified DNA fragments (amplicons), *(iii)* hybridization and *(iv)* quantitation of hybridized DNA. This part is not aimed to give a comprehensive review of the topic but to illustrate some questions of practical importance. Next, two experiments are described. The first experiment (Protocol 1) shows a hybridization reaction in a microplate using different concentrations of both capture probe and a single-stranded target DNA. It is demonstrated that the reaction exhibits a nonlinear concentration dependence resulting in a plateau at high concentrations of target DNA. Practical implications of this finding are pointed out. The second experiment (Protocol 2) is an example in which a heterogeneous hybridization assay is used as part of a reverse transcription polymerase chain reaction (RT-PCR) method to assess the mRNA expression of a gene of interest. Here, the quantitation of dopamine D_2 receptor (D2R) mRNA is shown.

BACKGROUND

In the last few years, an increasing number of protocols have been designed for the quantitative detection of PCR products in microplates using DNA hybridization assays. Microplate methods are most advantageous if large amounts of samples have to be analyzed, because the protocols involved are usually simple and rapid, the equipment required is found in many laboratories, and the procedures performed are amenable to automation.

In DNA hybridization assays, a reasonable combination of specificity control and product determination is achieved by hybridizing the PCR products with one or two probes directed to the amplicons of interest before the quantitative detection takes place. Several test formats have been proposed, but typically a hybridization assay includes four basic steps (20) (Figure 1): *(i)* DNA binding to microplate, *(ii)* strand separation of the amplicon, *(iii)* hybridization of target DNA with capture and detection probe(s) and *(iv)* quantitative detection of hybridized PCR products.

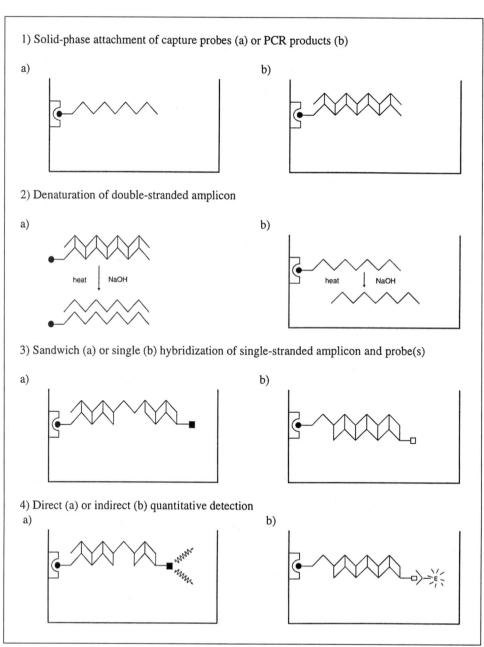

Figure 1. Basic steps of a DNA hybridization assay.

Binding of DNA to Microplates

The immobilization step is crucial for the analytical sensitivity of an assay because it is a main determinant of the signal-to-noise ratio (10). In heterogeneous hybridization assays (see below), the DNA binding to the microplate includes two different approaches: *(i)* the attachment of the target DNA of interest or *(ii)* the attachment of the capture probe designed to remove the target DNA from the amplification reaction mixture. In the homogeneous assay, the hybrid of probe and target DNA is bound to the microplate.

Small amounts of DNA can be fixed to microplates by passive adsorption, which can be increased in the presence of salts and by UV irradiation (10,15,24). More efficient strategies for DNA immobilization in hybridization assays are covalent binding and the use of microwells, the surfaces of which are specifically precoated.

Covalent DNA binding is achieved by a polyfunctional linker (mostly carbodiimide)-mediated condensation of DNA to aminated, carboxylated or oxysuccinimide-grafted microwells (16,18,24,28,34). Plates designed for covalent fixation of DNA are commercially available from a number of suppliers (e.g., Corning Costar [Cambridge, MA, USA], Genosys [The Woodlands, TX, USA] and Nalge Nunc International [Rochester, NY, USA]). Covalent binding increases the stability and capacity of the attachment, yielding ≥1 pmol of immobilized DNA (24) and results in a more defined orientation of the molecules than passive adsorption (27). Recently, carbodiimide and other cationic detergents were found to mediate another type of DNA immobilization presumably based on hydrophobic interactions of solid-phase and detergent-associated DNA (24). This procedure is rather simple and independent on special plates or functional groups of nucleic acids with a yield of immobilized DNA of about 1 pmol per well, which is comparable to that of covalent binding.

Improved DNA attachment can also be accomplished by physical or chemical modifications of the microplates. For example, UV irradiation (10), organosilanes (2) and poly(lys/phe) (28) were shown to enhance binding of unmodified DNA to microwells. Surface coating with antibodies against digoxigenin (23) and biotin has been used for the immobilization of digoxigenin- or biotin-labeled DNA, respectively. Also, DNA-binding proteins can be applied to capture PCR products (17). However, the most popular coating protein so far is streptavidin or avidin. Because of its affinity constant of 10^{-15} M, streptavidin binds biotinylated DNA very tightly. This interaction withstands alkali denaturation of double-stranded DNA and stringent wash conditions, a feature that favors the specific detection of the amplicon of interest. Streptavidin-coated plates bind 1–10 pmol of biotinylated nucleic acid within minutes. The system is easy to handle and seems to work with virtually any microplate. These advantages compensate for the need for biotinylated DNA.

Strand Separation of the Amplicon

The interaction of PCR products with specific probes requires strand separation of the amplicon. Amplified DNA can be simply denatured by heating or by incubating with NaOH, but an exonuclease reaction has also been used to remove one of the amplicon's strands (12). An easy way to obtain single-stranded DNA is by programming a heat denaturation step following the amplification cycles. However, with regard to quantitative determinations, the competition of opposite strand and probe during post-PCR procedures can be problematic. This difficulty can be circumvented by removing the strand competing with the probe. Strand removal is usually accomplished by phase separation; i.e., one of the strands of the denatured DNA is rapidly fixed to a solid phase while the other is brought into or left in solution. Solid phases most frequently used for strand separation are magnetic beads or simply the microplates used in the assay.

Hybridization of Target DNA with Capture and Detection Probes

Hybridization reactions can take place with both reactants held in solution (homogeneous hybridization) or with one of the participants kept immobilized to a solid phase (heterogeneous hybridization). Solution hybridization was reported to be advantageous in that it is faster and nearly complete and lacks steric hindrance (4,22,30), but many applications have successfully used solid-phase hybridization too. Indeed, the drawbacks of hybridization with immobilized DNA in microplates appear rather modest because the reaction can be saturated within a few minutes to 1 h (2,7,15), and it occurs with an efficiency of >75% using single-stranded DNA (7). Importantly, only half of that figure is achieved using double-stranded target DNA (7,15), suggesting that removal of the opposite strand is a major determinant for the test results. Hybridization assays can incorporate one or two probes [sandwich hybridization (25)] directed to the target sequence. Even if the additional probing step within sandwich hybridization enhances the specificity of detection, a single hybridization assay can reliably identify point mutations in an amplicon (3).

In setting up a hybridization assay, careful choice of the probes is recommended. Not only should secondary structures and self-complementarity be avoided, but also care should be taken in determining the probe length. Hybridization signals are higher when longer probes are used (2,15), presumably because of enhanced stability of hybrids during washing steps. On the other hand, if differential hybridization is used to discriminate, for example, between internal PCR standard and target amplicon, a probe length that is optimal for both stability of hybridization and discrimination by stringent washing steps should be chosen. In our experience, 18-mer oligonucleotides are well-suited to distinguish target sequences of about 100 bp that differ by a single base. To achieve this discrimination, stringent washing is required because mismatches of up to 10% have no effects on the hybridization efficiency under appropriate reaction conditions (32).

As reviewed by Wetmur (32), hybridization efficiency is mainly influenced by temperature, ion strength and viscosity of the reaction medium, but it is independent of the percentage of CG of the strands. Salt concentration in the media should be above 0.25 M. Dextran sulfate can accelerate the hybridization because of an exclusion of reaction volume, thereby increasing the concentration of hybridizing strands while addition of denaturing solvent reduces hybridization rates [e.g., at 1.1% per each percent of formamide (32)].

Quantitation of Hybridized PCR Products

A driving force for the development of DNA hybridization assays in microplates has been the analogy to enzyme immunotechniques that allows for the use of established equipment and methods. This feature is most evident in the detection step, in which the quantitative determination of hybridized DNA takes advantage of almost all the strategies used in enzyme-linked immunosorbent assay (ELISA) applications

(10,19,31,32). Today, many refined detection protocols and reagents designed for colorimetric, fluorometric or luminometric assays are available.

Detection formats can be divided into direct and indirect systems. In the direct mode, hybridized DNA already contains the label producing the signal to be measured [e.g., isotopes, fluorophores (11), luminogens (1) or probe-coupled enzymes (30)]. The indirect detection mode requires an additional reaction step to introduce a measurable label into the target DNA. Enzyme conjugates that are not directly coupled to nucleic acid fall into this category (e.g., those containing streptavidin and digoxigenin to trace biotin and digoxigenin, respectively). Also, antibodies directed against double-stranded DNA (21) or DNA–RNA hybrids were used for indirect detection protocols (5,29).

PROTOCOLS

Protocol 1: Concentration Dependencies Relevant for Quantitation

Background

Following amplification reactions, the concentration of target DNA present in a hybridization assay can vary over a broad range. Therefore, knowledge of the dependence of the signal intensity from the concentration of the target DNA and hybridization probe is a prerequisite for a correct interpretation of the test results. By analogy to the enzyme immunoassay, relevant information can be gained from a standard curve obtained by hybridizing the capture probe with different concentrations of an external standard. This is illustrated by the following experiment in which a synthetic target DNA representing a sequence from exon 1 of the K-*ras* protooncogene is hybridized with a solid phase-fixed capture probe.

Materials and Reagents

Microplates, capture probes and target DNA

- Streptavidin-coated microplates (black) (Boehringer Mannheim, Mannheim, Germany)

  Note: Streptavidin-coated plates are expensive. Refer also to several noncommercial protocols to coat surfaces with streptavidin (6,13,14,33).

- Capture probes: 5′ biotinylated capture probes (5′-biotin-gtt-gga-gct-ggt-ggc-gta-3′) (BioTeZ, Berlin, Germany)
- Target DNA: 5′ digoxigenin-labeled target DNA (5′-digoxigenin-agc-tgt-atc-gtc-aag-gca-ctc-ttg-cct-acg-cca-gct-cca-act-acc-aca-agt-tta-tat-tca-gtc-3′) (BioTeZ)

Buffers and reagents

- Phosphate-buffered saline (PBS) (1×): 58 mM Na_2HPO_4, 17 mM NaH_2PO_4, 68 mM NaCl, pH 7.3
- PBST: 1× PBS plus 0.2% (vol/vol) Tween® 20
- Hybridization buffer: 250 mM NaH_2PO_4, 1 mM Na_2 EDTA, 7% sodium dodecyl sulfate (SDS), 1% blocking reagent at pH 7.2 (Tropix, Bedford, MA, USA)

- Wash solution (0.1× standard saline citrate [SSC], 0.1% SDS): 15 mM NaCl, 1.5 mM sodium citrate, 1 mL/L SDS
- Conjugate buffer: 100 mM maleic acid, 150 mM NaCl, 10 mL/L blocking reagent at pH 7.5 (Boehringer Mannheim)
- Assay buffer: 100 mM DEA, 1 mM $MgCl_2$, 10% (vol/vol) Emerald™ enhancer (Tropix), 0.02% (wt/vol) sodium azide at pH 7.0

Instruments

- Luminoskan luminometer (Labsystems Oy, Helsinki, Finland)

Procedures

Attachment of the capture probe

1. Dilute the capture probe in 1× PBS to the appropriate concentration (see Results and Discussion).
2. Wash the wells of streptavidin-coated microplates once with 50 µL of 1× PBST.
   Note: This step reduces unspecific binding of enzyme conjugates to streptavidin.
3. Add 50 µL of the capture probe dilution per well and incubate for 1 h at 37°C.
4. Wash three times with 50 µL PBST.

Hybridization with target DNA and stringent washing

1. Dilute the target DNA in hybridization buffer to the desired concentration (see Results and Discussion).
2. Add 50 µL per well to capture probe-coated microplates and incubate at 42°C for 45 min.
   Note: The hybridization time should be empirically determined for each system of target DNA and probe.
3. Aspirate the hybridization buffer and wash the wells three times with 0.1× SSC, 0.1% SDS each at 48.5°C for 10 min.
   Notes: (i) The washing buffer must be prewarmed to 48.5°C, and *(ii)* the wash conditions should be empirically optimized.

Incubation with enzyme conjugate and measurement

1. Wash each well once with maleic acid buffer plus 1% blocking reagent for 5 min at room temperature.
   Note: This is a preincubation step to reduce unspecific binding of enzyme conjugate.
2. Dilute antidigoxigenin–alkaline phosphatase enzyme conjugate 1:5000 in conjugate buffer and add 50 µL to each hybridization sample. Incubate for 30 min at room temperature on a mircoplate shaker.
3. Remove the buffer and wash the wells three times with PBST. Transfer the microplates into the luminometer and inject 50 µL of assay buffer containing CSPD® (Tropix) plus Emerald luminescence enhancer per well.
   Note: There are other enzyme conjugates and substrates available that make it easy to switch to colorimetric or fluorometric tests based on the protocol given.

Table 1. Parameters Describing the Concentration Dependencies of Hybridization Assays Using the Logistic Model According to the Protocol of Raab (26)

Parameter	Capture Probe (fmol/well)					
	5000	2000	1000	500	100	50
R_{max}	11064	9076	8747	7962	3604	2259
$D_{0.5}$	1966	1932	1338	1062	2812	2687
n	1.09	1.26	1.23	1.12	1.25	1.26

R_{max}: maximal signal in relative light units (RLU)
$D_{0.5}$: probe concentration yielding R_{max} (fmol)
n: coefficient of sigmoidity

For example, using *p*-nitrophenyl phosphate as substrate and transparent microplates instead of black ones, one can detect the products colorimetrically at 405 nm in a conventional reader.

4. Measure the luminescence with an integration time of 5 s per well.

Results and Discussion

The hybridization assay was run using capture probe concentrations from 50 to 5000 fmol per well and varying the amount of target DNA from 1 to 2000 fmol per well. As shown in Figure 2, the target DNA exhibits sigmoid concentration dependence at all capture probe concentrations tested. This feature equals that of enzyme immunoassays. The standard curves shown can be well-described by mathematical models [e.g., by the logistic model or mass-action model (26)]. Figure 2 demonstrates experimental data fitted with nonlinear regression applying the logistic model. The nonlinear concentration dependence of the hybridization assay has important practical implications because PCR product quantitation leads to false results if the concentration of target DNA is in the plateau region. There are two ways to face the problem. One is to include a calibration curve of an external standard into each assay, as is done in ELISA, to check for the range in which the target DNA can be quantified. The other is to run each assay with serial dilutions (e.g., 1/5 steps) of the amplicons. At the concentration range applicable for quantitation, product dilutions will result in the respective decrease of the measured signals. Also, if an internal PCR standard is used, the concentration differences between sample and standard will be mirrored by the different dilution factors required to produce equal signal intensity.

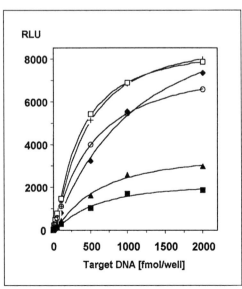

Figure 2. Concentration dependence of target DNA at different amounts of capture probe. Experimental data (black; mean values of three experiments) and calculated values (white) are presented.

Table 2. Sequences of the Primers and Capture Probes Used in Protocol 2

Oligonucleotide	Sequence
forward primer	5′-biotin-cag-acc-atg-ccc-aat-ggc-3′
reverse primer	5′-digoxigenin-cac-acc-gag-aac-aat-ggc
sample capture probe	5′-atg-agc-cgc-aga-agg-ctc-tcc-cag-cag-aag-gag-3′
internal standard capture probe	5′-atg-agc-cgc-aga-cct-tca-gag-cag-cag-aag-gag-3′

Protocol 2: Quantitation of Dopamine Receptor mRNA

Background

To study developmental changes in gene expression of D2R mRNA in cell cultures, a method composed of nonradioactive competitive RT-PCR and heterogeneous hybridization assay in microplates was worked out. Its main features are as follows (Figure 1): *(i)* target RNA and an internal standard cRNA (IS RNA) containing a stretch of nine mutations are reverse-transcribed and amplified with the same primer pair, yielding amplicons of identical length; *(ii)* PCR products are labeled by incorporating biotin- and digoxigenin-labeled primers; *(iii)* following strand separation using NaOH, biotinylated DNA is fixed to streptavidin-coated magnetic particles; *(iv)* the single-stranded amplicons of sample and internal standard are distinguished by differential hybridization with specific capture probes immobilized to microplates followed by *(v)* quantitation by a colorimetric assay using an anti-digoxigenin–peroxidase enzyme conjugate.

In this chapter, the procedures associated with the microplate assay are stressed, whereas RT and PCR, which were documented elsewhere (9), are only briefly outlined.

Materials and Reagents

RNA preparation reagents

- InViSorb™ RNA Preparation Kit II (InViTek GmbH, Berlin, Germany): mineralic carrier (adsorbin), isopropanol and a guanidinium salt-containing lysis solution
- Reagents for in vitro transcription of the IS RNA: transcription buffer, 1.5 mM each rNTP, 40 U RNase inhibitor, 40 U T7 RNA polymerase and 5 µg plasmid DNA for the IS RNA (Boehringer Mannheim)

Reverse transcription and DNA amplification reagents

- RT reagents: 10 mM Tris-HCl, pH 8.3, 50 mM KCl, 5 mM $MgCl_2$, 1 mM dNTPs, 25 U RNasin® (Promega, Madison, WI, USA), 2 pmol/µL random hexamer primers, 4 U/µL Moloney murine leukemia virus (MMLV) reverse transcriptase, 10 mM dithiothreitol, 0.01 g/L gelatin and nuclease-free water (Boehringer Mannheim)
- PCR reagents: 10 mM Tris-HCl, pH 8.3, 50 mM KCl, 0.01 g/L gelatin, 0.4 mM dNTP, 0.4 pmol/µL each of the forward and reverse primers, 2.5 mM $MgCl_2$, 1.5 U *Taq* DNA polymerase and 10 µL RT product (cDNA) (Boehringer Mannheim)
- Primers: 5′ digoxigenin- and 5′ biotin-labeled primers (Table 2) (BioTeZ)

Note: The D2R protein has short and long isoforms (D2Rs and D2Rl) originating from alternative splicing of the same gene (8). The PCR primers recognize regions common to each transcript of the two D2R forms and generate a single PCR product of 119 bp. The product of genomic DNA includes an intron.

Microplates, capture probes and probe-binding reagents

- MaxiSorp™ microplates (Nalge Nunc International)
- The sequences of target- and internal standard-specific capture probes are shown in Table 2.
- EDC: 10 mM 1-ethyl-3-(3-dimethylaminopropyl)carbodiimide hydrochloride solution (Sigma Chemical, Deisenhofen, Germany)
- MaxiSorp wash solution (WS1): 10 mM Tris-HCl, 150 mM NaCl, 0.05% Tween 20, pH 7.5 (Nalge Nunc International)

Magnetic beads

- Streptavidin-coated magnetic particles (10 mg/mL; Boehringer Mannheim)
- Wash buffer for magnetic particles (2× WS2): 10 mM Tris-HCl, 1 mM Na_2 EDTA × $2H_2O$, 2 M NaCl, pH 7.5 (Boehringer Mannheim)

Hybridization and detection reagents

- Hybridization buffer: 1.6 g/100 mL I-Block™ reagent (Tropix), 10% SDS, 0.16 mM Na_2 EDTA, 40 mM Na_2HPO_4 (Tropix)
- Wash solution (WS3): 0.1× SSC plus 0.1% (vol/vol) SDS (1× SSC = 150 mM NaCl, 15 mM sodium citrate)
- Anti-digoxigenin–peroxidase conjugate (Boehringer Mannheim)
- Maleic acid solution: 100 mM maleic acid, 150 mM NaCl, pH 7.5 (Boehringer Mannheim)
- Blocking buffer: maleic acid solution with 1% blocking reagent (Boehringer Mannheim)
- Wash solution (WS4): 137 mM NaCl, 2.7 mM KCl, 1.5 mM KH_2PO_4, 7.75 mM Na_2HPO_4, 0.1% Tween 20, pH 7.4 (Boehringer Mannheim)
- ABTS (2,2′ azino-di-3-ethylbenzthiazoline sulfonate) substrate solution (1 mg/mL; Boehringer Mannheim)

Instruments

- Thermal cycler: GeneAmp® PCR System 9600 (Perkin-Elmer, Weiterstadt, Germany)
- Magnetic Particle Concentrator (MPC®-10; Dynal, Oslo, Norway)
- Model 5436 Thermomixer (Eppendorf-Netheler-Hinz GmbH, Hamburg, Germany)
- Microplate reader: Spectra II (Tecan, Crailsheim, Germany)

Procedures

Preparation of total RNA and IS RNA

1. Total RNA was isolated using the InViSorb RNA Preparation Kit II according to

the manufacturer's instructions. The RNA was quantified by measuring the absorbances at 260 and 280 nm and stored at -80°C in diethyl pyrocarbonate (DEPC)-treated water.

2. IS RNA was synthesized using a DNA template containing the T7 promotor and the target sequence harboring 9 mutations. Purified IS RNA was directly used in the RT-PCR. Aliquots were stored at -80°C in DEPC-treated water for a maximum of 4 weeks.

Reverse transcription and polymerase chain reaction

1. A fixed amount of IS RNA was reverse-transcribed and amplified with different amounts (0.05, 0.075, 0.1, 0.25, 0.3 µg) of total sample RNA. RNA, IS RNA and primers were preincubated at 70°C for 10 min, and the reaction mixture was added and the samples incubated for 10 min at 25°C, 60 min at 37°C and 10 min at 70°C. The reaction was finished at 4°C and the products immediately used in the PCR or stored overnight at -20°C.

2. After 5 min at 94°C, the PCR was carried out with 30 cycles of amplification, each for 20 s at 94°C, 20 s at 57°C and 30 s at 72°C. After a final step for 10 min at 72°C, the PCR products were stored at 4°C. Amplicons were examined on agarose gels with ethidium bromide staining to verify the expected size of the amplified fragment.

Quantitation of the PCR products using microplates

(A) Binding of the capture probes to microwells:

1. Dilute capture probes in freshly prepared 10 mM EDC solution to a final concentration of 2 µM.

    Notes: *(i)* The carbodiimide functional group of EDC is moisture-sensitive, being relatively rapidly hydrolyzed, and therefore, freshly prepared EDC solutions should be used; and *(ii)* prepare four wells (each two with sample and standard capture probe) to measure one PCR sample.

2. Add 50 µL of this solution per well and incubate at room temperature overnight.

3. Wash the plates 3 times using WS1 immediately before the addition of the single-stranded amplicons.

(B) Strand separation of the amplicons:

1. Shake carefully the magnetic particle stock solution, pipet 3 µL of magnetic particles into an Eppendorf® tube (Brinkmann Instruments, Westbury, CT, USA) and wash it 2 times with 1× WS2.

2. Remove the supernatant and add 3 µL of 2× WS2, 3 µL of the amplicons and 44 µL of 1× WS2.

3. Mix at room temperature for 20 min to bind the double-stranded amplicons to the magnetic particles.

4. Remove the supernatant using the MPC and wash the magnetic particles once with 100 µL of 1× WS2.

5. Add to the tube containing the magnetic particles 35 µL 1× WS2 and denature the amplified product by a 5-min incubation of the sample in the Eppendorf Thermomixer at 94°C.

6. Transfer the tube for 1 min on ice, add 35 µL of 0.2 M NaOH and incubate for one more minute on ice.

7. Using the MPC, separate the supernatant and transfer it into a 2-mL Eppendorf tube.

8. Neutralize the single-stranded amplicons by adding 35 µL of 0.2 M HCl.

(C) Hybridization and stringent washing:

1. Add 100 µL of prewarmed (50°C) hybridization buffer to an Eppendorf tube containing the digoxigenin-labeled amplicons.

2. Pipet aliquots of 40 µL of the hybridization mixture to each two wells coated with capture probe for the target DNA and internal standard. Incubate the samples at 42°C for 30 min in a thermal shaker.

   Notes: (i) Because the standard and sample differ in composition by more than 10%, similar hybridization efficiencies of both should be verified in pilot experiments; *(ii)* include one blank composed of 1 part 1× WS2, 1 part 0.2 M NaOH, 1 part 0.2 M HCl and 3 parts of hybridization buffer per microplate.

3. To remove unspecific hybridization products, wash the microplate 3 times with 100 µL of preheated WS3 solution, each at 48°C for 5 min.

(D) Incubation with enzyme conjugate and measurement:

1. Wash each well with 100 µL maleic acid plus blocking buffer for 5 min at room temperature.

2. Add anti-digoxigenin–peroxidase conjugate (diluted 1:1000 in maleic acid blocking buffer) and incubate for 30 min at room temperature.

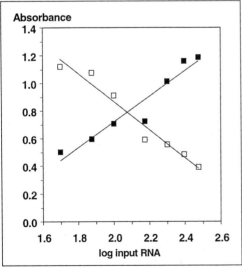

Figure 3. **Quantitation of D2R mRNA in the striatum of adult rats.** The following input RNA amounts (in ng/tube) were used for the RT-PCR: 50, 75, 100, 150, 200, 250, 300. A constant amount of 300 fg of D2R IS RNA was added to each tube. The point of equivalence was calculated from the regression equations.

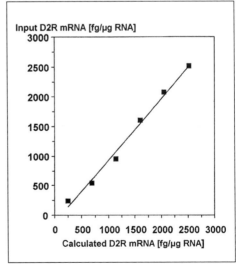

Figure 4. **Relationship between the theoretical and the experimental results of the recovery determined in Protocol 2.**

Note: After this step, the microplate can be stored tightly covered at 4°C overnight.

3. Wash the microplate 3 times with WS4 and add 50 µL of ABTS solution per well.
4. Measure the absorbance at 492 nm and calculate the mRNA/µg RNA content according to the procedure explained in Figure 4.

Results and Discussion

The present method was used to quantify the amount of specific D2R mRNA in the striatal tissue of adult rats and in primary cultures of mesencephalic cells from fetal rats. The PCR gives rise to only one discrete band of the predicted size because standard and target produce amplicons of identical length (not shown). In the quantitation assay, the competition between the target and IS RNA results in two linear curves of absorbance vs. total RNA (Figure 3), the intersection of which indicates equal amounts of D2R mRNA. In the example shown, 119 µg of total striatal RNA corresponds to 300 fg of IS RNA. Based on this estimation, a content of 2514 fg D2R mRNA per µg RNA in the adult striatum was calculated. To determine the D2R mRNA content in mesencephalic cell cultures, a smaller amount of IS RNA (50 fg) was used, and the mRNA content calculated was 246 fg of D2R mRNA per µg of total RNA (i.e., one order of magnitude below the adult level).

The performance of the RT-PCR method presented was evaluated by experiments determining its accuracy and precision. Total RNA from mesencephalic cell cultures containing a low level of D2R mRNA and from striata of adults rats containing a higher level of D2R mRNA was mixed in defined ratios (1:5, 2:5. 3:5, 4:5 striatum RNA and cell culture RNA, respectively) and quantified as described. The regression line between theoretical and experimental results of the recovery experiment demonstrates an accurate quantitation of D2R mRNA by the applied approach (Figure 4). The variation coefficient of 12 independent experiments using striatal tissue was 10% (mean = 2384 fg, standard deviation = 240 fg).

ACKNOWLEDGMENT

We wish to thank Marion Bebenroth, Iris Müller and Kerstin Oehlschlegel for skilled technical assistance.

REFERENCES

1. **Arnold, L.J., P.W. Hammond, W.A. Wiese and N.C. Nelson.** 1989. Assay formats involving acridinium-ester-labeled DNA probes. Clin. Chem. *35*:1588-1594.
2. **Berndt, C., M. Bebenroth, K. Oehlschlegel, F. Hiepe and W. Schößler.** 1995. Quantitative polymerase chain reaction using a DNA hybridization assay based on surface-activated microplates. Anal. Biochem. *225*:252-257.
3. **Berndt, C., G. Wolf, G. Schröder, M. Bebenroth, K. Oehlschlegel, T. Hillebrand and J. Zanow.** 1996. A microplate assay for K-ras genotyping. Eur. J. Clin. Chem. Clin. Biochem. *34*:837-840.
4. **Bünemann, H.** 1982. Immobilization of denatured DNA to macroporous supports: II. Steric and kinetic parameters of heterogeneous hybridization reactions. Nucleic Acids Res. *10*:7181-7196.
5. **Coutlee, F., L. Bobo, K. Mayur, R.H. Yolken and R.P. Viscidi.** 1989. Immunodetection of DNA with biotinylated RNA probes: a study of reactivity of a monoclonal antibody to DNA-RNA hybrids. Anal. Biochem. *181*:96-105.
6. **Galvan, B., T.K. Christopoulos and E.P. Diamandis.** 1995. Detection of prostate specific antigen mRNA by reverse transcription polymerase chain reaction and time-resolved fluorometry. Clin. Chem. *41*:1705-1709.

7. **Gingeras, T.R., D.Y. Kwoh and G.R. Davis.** 1987. Hybridization properties of immobilized nucleic acids. Nucleic Acids Res. *15*:5373-5390.
8. **Giros, B., P. Sokoloff, M. Martres, J.F. Riou, L.J. Emorine and J.C. Schwartz.** 1989. Alternative splicing directs the expression of the two D_2 dopamine receptor isoforms. Nature *342*:923-926.
9. **Gross, J., I. Müller, C. Berndt, U. Ungethüm, J. Heldt, M. Elizalde, L.E. Peters and K. Andersson.** Quantitation of dopamine D_2 receptor mRNA in a mesencephalic cell culture using a nonradioactive competitive reverse transcription polymerase chain reaction method. J. Neurosci. Methods (In press).
10. **Guesdon, J.L.** 1992. Immunoenzymatic techniques applied to the specific detection of nucleic acids. J. Immunol. Methods *15*:33-49.
11. **Heller, M.J.** 1994. Fluorescent detection methods for PCR analysis, p. 134-141. *In* K.B. Mullis, F. Ferré and R.A. Gibbs (Eds.), The Polymerase Chain Reaction. Birkhäuser, Boston.
12. **Holloway, B., D.D. Erdman, E.L. Durigon and J.J. Murtagh.** 1993. An exonuclease-amplification coupled capture technique improves detection of PCR product. Nucleic Acids Res. *21*:3905-3906.
13. **Holmstrom, K., L. Rossen and O.F. Rasmussen.** 1993. A highly sensitive and fast nonradioactive method for detection of polymerase chain reaction products. Anal. Biochem. *209*:278-283.
14. **Holodniy, M., M.A. Winters and T.C. Merigan.** 1992. Detection and quantification of gene amplification products by a nonisotopic automated system. BioTechniques *12*:36-39.
15. **Kawai, S., S. Maekawajiri and A. Yamane.** 1993. A simple method of detecting amplified DNA with immobilized probes on microtiter wells. Anal. Biochem. *209*:63-69.
16. **Keller, G.H., D.P. Huang and M.M. Manak.** 1989. A sensitive nonisotopic hybridization assay for HIV-1 DNA. Anal. Biochem. *177*:27-32.
17. **Kemp, D.J., D.B. Smith, S.J. Foote, N. Samaras and M.G. Peterson.** 1989. Colorimetric detection of specific DNA segments amplified by polymerase chain reactions. Proc. Natl. Acad. Sci. USA *86*:2423-2427.
18. **Kohsaka, H., A. Taniguchi, D.D. Richman and D.A. Carson.** 1993. Microtiter format gene quantification by covalent capture of competitive PCR products: application to HIV-1 detection. Nucleic Acids Res. *21*:3469-3472.
19. **Kricka, L.J.** 1994. Selected strategies for improving sensitivity and reliability of immunoassays. Clin. Chem. *40*:347-357.
20. **Lassner, D.** 1995. Quantitation of mRNA by the ELOSA technique using external standards, p. 117-123. *In* T. Köhler, D. Lassner, A.K. Rost, B. Thamm, B. Pustowoit and H. Remke (Eds.), Quantitation of mRNA by Polymerase Chain Reaction. Springer Verlag, Berlin.
21. **Mantero, G., A. Zonaro, A. Albertini, P. Bertolo and D. Primi.** 1991. DNA enzyme immunoassay: general method for detecting products of polymerase chain reaction. Clin. Chem. *37*:422-429.
22. **Matthews, J.A. and L.J. Kricka.** 1988. Analytical strategies for the use of DNA probes. Anal. Biochem. *169*:1-25.
23. **Nainan, O.V., T.L. Crommeans and H.S. Margolis.** 1996. Sequence-specific, single-primer amplification of PCR products for identification of hepatitis viruses. J. Virol. Methods *61*:127-134.
24. **Nikifirov, T.T. and Y.H. Rogers.** 1995. The use of 96-well polysterene plates for hybridization-based assays: an evaluation of different approaches to oligonucleotide immobilization. Anal. Biochem. *227*:201-209.
25. **Polsky-Cynkin, R., G.H. Parsons, L. Allerdt, G. Landes, G. Davis and A. Rashtchian.** 1985. Use of DNA immobilized on plastic and agarose supports to detect DNA by sandwich hybridization. Clin. Chem. *31*:1438-1443.
26. **Raab, M.** 1983. Comparison of a logistic and a mass-action curve for radioimmunoassay data. Clin. Chem. *29*:1757-1761.
27. **Rasmussen, S.R., M.R. Larsen and S.E. Rasmussen.** 1991. Covalent immobilization of DNA onto polysterene microwells: the molecules are only bound at the 5′ end. Anal. Biochem. *198*:138-142.
28. **Running, J.A. and M.S. Urdea.** 1990. A procedure for productive coupling of synthetic oligonucleotides to polystyrene microtiter wells for hybridization capture. BioTechniques *8*:276-279.
29. **Stollar, B.D. and A. Rashtchian.** 1987. Immunochemical approaches to gene probe assays. Anal. Biochem. *161*:387-394.
30. **Tu, E. and E. Jablonski.** 1994. Enzyme-labeled oligonucleotides, p. 142-150. *In* K.B. Mullis, F. Ferré and R.A. Gibbs (Eds.), The Polymerase Chain Reaction. Birkhäuser, Boston.
31. **Tullis, R.H.** 1994. Ultrasensitive nonradioactive detection of PCR reactions: an overview, p. 123-159. *In* K.B. Mullis, F. Ferré and R.A. Gibbs (Eds.), The Polymerase Chain Reaction. Birkhäuser, Boston.
32. **Wetmur, J.G.** 1991. DNA probes: applications of the principles of nucleic acid hybridization. Crit. Rev. Biochem. Mol. Biol. *26*:227-259.
33. **Whitby, K. and J.A. Garson.** 1995. Optimisation and evaluation of a quantitative chemiluminescent polymerase chain reaction assay for hepatitis C virus RNA. J. Virol. Methods *51*:75-88.
34. **Zammatteo, N., C. Girardeaux, D. Delforge, J.J. Pireaux and J. Remacle.** 1996. Amination of polystyrene microwells: application to the covalent grafting of DNA probes for hybridization assays. Anal. Biochem. *236*:85-94.

Section III

Cloning Differentially Expressed cDNAs

11 Improvements to Differential Display

Timothy Bos and Martin Hadman
Department of Microbiology and Immunology, Eastern Virginia Medical School, Norfolk, VA, USA

OVERVIEW

The differential display technique has proven to be a valuable tool in the study of differential gene expression. Differential display has gained popularity because it allows direct and simultaneous comparison of multiple samples for both positive and negative changes in gene expression. The procedure uses only small amounts of RNA and is therefore a perfect method for studying changes in gene expression from scarce biological materials. In principle, a comparative differential display profile can be generated in a few days. However, because of the large number of false positives that are generated, the process of screening can be very labor-intensive. This chapter focuses on the critical steps in the differential display procedure that have been targeted for modification in an effort to reduce the number of false positives. In addition, we include a detailed protocol that incorporates a number of these modifications.

BACKGROUND

A common approach to studying the mechanisms of complex biological processes is to identify differences that exist at the level of gene expression in response to various stimuli, treatments or conditions. Several methods including cDNA library subtraction, differential screening and differential display allow this type of comparison.

Differential display was originally developed by Liang and Pardee (13). As the name implies, the differential display technique results in a visual display of differences between two or more test populations of RNA expression products. To generate a display, RNA from two or more cell populations is reverse-transcribed into cDNA using an anchored oligo(dT) primer. The cDNAs are then tagged and amplified by polymerase chain reaction (PCR) using an arbitrary 5′ primer in combination with the same anchored 3′ oligo(dT) primer. These PCR products are separated on a high-resolution gel on which direct comparisons can be made. Bands that appear differential are then excised from the gel, reamplified by PCR and examined by sequence and Northern blot analysis.

The two main advantages of this method over subtraction and differential screening are that a number of different samples can be compared simultaneously, and both positive and negative differences can be visualized from the same set of reactions. This makes differential display an ideal method for studying changes in gene expression profiles associated with multistep biological processes such as development, differentiation and tumor progression.

The major limitation with differential display is the large number of false positives

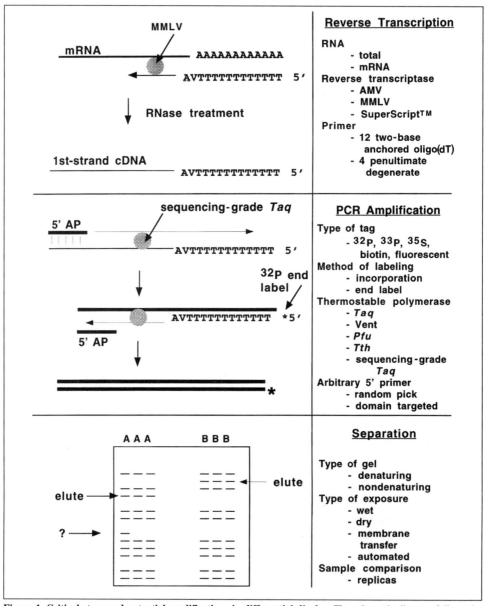

Figure 1. Critical steps and potential modifications in differential display. The schematic diagram follows the procedure outlined in this chapter. The modifications listed to the right demonstrate the potential for large numbers of permutations in the differential display protocol.

that are generated. Although the technique itself is rapid and can generally produce a comparative profile in a matter of days, the process of screening potential differential cDNAs can be an enormous task. In an attempt to cut down on the time required for screening, several modifications to the original procedure have surfaced. The major impetus for each modification has been to reduce the proportion of false-positive bands. Out of this quest for a more refined procedure have arisen a number of similar protocols that are distinguished mainly by the choice of primers used for amplification. The original procedure, differential display, uses an anchored 3′ oligo(dT) primer and an arbitrary 5′ primer (13). RNA fingerprinting by arbitrarily primed PCR (RAP-PCR) uses one or two arbitrary primers rather than an anchored oligo(dT) primer (30). Domain-specific differential display uses primers based on specific sequence motif information as a way to target specific types of genes (9). Each of these techniques, however, uses the same overall concept to display differences in gene expression patterns.

The purpose of this chapter is to focus attention on some of the difficulties associated with differential display and the multitude of modifications that have been tried to circumvent the problems and increase the usefulness of the procedure. We have included a detailed protocol that we have used and highlighted its differences from the original procedure. In addition, we have provided a detailed reference list in which the reader can access a number of excellent technical reviews and technique-oriented papers that discuss some of the fine details of differential display implementation.

Targeted Modifications

The differential display technique can be divided roughly into five basic steps: *(i)* reverse transcription (RT), *(ii)* PCR, *(iii)* separation, *(iv)* recovery and *(v)* screening. Figure 1 diagrams schematically the steps required to generate a display and outlines many of the critical parameters that have been targeted for modification.

Reverse Transcription: Generation of Primary cDNA Pools

One of the most critical steps in differential display is RT. It is at this step that representative cDNA pools are generated. Several factors are important in this first step including choice of RNA, choice of enzyme and choice of primer.

<ins>Choice of RNA</ins>

Both total RNA and mRNA have been used successfully in differential display. The most important element at this step is to isolate an undegraded RNA population free of DNA. RNase-free DNase I is often used to ensure that no genomic DNA sequences have been retained in the RNA sample. Differential display bands derived from total RNA pools are often derived from nonmessenger fractions. If the number or intensity of these bands is high, it can complicate band visualization and interpretation. An obvious remedy to this problem is to use poly(A)-selected RNA as the starting material. This, however, has lead to problems of its own. For instance, differential displays generated from mRNA are often highly smeared because of contamination with small amounts of oligo(dT) (from the mRNA isolation), which results in priming at multiple points along the poly(A) tail (6,11).

Choice of enzyme

The enzyme used during the RT reaction can affect the quality of the first-strand cDNA generated. Most protocols use the RNase H-negative Moloney murine leukemia virus (MMLV) reverse transcriptase. Avian myeloblastosis virus (AMV) reverse transcriptase, which has significant RNase H activity, generally results in smearing (24). Recently, the thermally stable *Tth* enzyme has been used successfully (23). The advantage of this enzyme is that the RTs and PCRs can be done in the same tube with the same enzyme.

Choice of primer

The choice of primer has received much attention. The original procedure (13) called for 12 different anchored oligo(dT) primers, each with a different combination at the ultimate and penultimate residues. The theory was that the number and therefore resolution of bands generated by differential display would be reduced if the starting population were divided into discrete populations based on the RT reaction. Subsequent trials have used primers that are degenerate for G, A and C at the penultimate base so that only 4 RT reaction pools will be generated. This cuts down on the number of different differential display reactions that need to be performed. However, because base pairing at the degenerate site can favor one nucleotide over another, it has been suggested that these degenerate primers result in uneven pools (2). One method to circumvent this problem has been the use of inosine, which pairs at the penultimate position with equal affinity to all four bases (21).

Polymerase Chain Reaction: Amplification of RNA Pools

Once the RNA has been reverse-transcribed into first-strand cDNA, the next step is to amplify these products by PCR with the intent of generating amplification products that can be resolved by electrophoresis. To accomplish this, the RT products are used as template for PCR. At this step, the 3′ primer is generally the same one used during the RT reaction. The 5′ primer is a small arbitrary primer. Theoretical calculations have suggested that total representation of RNA expression pools requires 24–26 arbitrary 5′ primers and twelve 3′ primers (2). This equates to 288–312 PCRs from each sample. Obviously, reactions carried out on this scale would require significant effort and resources. In practice, many investigators have discovered true differential clones using many fewer primer combinations.

A number of variables in the PCR amplification step can be incorporated that will ultimately affect implementation of subsequent steps. These include: *(i)* choice and method of incorporation of a tag, *(ii)* source of thermostable polymerase and *(iii)* design of 5′ arbitrary primers.

Choice and method of incorporation of tags

The tag used for subsequent visualization of the displayed PCR products can vary tremendously. The original procedure called for incorporation during PCR of ^{35}S-labeled dCTP (13). This procedure resulted in highly labeled bands that could easily be visualized. However, incorporation of label during synthesis has a number of disadvantages. The intensity of a specific band depends not only on its abundance but also on its length and relative content of C residues. In addition, a significant portion of PCR products are derived from priming between two 5′ arbitrary binding sites rather than from the 3′ and 5′ primer combinations. A simple method to avoid these problems is to tag the 3′ primer with a label (6). Here, the band intensity should be proportional to the amount of product rather than length or nucleotide content. In addition, only bands generated with the 3′ primer on one end will be visualized. Because the 3′ primer is mostly T residues, significant amplification between two 3′ primers is not generally a problem (6). A positive side effect of using a labeled 3′ primer is that the smearing often seen when using mRNA as template is eliminated (6). This is because priming from nonanchored oligo(dT) primers carried over from the mRNA isolation will not be labeled and therefore not visualized.

The specific tag used to label the PCR product will dictate implementation of subsequent steps (see below). ^{35}S, ^{33}P, ^{32}P, biotin, digoxigenin and fluorescent tags have all been used successfully (1,4,6,8,13,16,22,26).

Source of thermostable polymerase

A number of different thermostable polymerases have been used during the PCR amplification step. These include Vent®, Deep Vent™ (both from New England Biolabs, Beverly, MA, USA), *Pfu*, *Tth*, *Taq* and sequencing-grade *Taq*. Choice of polymerase might be important depending on subtle changes in procedure. For instance, when using end-labeled primers, it is important to use a polymerase that does not have significant levels of 5′ to 3′ exonuclease activity because this will result in loss of the label (6). Sequencing-grade *Taq* DNA polymerase is modified to eliminate this activity and is therefore a good choice under these conditions. Likewise, modifications in primer design that specify an increased potential for long PCR products will benefit from the use of enzymes or enzyme combinations with proofreading activities such as Vent, Deep Vent and *Pfu* (5).

Design of arbitrary primers

The design of the 5′ arbitrary primers has been a focus of many investigators. The size and sequence of these arbitrary primers can vary dramatically. A number of investigators have generated nonrandom arbitrary primers that will decrease the number of bands derived from unwanted RNA populations (5). One paper tested arbitrary primer sequences for low identity to ribosomal and mitochondrial RNA populations (7). Another tested primers for sequences found at high frequency in mammalian cDNAs (5). Likewise, primers can be designed to target specific RNA populations such as those that can encode zinc finger motifs (9). In most cases, 5′ primers of 10–12 bases are used. However, because of their small size, low annealing temperatures must be used during the PCRs. These low annealing temperatures often result in mismatched priming that can result in variable amplification patterns. Several investigators have dramatically increased the size of their 5′ and 3′ primers to be able to use higher annealing temperatures and to increase specificity (3).

Separation: Display of Differences

This step forms the foundation from which differences in expression from different RNA pools are visualized. In the original protocol, ^{35}S-labeled PCR amplification products were separated on denaturing sequencing gels. The gel was then dried and autoradiographed. Differences could be resolved in the range below about 500–600 bases. Although this step is fairly straightforward, a number of variations have arisen. Most of these involve the process and method used for visualization.

The process for visualization depends on the specific tag used. Low-energy isotopes such as ^{35}S and ^{33}P require the gel to be dried first before autoradiography. This allows a fine level of resolution but requires the extra step of drying. If a high-energy isotope such as ^{32}P is used, drying of the gel is optional, and exposure times are reduced (6). Gels run with biotin- or digoxigenin-labeled primers need to be transferred to a membrane (4). Direct visualization involves use of a substrate that will develop directly on the membrane. The band can therefore be excised with great precision (4). Indirect visualization using chemiluminescence is more sensitive but requires X-ray film and subsequent film alignment over the bands on the membrane. Fluorescence-labeled primers allow high-throughput, high-resolution analysis on automated DNA sequencers (8,16). However, because the signal is read during the gel run, there is not an easy way to recover specific bands. Band recovery requires the running of an additional gel using one of the other methods for band labeling (8,16).

In addition to using different gel-processing protocols, a number of simple things can be done that greatly enhance the confidence with which differential bands are later selected for analysis. An obvious but often neglected one is to run reactions in duplicate or triplicate (20,25). Ideally, these reactions will originate from completely independent RNA isolations and therefore represent true repetitions. If a differential band occurs at the same place in equivalent but independent RNA preparations, it is likely that the band itself does not result as an artifact of a particular reaction. Likewise, if a band is real, one would expect to be able to regenerate it in a completely independent experiment. These redundant samples serve as important internal controls that can greatly reduce the number of false positives screened in later steps.

Recovery: Elution and Cloning

Recovery involves several steps. Autoradiographs and gels must be precisely aligned so that a band can be excised. The DNA in the band is then eluted and reamplified by PCR. Several modified elution protocols have been described (3,18,28). Once reamplified, most DNA preparations will contain more than one distinct DNA in significant concentration. The reamplified DNAs are cloned using any one of a variety of TA cloning vectors, and subsequent individual clones are assayed in the subsequent screening step.

The recovery step requires precision and patience and is often frustrating. It is not unusual to consistently find differential bands that can not be reamplified. The reason for this is not entirely clear. Possibilities include contaminating gel material that inhibits the PCR and misalignment of the autoradiograph with the original gel.

Screening: Determining Which Clones Are Really Differential

The recovery and elution step really marks the end of the differential display technique. Once clones have been generated, the real challenge of determining if they are truly differential begins. The modifications mentioned up to this point have been geared towards increasing the specificity of the display to decrease the number of nonreproducible or artifactual bands.

A common and accepted method to confirm the differential nature of a specific clone is to examine it by Northern blot analysis. This has been difficult for clones isolated by differential display for a variety of reasons. Most of the clones isolated are short (100–300 bases long). These short fragments are difficult to label using common random-primed methods. Riboprobes can be generated, but this requires some knowledge of the orientation of the clone, which usually necessitates the additional step of sequencing. A further disadvantage of the Northern blot assay is its sensitivity, which precludes analysis of low-abundance RNAs. The most serious limitation, however, is that only one clone at a time can be assayed. Because differential display tends to generate a large number of differential bands and each band generates a number of distinct clones during recovery, screening each one individually is an extremely inefficient process. An additional disadvantage to this approach to screening is that a large pool of RNA is required for generation of multiple Northern blots. Because differential display is often performed from small amounts of RNA isolated directly from tissue biopsies and primary cultures, large amounts of RNA are often not available. In an effort to overcome some of these hurtles, a number of alternate approaches have been developed that allow quicker screening of multiple differential clones using a minimum amount of sample RNA. These screening techniques are not unique to differential display and have been used as a way to screen differences in gene expression generated by a variety of other methods.

A number of reports have skipped the screening step and proceeded directly to sequencing of potential clones (29,33). The thought is that enough sequence information can be obtained to perform database searches to determine if the clone in question is a known or interesting gene. Although this approach is suitable for automation and high-throughput analysis, for it to yield relevant information, the number of false positives generated from the display must be significantly reduced.

Another method that deserves mention is called affinity capture, in which labeled bands obtained after gel elution (before reamplification) are used to probe a Northern blot or dot blot of differential clones (10). The differential bands are excised from the Northern blot, eluted and cloned. Because the amount of label on the original gel fragment recovered from the sequencing gel is low, the sensitivity of this technique limits its utility.

Perhaps the most powerful method used for screening after differential display is the reverse Northern blot procedure (2,3,15,17–19,27,28,31). In this method, DNA from differential clones is fixed to a membrane and screened with probes generated from the different original populations of RNA. The probes can be made by a variety of methods ranging from labeling cDNA from RNA or cDNA templates (17,28,31) using the original primary labeled differential display product (3,27), generating labeled amplified RNA from cDNA (19) and labeling freshly synthesized RNA from nuclear runon reactions (2). The source and amount of the starting RNA will dictate which method will be most efficient in specific applications. For instance, if a very small tissue section is used for differential display, the amount of RNA that can be easily obtained will be minimal. In this case, use of an amplified RNA or cDNA probe generated from a cDNA template would provide an abundant source of probe without the need for further RNA isolations. The primary advantage of the reverse Northern blot method is that it allows screening of large numbers of potential differential clones simultaneously while using only a small amount of sample material as a source for probe generation. Clones that are shown to be different by this differential screen can then be further analyzed by Northern blot, quantitative RT-PCR (7) or sequence analysis.

PROTOCOLS

The following method has been used successfully in our laboratory. Several of the modifications mentioned in preceding sections are incorporated into this protocol. Many of the papers referenced in this chapter contain detailed protocols in which very specific modifications have been introduced. In addition, there are a number of excellent technical reviews that detail the traditional differential display protocol (2,11,12,14,20). The method presented below is described only as far as the actual display and recovery of bands. Screening is largely a matter of preference that will be dictated by the availability of large amounts of RNA and numbers of clones generated. We direct the reader to references cited in the section above on screening for detailed protocols appropriate for different applications.

The procedure detailed below (6) contains several modifications from the original procedure of Liang and Pardee (13). First, mRNA is used as the starting template material. This allows a greater representation of expressed messages without the complication of bands generated from ribosomal and transfer RNAs. Second, ^{32}P rather than ^{35}S is used to label the display products. This allows direct exposure of display gels without the need for drying. Although the resolution is not as good as with ^{35}S, we have not found this to be a major limitation. This is largely because several of the other modifications have cut down on the total number of bands generated. Third, we have end-labeled the 3' anchored oligo(dT) primer rather than incorporate label during PCR. We have found that addition of the 3' primer alone does not result in significant banding patterns. Therefore, bands generated with a labeled 3' primer are representative of amplification between the arbitrary 5' primer and the 3' primer. This has also eliminated the smearing often seen when using mRNA as a template. An example of a display generated with this protocol is shown in Figure 2.

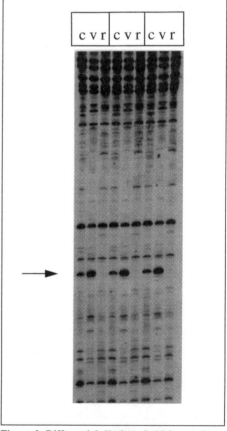

Figure 2. Differential display of chicken embryo fibroblasts (CEF) expressing c-Jun (c), v-Jun (v) or vector (r). The display was generated using the procedure outlined in the Protocols section. The arrow points to a differential band found in both c-Jun and v-Jun expressing CEF but not in vector-infected CEF.

Preparation of RNA

Isolation of mRNA

Prepare mRNA from cells or tissue using one of the many mRNA isolation kits commercially available. We have experience with the FastTrack® Kit (Invitrogen, Carlsbad, CA, USA) and Oligotex™ (Qiagen, Chatsworth, CA, USA).

DNase treatment of mRNA

It is important to treat the RNA sample with DNase to avoid any amplification that might occur as a result of contaminating genomic DNA.

Materials and Reagents

- DNA digestion mixture: 10 µL 100 mM MgCl$_2$, 10 mM dithiothreitol (DTT), 1 µL RNasin® (Promega, Madison, WI, USA), 1 µL RNase-free DNase (20 U/µL; Promega), TE buffer (10 mM Tris, pH 7.4, 1 mM EDTA) to 50 µL
- DNase stop mixture: 50 mM EDTA, 1.5 M sodium acetate, 1% sodium dodecyl sulfate (SDS)
- Diethyl pyrocarbonate (DEPC)-treated water

Procedure

1. Add 50 µL of DNA digestion mixture to 50 µL of the mRNA sample and incubate for 15 min at 37°C.
2. Add 25 µL DNase stop mixture.
3. Extract once with phenol–chloroform–isoamyl alcohol and once with chloroform–isoamyl alcohol.
4. Precipitate the RNA with 3.5 vol of ethanol at -70°C for 20–30 min.
5. Resuspend in DEPC-treated water.

Reverse Transcription

This setup is for one mRNA template extended by one of four unlabeled 3′ primers. Prepare a master mixture for the appropriate number of templates.

Materials and Reagents

	Final Concentration
• DEPC-treated water	
• 3 µL 5× RT buffer	1×
• 2 µL 100 mM DTT	10 mM
• 2 µL 5 mM dNTP	0.5 mM
• 0.5 µL 0.5 µg/µL actinomycin D	500 ng

- 1 µL 20 U/µL RNasin   20 U
- 1 µL 25 pmol/µL 3′ primer   25 pmol*
- mRNA template   0.2–0.4 µg*
- 1.5 µL 200 U/µL MMLV reverse transcriptase   300 U

*Omit these ingredients when preparing the master mixture because templates and primers will vary.

Note: 5× RT buffer contains 250 mM Tris-HCl, pH 8.3, 40 mM $MgCl_2$, 50 mM DTT.

Procedure

1. In a sterile RNase-free tube, mix 0.2–0.4 µg of mRNA with 25 pmol of 3′ primer. Add DEPC-treated water to a volume of 6 µL. Incubate at 65°C for 5–10 min to denature the mRNA. Anneal the primer and mRNA at 30°C for 15–30 min. Spin briefly. Chill on ice until ready.
2. Prepare the master mixture on ice without mRNA and primer.
3. Add mixture to mRNA–primer reaction on ice.
4. Incubate at 35°–37°C for 45 min to 1 h. Inactivate the reverse transcriptase by heating at 90°C for 3 min. Spin briefly. Chill on ice until ready.
5. Bring the reaction to room temperature. Add 1 µL DNase-free RNase A. Incubate at 37°C for 15 min.
6. Purify the cDNA on a CHROMA SPIN™-100 column (CLONTECH Laboratories, Palo Alto, CA, USA) to remove unincorporated dNTP and unlabeled primers. Store reaction mixtures at -20°C.

Primer Labeling Reaction

Materials and Reagents

- Kinase reaction mixture (21-µL volume): 11.3 µL water, 2 µL 10× kinase buffer (Promega), 1.7 µL 100 pmol/µL of the 3′ primer (5′-TTTTTTTTTTTTVA-3′ [V = A, C or G], 5 µL [γ-^{32}P]ATP (3000 Ci/mmol; 10 mCi/mL), 1 µL 8 U/µL T4 polynucleotide kinase

Procedure

1. Prepare kinase reaction on ice. Incubate at 37°C for 20–30 min.
2. Inactivate the kinase by heating at 90°C for 2–3 min.
3. Place on ice until ready to use or store at -20°C.

Note: There is no need to purify the kinase reaction.

Polymerase Chain Reaction

This setup is for one PCR. Be sure that the 3′ labeled primer used is identical to the 3′ primer used in the RT reaction. Prepare a master mixture for the appropriate number of templates.

Materials and Reagents

	Final concentration
• 3.8 µL DEPC-treated water	
• 2 µL 10× Mg-free *Taq* buffer (Promega)	1×
• 0.8 µL 25 mM MgCl$_2$	1 mM
• 4 µL 1 mM dNTP	0.2 mM
• 1 µL 20 pmol/µL 5′ primer (CTTGATTGCC)	20 pmol
• 2.4 µL 8.5 pmol/µL 3′ labeled primer	20 pmol
• 5 µL cDNA template	ca. 0.0125 pmol
• 1 µL 5 U/µL sequencing-grade *Taq*	2.5 U

Note: Final volume is 20 µL.

Procedure

1. Prepare the PCR mixtures on ice.
2. Use the following PCR amplification profile: denaturation at 90°C for 45 s, annealing at 43°–45°C for 1–2 min, extension at 70°C for 1 min for 30–35 cycles, followed by a final extension at 70°C for 5 min.
3. Soak at 4°C until ready.

Separation

Separate the labeled differential display products on a standard 6% denaturing sequencing gel. For loading, add 2 parts formamide stop dye to 3 parts PCR master mixture. Load 3–5 µL on a prerun 6%–7% denaturing sequencing gel at 80 W constant power. Run the gel until the xylene cyanol runs off the bottom (about 75 bp).

After sequencing, mark the gel with fluorescent tape and expose for autoradiography. Because ^{32}P was used as the label, there is no need to dry the gel.

Elution and Reamplification

Once bands have been identified and excised from the gel, their DNA can be amplified using the following procedure.

Procedure

1. Incubate the gel slice in 100 µL TE buffer at 37°C overnight (16–18 h) or boil the gel slice in TE buffer for 10 min and chill on ice.
2. Spin the sample briefly and remove the supernatant to a new tube. Store at −20°C.
3. Use 1–10 µL of eluate in a 100-µL PCR. Use the same primers (do not label the 3′ primer this time) and protocol as used for the differential display reaction. Generally, one round of 30–35 cycles will yield DNA.
4. Check the DNA by running 10–15 µL on a 5% polyacrylamide gel in TBE buffer; stain with ethidium bromide.

REFERENCES

1. **An, G., G. Luo, R.W. Veltri and S.M. O'Hara.** 1996. Sensitive, nonradioactive differential display method using chemiluminescent detection. BioTechniques *20*:342-346.
2. **Bauer, D., P. Warthoe, M. Rohde and M. Strauss.** 1994. Detection and differential display of expressed genes by DDRT-PCR. PCR Methods Appl. *4*:97-108.
3. **Callard, D., B. Lescure and L. Mazzolini.** 1994. A method for the elimination of false positives generated by the mRNA differential display technique. BioTechniques *16*:1096-1103.
4. **Chen, J.J.W. and K. Peck.** 1996. Non-radioisotopic differential display method to directly visualize and amplify differential bands on nylon membrane. Nucleic Acids Res. *24*:793-794.
5. **Diachenko, L.B., J. Ledesma, A.A. Chenchik and P.D. Siebert.** 1996. Combining the technique of RNA fingerprinting and differential display to obtain differentially expressed mRNA. Biochem. Biophys. Res. Commun. *219*:824-828.
6. **Hadman, M., B.-L. Adam, G.L. Wright, Jr. and T.J. Bos.** 1995. Modifications to the differential display technique reduce background and increase sensitivity. Anal. Biochem. *226*:383-386.
7. **Ikonomov, O.C. and M.H. Jacob.** 1996. Differential display protocol with selected primers that preferentially isolates mRNAs of moderate- to low-abundance in a microscopic system. BioTechniques *20*:1030-1042.
8. **Ito, T., K. Kito, N. Adati, Y. Mitsui, H. Hagiwara and Y. Sakaki.** 1994. Fluorescent differential display: arbitrarily primed RT-PCR fingerprinting on an automated DNA sequencer. FEBS Lett. *351*:231-236.
9. **Johnson, S.W., N.A. Lissy, P.D. Miller, J.R. Testa, R.F. Ozols and T.C. Hamilton.** 1996. Identification of zinc finger mRNAs using domain-specific differential display. Anal. Biochem. *236*:348-352.
10. **Li, F., E.S. Barnathan and K. Kariko.** 1994. Rapid method for screening and cloning cDNAs generated in differential display: application of Northern blot for affinity capturing of cDNAs. Nucleic Acids Res. *22*:1764-1765.
11. **Liang, P., L. Averboukh and A.B. Pardee.** 1993. Distribution and cloning of eukaryotic mRNAs by means of differential display: refinements and optimization. Nucleic Acids Res. *21*:3269-3275.
12. **Liang, P., D. Bauer, L. Averboukh, P. Warthoe, M. Rohrwild, H. Muller, M. Strauss and A.B. Pardee.** 1995. Analysis of altered gene expression by differential display. Methods Enzymol. *254*:304-321.
13. **Liang, P. and A.B. Pardee.** 1992. Differential display of eukaryotic messenger RNA by means of the polymerase chain reaction. Science *257*:967-971.
14. **Liang, P. and A.B. Pardee.** 1995. Recent advances in differential display. Curr. Opin. Immunol. *7*:274-280.
15. **Liu, C. and K.G. Raghothama.** 1996. Practical method for cloning cDNAs generated in an mRNA differential display. BioTechniques *20*:576-580.
16. **Luehrsen, K.R., L.L. Marr, E. van der Knaap and S. Cumberledge.** 1997. Analysis of differential display RT-PCR products using fluorescent primers and GENESCAN™ software. BioTechniques *22*:168-174.
17. **Martin-Laurent, F., P. Franken and S. Gianinazzi.** 1995. Screening of cDNA fragments generated by differential RNA display. Anal. Biochem. *228*:182-184.
18. **Mou, L., H. Miller, J. Li, E. Wang and L. Chalifour.** 1994. Improvements to the differential display method for gene analysis. Biochem. Biophys. Res. Commun. *199*:564-569.
19. **Poirier, G.M.C., J. Pyati, J.S. Wan and M.G. Erlander.** 1997. Screening differentially expressed cDNA clones obtained by differential display using amplified RNA. Nucleic Acids Res. *25*:913-914.
20. **Reuber, T.L. and F.M. Ausubel.** 1995. Differential mRNA display. Methods Cell Biol. *49*:431-440.
21. **Rohrwild, M., R.S. Alpan, P. Liang and A.B. Pardee.** 1995. Inosine-containing primers for mRNA differential display. Trends Genet. *11*:300.
22. **Røsok, Ø., J. Odeberg, M. Rode, T. Stokke, S. Funderud, E. Smeland and J. Lundeberg.** 1996. Solid-phase method for differential display of genes expressed in hematopoietic stem cells. BioTechniques *21*:114-121.
23. **Schwarz, H.** 1996. Rapid, nonradioactive differential display using *Tth* polymerase. Trends Genet. *12*:396-397.
24. **Shoham, N.G., T. Arad, R. Rosin-Abersfeld, P. Mashiah, A. Gazit and A. Yaniv.** 1996. Differential display assay and analysis. BioTechniques *20*:182-184.
25. **Sompayrac, L., S. Jane, T.C. Burn, D.G. Tenen and K.J. Danna.** 1995. Overcoming limitations of the mRNA differential display technique. Nucleic Acids Res. *23*:4738-4739.
26. **Tokuyama, Y. and J. Takeda.** 1995. Use of ^{33}P-labeled primer increases the sensitivity and specificity of mRNA differential display. BioTechniques *18*:424-425.
27. **Vögeli-Lange, R., N. Bürckert, T. Boller and A. Wiemken.** 1996. Rapid selection and classification of positive clones generated by mRNA differential display. Nucleic Acids Res. *24*:1385-1386.
28. **Wadhwa, R., E. Duncan, S.C. Kaul and R.R. Reddel.** 1996. An effective elimination of false positives isolated from differential display of mRNAs. Mol. Biotechnol. *6*:213-217.
29. **Wang, X. and G.Z. Feuerstein.** 1995. Direct sequencing of DNA isolated from mRNA differential display. BioTechniques *18*:448-453.

30. **Welsh, J., K. Chada, S.S. Dalal, R. Cheng, D. Ralph and M. McClelland.** 1992. Arbitrarily primed PCR fingerprinting of RNA. Nucleic Acids Res. *20*:4965-4970.
31. **Zhang, H., R. Zhang and P. Liang.** 1996. Differential screening of gene expression difference enriched by differential display. Nucleic Acids Res. *24*:2454-2455.
32. **Zhao, S., S.L. Ooi and A.B. Pardee.** 1995. New primer strategy improves precision of differential display. BioTechniques *18*:842-850.
33. **Zhao, S., S.L. Ooi, F.-C. Yang and A.B. Pardee.** 1996. Three methods for identification of true positive cloned cDNA fragments in differential display. BioTechniques *20*:400-404.

12. DD/AP-PCR: Differential Display and Arbitrarily Primed PCR of Oligo(dT) cDNA

Cynthia B. Rothschild[1], Donald W. Bowden[1] and Gregory S. Shelness[2]

Departments of [1]Biochemistry and [2]Comparative Medicine, Bowman Gray School of Medicine, Wake Forest University, Winston-Salem, NC, USA

OVERVIEW

We have previously described the comparison of differential display (DD) and arbitrarily primed polymerase chain reaction (AP-PCR) amplification of oligo(dT)-primed cDNA. Our results indicated that both of these widely used RNA fingerprinting techniques have their respective advantages and limitations and that both techniques can use cDNA synthesized using a generic anchored oligo(dT) primer $dT_{12}VN$, where V is equimolar dA, dC and dG. This efficiently selects for poly(A$^+$) sequences from total RNA and significantly reduces the number of cDNA preparations required per experiment. In addition, because of the problems inherent to the isolation of undegraded mRNA from clinical samples, we have developed a simple protocol (S100/S101 assay) that serves as a reliable indicator of whether cDNA prepared from a sample is suitable for DD or AP-PCR. Because the RNA is usually the limiting resource, maximum utilization can be achieved by generating a single pool of $dT_{12}VN$-primed cDNA, testing the cDNA by S100/S101 analysis and then performing AP-PCR, DD or a combined approach (DD/AP-PCR).

INTRODUCTION

The detection of differentially expressed genes is of interest in molecular analysis of biological processes including differentiation, development and carcinogenesis. Previous techniques such as subtractive or differential hybridization were laborious and required significant amounts of material. Recently, two PCR-based techniques, DD and AP-PCR, have been developed as methods for the identification and isolation of mRNA specific to a cell type.

AP-PCR enables the generation of cell-specific fingerprints from DNA (9) or RNA (4,8). Traditionally, in RNA fingerprinting by AP-PCR (4,8), cDNA is synthesized using a specific but arbitrarily chosen primer (20-mer). The cDNA is then PCR-amplified using the same arbitrary primer, with one low-stringency cycle to promote incorporation of the primer into the second strand of cDNA, followed by several cycles of high-stringency amplification. AP-PCR products are radiolabeled during PCR and resolved by gel electrophoresis, generating a fingerprint that is specific to the RNA population and arbitrary primer used.

DD (1,2) uses 3′-anchored oligo(dT) primers for the synthesis of cDNA subsets representative of the poly(A$^+$) RNA population. Oligo(dT) primers that are anchored with two specific nucleotides (e.g., $dT_{12}AG$) (2), degenerate at the penultimate base (e.g., $dT_{12}VG$, where V is equimolar dA, dG and dC) (1) or anchored with only one base (e.g., $dT_{12}G$) (3), have been used successfully for DD. The individual cDNA pools are then amplified in the presence of a radiolabel using the same anchored oligo(dT) primer and a 10-mer oligonucleotide. For both AP-PCR and DD, bands that appear to be specific to the cell population of interest can be excised from the gel and the DNA eluted and reamplified for subsequent cloning and characterization.

We were interested in applying either DD or AP-PCR to the analysis of changes in gene expression occurring in human prostate carcinogenesis. However, because of the difficulty in obtaining human prostate tumor that had been processed quickly enough to obtain relatively intact mRNA, we wanted to optimize our limited samples. Thus, we decided to compare DD and AP-PCR using an animal model system to determine whether either approach would be better for the analysis of human samples (5). In the course of this work, we also developed a simple AP-PCR protocol using primers (S100 and S101) from the signal peptidase gene to analyze whether a cDNA pool is of suitable quality for DD or AP-PCR.

Both DD and AP-PCR can be used to generate RNA fingerprints from oligo(dT)-generated cDNA. Both techniques are straightforward and reproducible, each with its respective advantages and limitations. DD produces profiles that are specific to the oligo(dT) primer used (e.g., $dT_{12}VG$ vs. $dT_{12}VC$) for amplification and is suited for isolation of A-rich and 3′ mRNA sequences. AP-PCR displays significant redundancy in profiles generated from various oligo(dT) cDNA pools but is not as biased to the isolation of AT-rich or 3′ sequences. Thus, in cases where mRNA might be a limiting factor, maximum utilization can be achieved by generating a single pool of $dT_{12}VN$-primed cDNA, testing the cDNA by S100/S101 analysis and then performing AP-PCR, DD or a combined approach (DD/AP-PCR).

PROTOCOLS

Primers

Primers used for DD or AP-PCR were either 20-mers randomly chosen from primers used in our laboratory, 10-mers comprising the 3′ end of the 20-mers or anchored oligo(dT) primers. Primers synthesized by Operon Technologies (Alameda, CA, USA) were 20-mers MFD95GT (5′-CTTTATCTTCACACAGCTTC-3′) and PPGB (5′-GATGAGAAGACCCTTCAACC-3′) (5). Primers synthesized by the DNA Synthesis Core Laboratory of the Comprehensive Cancer Center of Wake

Forest University were MFD-3' (5'-ACACAGCTTC-3'), PPGB-3' (5'-CCCTTCA-ACC-3') and the anchored oligo(dT) primers $dT_{12}VT$, $dT_{12}VA$, $dT_{12}VG$ and $dT_{12}VC$ (V = equimolar dA, dC and dG).

The primers used for evaluation of the cDNA quality for use in DD or AP-PCR are S100 (5'-gagtcgacATGGARCCNGCNTTYCA-3') (R = A or G, Y = C or T, N = A, C, G or T) and S101 (5'-cagaattcRTTRTTRTCNCCYTTNGT-3'). The primer sequences shown in upper case are specific to two regions corresponding to nucleotides 207–223 (S100) and 374–392 (S101) of canine signal peptidase (GenBank® Accession No. J05466) (6), known to be highly conserved from yeast to humans. The primers are degenerate at several positions to enable amplification across species and are extended on the 5' ends (lower case) with restriction endonuclease sites (SalI and EcoRI, respectively).

Isolation of RNA and cDNA Synthesis

Rat tissue (liver, spleen and brain) is snap-frozen in liquid nitrogen and stored at -70°C. Total RNA is isolated using TRIZOL® Reagent (Life Technologies, Gaithersburg, MD, USA) and treated with DNase I (Life Technologies) as recommended by the manufacturer. Poly(A+) RNA is isolated using the Micro-FastTrack™ Kit (Invitrogen, Carlsbad, CA, USA). cDNA is synthesized from mRNA (100 ng mRNA/20-µL reaction) or DNA-free total RNA (1 µg/20-µL reaction) using the cDNA Cycle® Kit (Invitrogen) or SUPERSCRIPT™ II RNase H- Reverse Transcriptase (Life Technologies). cDNA (2 µL) is used in AP-PCR or DD reactions as described below.

Arbitrarily Primed PCR

cDNA is synthesized using oligo(dT) (20 ng/µL) (cDNA Cycle Kit) or anchored oligo(dT) primers (20 ng/µL). cDNA pools are amplified by AP-PCR using 20-mers with two low-stringency cycles (5 min at 94°C, 5 min at 40°C and 5 min at 72°C) for incorporation of primer into initial amplification products), followed by 40 high-stringency cycles (1 min at 94°C, 1 min at 58°C and 2 min at 72°C). Amplification reactions (15 µL) contain 2 µL cDNA, 50 mM KCl, 10 mM Tris-HCl, pH 8.3, 0.01% bovine serum albumin (BSA), 2.5 mM $MgCl_2$, 250 ng specific 20-mer (3.3 µM), 100 µM dNTPs, 2.5 µCi [α-^{32}P]dCTP (3000 Ci/mM; 10 mCi/mL) and 0.75 U Taq DNA polymerase (Boehringer Mannheim, Indianapolis, IN, USA). In some experiments, a two-step amplification protocol (9) was used in which cDNA was amplified for 2 low-stringency and 10 high-stringency cycles using increased primer (10 µM) and $MgCl_2$ (4 mM), followed by 30 high-stringency cycles with reduced primer (1 µM) and $MgCl_2$ (1.5 mM) and 5 µCi [α-^{32}P]dCTP. Results were essentially the same using either protocol.

Differential Display

cDNA is synthesized using either oligo(dT) or the anchored oligo(dT) primers as described for AP-PCR. DD reactions (15 µL) contain 2 µL cDNA, 50 mM KCl, 10 mM Tris-HCl, pH 8.3, 1.5 mM $MgCl_2$, 0.01% BSA, 5 µM dNTPs, 5 µCi [α-^{32}P]dCTP, 100 ng of the primer used for cDNA synthesis, 50 ng of a specific 10-mer (e.g., PPGB-3') and 2.5 U Taq. Alternatively, DD products can be radiolabeled by incorporation of an anchored oligo(dT) primer (e.g., $dT_{12}VC$) that had been end-labeled using [γ-^{32}P]ATP; in these amplifications, 100 µM dNTPs are used. Samples are heated to 95°C for 1 min, followed by 40 cycles at 94°C for 45 s, 40°C for 90 s and 72°C for 30 s with a 2-s extension per cycle, followed by 5 min at 72°C.

Cloning of AP-PCR and DD Products

AP-PCR and DD products are resolved by electrophoresis on a 6% DNA sequencing gel. Bands of interest are excised from the gel and incubated in 100 µL TE buffer (10 mM Tris-HCl, pH 7.5, 1 mM EDTA) for 2 h at 60°C. An aliquot is then used for reamplification using the same primers and thermal cycling conditions used for original amplification (high stringency for AP-PCR). Reamplification reactions (100 µL) contain 7.5 µL eluted DNA, 10 mM Tris-HCl, pH 8.3, 50 mM KCl, 1.5 mM MgCl$_2$, 0.01% BSA, 200 µM dNTPs, 500 ng primer(s) and 2.5 U *Taq*. Successful reamplification of the fragment is verified by agarose gel electrophoresis. Products are ligated directly into the pCR®II TA Cloning® Vector (Invitrogen) for DNA sequencing.

RESULTS AND DISCUSSION

RNA Fingerprinting by Arbitrarily Primed PCR and Differential Display

The overall strategy for direct comparison of AP-PCR and DD is diagrammed in Figure 1. Rat liver was excised and immediately frozen in liquid nitrogen. Both total RNA and poly(A$^+$) RNA (mRNA) were isolated. cDNA pools from either total RNA

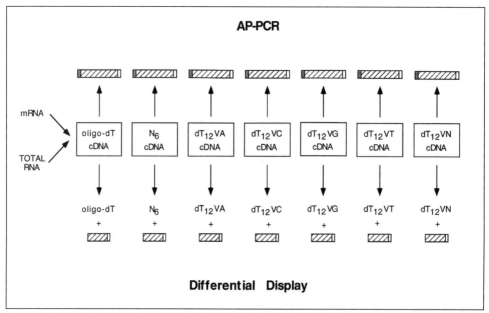

Figure 1. Strategy for comparison of AP-PCR and DD. Total RNA or mRNA was isolated from rat liver, and cDNA was generated by priming with either oligo(dT), random hexamers (N$_6$), anchored oligo(dT) primers (dT$_{12}$VA, dT$_{12}$VC, dT$_{12}$VG or dT$_{12}$VT) or a combination of the four anchored oligo(dT) primers (dT$_{12}$VN) as indicated. Individual cDNA pools were then amplified by AP-PCR using a specific 20-mer (upper) or by DD (lower) using both the primer used for cDNA synthesis and a 10-mer corresponding to the 3' half of the 20-mer used for AP-PCR.

or mRNA were generated using either oligo(dT), random hexamers or the anchored oligo(dT) primers $dT_{12}VA$, $dT_{12}VC$, $dT_{12}VG$, $dT_{12}VT$ (where V is degenerate as dA, dC or dG) or $dT_{12}VN$ (equimolar amounts of $dT_{12}VC$, $dT_{12}VG$, $dT_{12}VA$ and $dT_{12}VT$). In an effort to directly compare AP-PCR and DD as methods for generating RNA fingerprints, 10-mers used for DD corresponded to the 3′ half of 20-mers used for AP-PCR. For example, the cDNA pool generated with $dT_{12}VC$ was amplified either with MFD95GT (5′-CTTTATCTTC**ACACAGCTTC**-3′) for AP-PCR or $dT_{12}VC$ and MFD-3′ (5′-**ACACAGCTTC**-3′) for DD.

AP-PCR and DD profiles generated from rat liver total RNA are shown in Figure 2. cDNA was synthesized using either oligo(dT), random hexamers or the anchored oligo(dT) primers, and then aliquots were used for AP-PCR or DD. AP-PCR profiles generated with PPGB and MFD95GT are shown in Figure 2A. It can be seen that AP-PCR products generated from oligo(dT) (lanes 1), random hexamers (lanes 2) and $dT_{12}VN$ (lanes 7) cDNA pools have some overlap, although bands specific to each lane can be detected. cDNA generated with the anchored oligo(dT) primers ($dT_{12}VC$, $dT_{12}VG$, $dT_{12}VA$ or $dT_{12}VT$; lanes 3, 4, 5 and 6, respectively) generated profiles similar to each other (and $dT_{12}VN$). DD profiles generated using the oligonucleotide that was used for cDNA synthesis (e.g., $dT_{12}VC$ for $dT_{12}VC$-primed cDNA) and either PPGB-3′ or MFD-3′ are shown in Figure 2B. DD profiles generated from oligo(dT) cDNA (lanes 1) resulted in a smear as expected, because the 3′ end of the DD products will not be fixed. For random hexamers (lanes 2) and the anchored oligo(dT) primers ($dT_{12}VC$, $dT_{12}VG$, $dT_{12}VA$ or $dT_{12}VT$; lanes 3–6), however, the resulting profiles were specific to both the 10-mer and the anchored

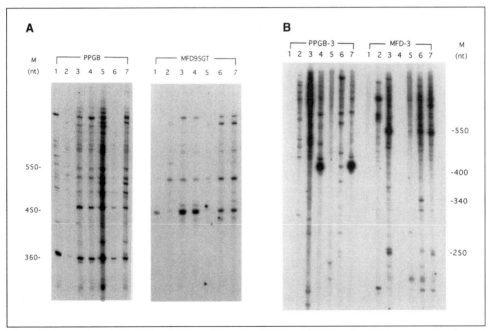

Figure 2. Comparison of AP-PCR and DD amplification of different cDNA pools. cDNA was generated from rat liver total RNA using oligo(dT) (lanes 1), random hexamers (lanes 2), anchored primers $dT_{12}VC$ (lanes 3), $dT_{12}VG$ (lanes 4), $dT_{12}VA$ (lanes 5), $dT_{12}VT$ (lanes 6) or $dT_{12}VN$ (equimolar mixture of the four anchored primers) (lanes 7). (A) AP-PCR with PPGB and MFD95GT. (B) DD with PPGB-3′ or MFD-3′ and the same primer used for cDNA synthesis. Approximate sizes are given in nucleotides (nt) and were determined by reamplification of bands and agarose gel electrophoresis.

oligo(dT) primer used for amplification. DD of $dT_{12}VN$-primed cDNA with $dT_{12}VN$ and a specific 10-mer resulted in a profile that contained some (but not all) of the bands seen in the individual anchored primer reactions, suggesting that a subset of the DD products were preferentially amplified.

Overall, although the complexity of the pattern generated in an individual lane was about the same with either technique (20–40 bands/lane), redundancy between cDNA pools was greater with AP-PCR than with DD. Similar results were found with other primer sets. It appears that RT with the anchored primers is not completely specific, and therefore AP-PCR gives redundant profiles, whereas DD enables an extra level of specificity because of the use of anchored $dT_{12}V\_$ primers during the amplification step. Overall, both AP-PCR and DD were qualitatively reproducible between experiments, although efficiencies of amplification for individual reactions were found to vary (Figure 2B; $dT_{12}VG$/MFD-3′).

When optimizing a new experimental system, AP-PCR and DD fingerprints generated from mRNA and total RNA should be compared. In our hands, both approaches (AP-PCR and DD) resulted in product bands of similar sizes and overall complexity, although the pattern was "cleaner" and showed less background with mRNA (7). For both AP-PCR and DD, there was significant overlap between fingerprints generated from mRNA versus total RNA, indicating that selection of poly(A+) sequences by oligo(dT)-primed cDNA synthesis was fairly efficient.

Use of $dT_{12}VN$-Primed cDNA for AP-PCR and DD

Typically, both AP-PCR and DD rely on the use of primer-specific cDNA synthesis for each analysis. For DD, individual cDNA pools are generated with each anchored oligo(dT) primer; each pool is then used for PCR with the same anchored oligo(dT) and a variety of 10-mers. For AP-PCR, the conventional method is to synthesize cDNA using the same 20-mer that is subsequently used for amplification. For both techniques, to ensure adequate representation of the transcripts present, multiple cDNA pools must be generated for each sample.

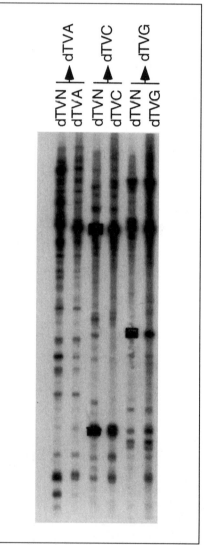

Figure 3. Use of $dT_{12}VN$-primed cDNA for DD. cDNA was synthesized from rat liver total RNA using either $dT_{12}VN$ or the specific anchored oligo(dT) (i.e., $dT_{12}VA$, $dT_{12}VC$ or $dT_{12}VG$) as indicated directly above the lane. The cDNA was then amplified by DD using the specific anchored oligo(dT) indicated (i.e., $dT_{12}VA$, $dT_{12}VC$ or $dT_{12}VG$) and the 10-mer MFD-3′. Boxed bands were eluted and the products reamplified (not shown).

We evaluated whether the composite set of anchored oligo(dT) primers ($dT_{12}VN$) could be used to generate a representative cDNA pool suitable for DD. In the experiment shown (Figure 3), cDNA synthesized using $dT_{12}VN$ was compared to cDNA synthesized using a specific anchored oligo(dT) primer (e.g., $dT_{12}VC$) to support DD amplification using the same specific anchored oligo(dT) primer (e.g., $dT_{12}VC$) and MFD-3′. It was found that cDNA synthesized with the mixed oligo(dT) set was an efficient substrate for DD. Thus, virtually identical DD profiles were generated using $dT_{12}VN$-primed cDNA and the cDNA primed with a specific anchored primer; this can reduce the number of cDNA syntheses required per DD experiment. This is distinct from the experiment shown in Figure 2, where $dT_{12}VN$ was used in the amplification step as well, resulting in only a subset of the cDNA pool being amplified. It was also found that cDNA synthesized using $dT_{12}VN$ resulted in AP-PCR fingerprints of similar complexity to those from cDNA synthesized using specific 20-mers (not shown). Thus, for some applications (e.g., when only total RNA as opposed to mRNA is available), it might be preferable to generate a representative cDNA pool using a composite set of anchored oligo(dT) primers [to select for poly(A^+) RNA], which can then be used for AP-PCR or DD amplification.

Comparison of Clones Isolated by AP-PCR and DD Amplification of $dT_{12}VN$-Primed cDNA

cDNA was generated from rat liver, spleen and brain total RNA and mRNA using the mixed anchored primers $dT_{12}VN$. These cDNA pools were amplified by AP-PCR using the 20-mer MFD95GT or by DD using the 10-mer MFD-3′. Negative controls (no RNA, no reverse transcriptase) were included to assess for potential false positives resulting from PCR artifacts or amplification of residual genomic DNA. Twenty-three representative (i.e., of differing sizes and intensities) AP-PCR and DD bands were excised from the gel, reamplified, subcloned into pCRII and sequenced.

It was found, as is typical of both AP-PCR and DD, that several of the bands consisted of more than one unique amplification product. In addition, there were two bands that could not be subcloned at all (5). For both AP-PCR and DD, several of the clones corresponded to known proteins. Thus, poly(A)-binding protein mRNA was detected using both AP-PCR and DD, whereas the Na^+/K^+ ATPase β subunit and dimethylglycine dehydrogenase were detected using AP-PCR, and intracisternal A particle element (IAPE), factor X, calcineurin and a membrane glycoprotein were detected by DD. For all AP-PCR subclones, potential flanking MFD-3′ amplification sites were identified; for DD subclones, flanking MFD-3′ amplification sites as well as internal or flanking T-rich sites were identified (5).

Evaluation of cDNA Integrity for Use in DD/AP-PCR

Although the initial characterization of DD/AP-PCR used an animal model system, we wanted to apply DD/AP-PCR to the analysis of changes in gene expression in tissue samples obtained from patients with prostate cancer. After radical prostatectomy, fresh unfixed prostate tissue samples were analyzed for the presence of carcinoma, and the tissue was frozen at -70°C for subsequent RNA isolation. However, the time required for evaluation of whether the tissue is normal or tumor (15–30 min at room temperature) can cause significant degradation of the sample RNA and therefore poor-quality cDNA. It was found that both DD and AP-PCR were inconsistent, working well for some samples but not for others. Thus, we developed a simple protocol that we found to be a highly reliable indicator of whether cDNA prepared from a sample is suitable for DD or AP-PCR. The assay uses PCR of the cDNA with primers designed to recognize a highly conserved ubiquitous protein, the SEC11 subunit of the microsomal signal peptidase complex. We reasoned that because signal peptidase is involved in protein processing, detection of SEC11 mRNA might be indicative of cellular protein synthesis. The primers are degenerate at several positions, enabling amplification of a 200-bp region of mRNA from species ranging from yeast to mammals (6). Successful amplification of this 200-bp region was found to be a highly reliable indicator of suitability of a cDNA template for DD. The assay, which we called the S100/S101 assay, uses only a small aliquot of the cDNA and is much less labor-intensive than processing cDNAs through an entire DD experiment.

For human prostate samples, tissue sections were frozen at -70°C and total RNA isolated using TRIZOL reagent. cDNA was synthesized from DNase I treated total RNA with $dT_{12}VC$, and after extraction with phenol–chloroform, an aliquot of the cDNA reaction mixture (2 µL) was used for testing and the remainder ethanol-precipitated. The S100/S101 assay reactions (20 µL) contain 2 µL of cDNA, 50 mM KCl, 10 mM Tris-HCl, pH 8.3, 2.5 mM $MgCl_2$, 0.01% BSA, 100 µM dNTPs, 100 ng S100 and S101 and 1 U *Taq*. Samples were amplified using 2 cycles at 94°C for 5

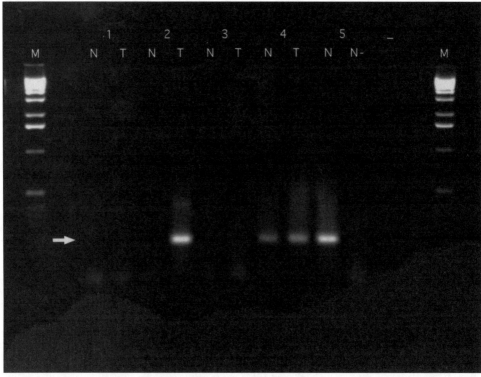

Figure 4. Evaluation of human prostate cDNA by S100/S101 PCR. $dT_{12}VC$ was used to prime cDNA synthesis using normal and tumor prostate total RNA from four patients (lanes 1–4: N and T, respectively) and from Patient 5 with (lane 5N) and without (lane 5N-) reverse transcriptase. Lane (-) corresponds to a buffer control (no RNA). An aliquot of the cDNA reaction mixture was removed and used for S100/S101 PCR, and the 200-bp product (arrow) was visualized by electrophoresis (1.8% agarose) and ethidium bromide staining. Lane M corresponds to molecular weight markers (1-kb DNA ladder; Life Technologies).

min, 40°C for 5 min and 72°C for 5 min, followed by 20 cycles at 94°C for 1 min, 55°C for 1 min and 72°C for 1 min, and 20 cycles at 92°C for 1 min, 55°C for 1 min and 72°C for 1 min. These cycling conditions were chosen to mimic the conditions used for DD or AP-PCR. Successful amplification of signal peptidase sequences from sample cDNA results in a 200-bp product detected by agarose gel electrophoresis and ethidium bromide staining.

Evaluation of a set of human prostate cDNAs using S100 and S101 is shown in Figure 4. cDNA was generated from normal and tumor tissue isolated from four different patients (lanes 1–4; N and T, respectively) and from normal tissue isolated from Patient 5 (5N). Lane 5N- (Patient 5 normal tissue RNA, no reverse transcriptase) and the last lane (-) (no RNA) are included to assess false positives resulting from PCR artifacts or amplification of residual genomic DNA. It can be seen that S100/S101 PCR was successful only for samples 2T, 4N, 4T and 5N.

The cDNAs were then used for DD with [^{32}P]dT$_{12}$VC and the 10-mer OPA-02 (5′-TGCCGAGCTG-3′; Operon Technologies). Only those cDNAs that supported amplification with the S100 and S101 primers were successfully amplified by DD (Figure 5). In addition, there was background amplification apparently because of residual genomic DNA (lane 5N-; no reverse transcriptase), although this is not seen for all samples.

To assess the quality of the DD amplification, the six bands shown in Figure 5 were cut from the gel, an aliquot of the eluted DNA was reamplified with dT$_{12}$VC and OPA-02, and 20 ng were subcloned into the pCRII TA Cloning vector. Four subclones were isolated for each band, and subclones that, upon digestion with Sau3A, appeared to be unique (7) were sequenced and compared to the GenEMBL and dBEST databases. All four subclones from band 1 (436 bp) were derived from human M2-type pyruvate kinase mRNA, subclones from band 2 (ca. 350 bp) were derived from human thymosin beta-10 mRNA, and subclones from band 4 (ca. 300 bp) were derived from human X123 mRNA. Band 3 subclones (325 bp) were identical in sequence to a cDNA (GenBank® Accession No. N52847) similar to transforming growth factor beta protein, and subclones from band 5 (ca. 270 bp) were homologous to an unknown cDNA (GenBank Accession No. R11323). Two different subclones were isolated from band 6 (ca. 240 bp), neither of which was homologous to sequences in the database. For each of the clones, flanking OPA-02 amplification sites and internal or flanking T-rich sites were identified. Thus, the DD amplification identified expressed genes and not ribosomal or genomic sequences.

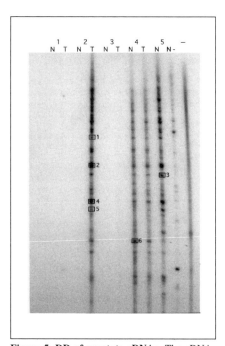

Figure 5. DD of prostate cDNAs. The cDNAs used for the S100/S101 PCR shown in Figure 1 were used for DD amplification with [^{32}P]dT$_{12}$VC and OPA-02. DD products shown in boxes were cut out and subcloned for further evaluation.

The use of a 3-step PCR protocol with low-stringency incorporation of the S100 and S101 primers into early amplification products resulted in more consistent amplification of the 200-bp region than when only high-stringency amplification (i.e., 55°C annealing) was used (not shown). In addition, cDNAs generated with individual anchored oligo(dT) primers or $dT_{12}VN$ were found to be suitable for the assay. As expected, cDNA generated with specific 20-mers unrelated to the signal peptidase gene did not generate the 200-bp product. To date, cDNAs that generate the 200-bp S100/S101 product have consistently worked for DD. This screening assay requires very little sample and can likely be adapted to virtually any tissue from any eukaryotic organism. We have used S100/S101 PCR to evaluate cDNAs generated from rat spleen, rat liver and human prostate, and we routinely use the assay to evaluate not only newly synthesized cDNAs but also cDNAs that have been stored (-70°C for several months).

SUMMARY

Both DD and AP-PCR generate RNA fingerprints of similar complexity and product size. By generating a cDNA pool with a mixed anchored oligo(dT) primer set ($dT_{12}VN$), both approaches can be used to select for poly(A^+) sequences even when total RNA is used as the template. DD selects for AT-rich sequences and thus preferentially identifies 3' regions, although the extent to which 5' sequences are identified by either approach is practically limited by the quality of the cDNA. Although the quality of an RNA preparation can be assessed by agarose gel electrophoresis and ethidium bromide staining, we have found the S100/S101 assay to be a highly reliable and less subjective indicator of whether cDNA (and therefore template mRNA) is suitable for subsequent DD or AP-PCR amplification. While in principle, any primers specific for a ubiquitous, highly conserved protein could be used, the studies described here indicate that the S100/S101 primer set specific for signal peptidase provides a facile, reliable and sensitive tool for routine evaluation of the suitability of mRNA for analysis by DD or AP-PCR. The S100/S101 PCR assay is simple to perform, and the results are easy to interpret. For many applications (e.g., the analysis of clinical biopsy specimens), it is difficult or impossible to isolate sufficient quantities of high-quality mRNA. In these cases, maximum utilization of the sample can be achieved by: *(i)* isolating total RNA, *(ii)* generating a large pool of cDNA using the degenerate anchored oligo(dT) primer ($dT_{12}VN$), *(iii)* evaluating the quality of the cDNA by S100/S101 analysis and *(iv)* performing both AP-PCR and DD as a combined strategy (DD/AP-PCR).

ACKNOWLEDGMENTS

This work was supported by an Individual Allocation to C.B.R. from an American Cancer Society Institutional Grant to Bowman Gray School of Medicine and NIH Grant No. CA46806. The authors are indebted to Dr. Ralph D. Woodruff of the Department of Pathology at BGSM for his aid in procuring prostate tumor samples.

REFERENCES

1.**Liang, P., L. Averboukh and A.B. Pardee.** 1993. Distribution and cloning of eukaryotic mRNAs by means of differential display: refinements and optimization. Nucleic Acids Res. *21*:3269-3275.

2. **Liang, P. and A.B. Pardee.** 1992. Differential display of eukaryotic messenger RNA by means of the polymerase chain reaction. Science *257*:967-971.
3. **Liang, P., W. Zhu, X. Zhang, Z. Guo, R.P. O'Connell, L. Averboukh, F. Wang and A.B. Pardee.** 1994. Differential display using one-base anchored oligo-dT primers. Nucleic Acids Res. *22*:5763-5764.
4. **McClelland, M., D. Ralph, R. Cheng and J. Welsh.** 1994. Interactions among regulators of RNA abundance characterized using RNA fingerprinting by arbitrarily primed PCR. Nucleic Acids Res. *22*:4419-4431.
5. **Rothschild, C.B., C.S. Brewer and D.W. Bowden.** 1997. DD/AP-PCR: combination of differential display and arbitrarily primed PCR of oligo(dT) cDNA. Anal. Biochem. *245*:48-54.
6. **Shelness, G.S. and G. Blobel.** 1990. Two subunits of the canine signal peptidase complex are homologous to yeast SEC11 protein. J. Biol. Chem. *265*:9512-9519.
7. **Shoham, N.G., T. Arad, R. Rosin-Abersfeld, P. Mashiah, A. Gazit and A. Yaniv.** 1996. Differential display assay and analysis. BioTechniques *20*:182-184.
8. **Welsh, J., K. Chada, S.S. Dalal, R. Cheng, D. Ralph and M. McClelland.** 1992. Arbitrarily primed PCR fingerprinting of RNA. Nucleic Acids Res. *20*:4965-4970.
9. **Welsh, J. and M. McClelland.** 1990. Fingerprinting genomes using PCR with arbitrary primers. Nucleic Acids Res. *18*:7213-7218.

13 Preferential Identification of Differentially Expressed mRNAs of Moderate to Low Abundance in a Microscopic System Using Selected Primers

Ognian C. Ikonomov[1] and Michele H. Jacob[2]
[1]*Department of Psychiatry, Wayne State University School of Medicine, Detroit, MI and* [2]*Department of Neuroscience, Tufts University School of Medicine, Boston, MA, USA*

OVERVIEW

The pool of transcribed mRNAs in a given cell, termed a transcriptome (27), underlies the cell phenotype. The transcriptome, which includes the expressed genes and their level of expression, changes during the cell cycle, growth, differentiation and death, and in response to extrinsic stimuli. Identifying the genes whose transcription is induced or silenced in the transition from one state to another is central to defining the precise molecular events that regulate cell type and function. Three distinct techniques are currently being applied to identify differentially expressed genes: *(i)* electronic subtraction (serial sequencing and comparison among a large number of clones isolated from cDNA libraries constructed from test and control cells), *(ii)* subtractive hybridization (subtracted cDNA library screening with population probes from test and control cells) and *(iii)* reverse transcription polymerase chain reaction (RT-PCR)-based technologies including differential display (RT-PCR amplification of RNA extracted from test and control cells and comparison of the multiple products). Comparison of these three techniques by applying them to the same starting polyadenylated RNA demonstrates that differential display provides many advantages, including independence of mRNA prevalence in the total RNA population, potentially enhancing the identification of genes of lower abundance, simultaneous detection of increases and decreases in mRNA steady-state levels, use of smaller amounts of RNA, greater speed and feasibility for a small laboratory (29). These features make RT-PCR-based differential screening extremely attractive. Modifications have been introduced to improve the original protocols. Not surprisingly, hundreds of successful applications have been reported over the past few years.

We describe here a modified differential display method that uses selected primers. This method dramatically increases the bias toward identifying moderate- to low-abundance transcripts that are differentially expressed, and it is applicable to microscopic biological systems. The preferential identification of these relatively rare transcripts is essential for the analysis of differential gene expression because these transcripts represent a high proportion (>90%) of the total cellular mRNA pool but only 50% of the mRNA mass because of the prevalence of a small number of highly abundant messages (1–3).

BACKGROUND

The underlying principle of RT-PCR-based methods for the identification of differentially expressed genes is the amplification of multiple mRNAs in each RNA sample isolated from test and control cells. Subsequently, the levels of the radiolabeled amplification products (RNA fingerprints) are compared in adjacent test and control lanes after band separation on the basis of size by polyacrylamide gel electrophoresis (PAGE). This pioneering approach was introduced as mRNA differential display by Liang and Pardee (17) and as RNA fingerprinting by arbitrarily primed PCR (RAP-PCR) by Welsh et al. (31). In both protocols, the amplification of multiple bands is accomplished by diminishing the selectivity of priming using a low-stringency PCR annealing temperature (40°–42°C) throughout or in the initial cycle. To display the products and achieve a high sensitivity of detection, PCR products are radiolabeled and separated using a 6% sequencing gel. Differentially appearing bands are selected, isolated, cloned and sequenced, and their discordant expression is confirmed by Northern blot analysis. The two original differential display protocols differ mainly in priming strategy. mRNA differential display uses oligo(dT)$_{12}$MN (M represents a degenerate base of A, C or G; N represents A, C, G or T) as antisense primer to target the 3' end of a subgroup of polyadenylated transcripts. An arbitrary 10-mer (50% GC-rich random sequence) is used as a sense primer. In contrast, in RAP-PCR, a single arbitrary 20-mer (50% GC content) is used as both sense and antisense primer. These two differential display protocols have gained widespread interest since their introduction in 1992 (for recent reviews, see References 8, 18, 19, 21 and 30). However, the practical implementation of both original protocols is accompanied by multiple problems, including: *(i)* low reproducibility of band pattern, *(ii)* the large number of false-positive cDNAs and *(iii)* frequent isolation of mRNA of higher abundance as well as ribosomal and mitochondrial transcripts (2,12,18,21,31).

To overcome these limitations, we have developed a modified differential display procedure that uses selected primers. Specifically, it preferentially identifies moderate- to low-abundance transcripts, provides good reproducibility of the multiple-band pattern with independent RNA extractions and is applicable to microscopic biological systems (12). This modified procedure combines the primer design (50% GC-rich 15–21-mers) and reaction conditions of specific RT-PCR with the sensitivity of amplification product separation and band detection on a sequencing gel as devel-

Preferential Identification of Differentially Expressed mRNAs

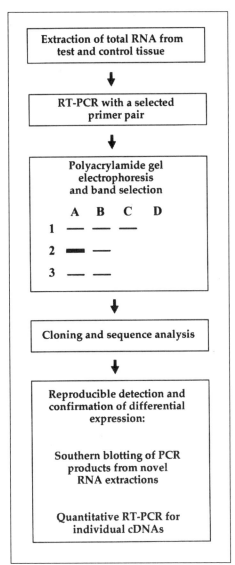

Figure 1. Flowchart of the RT-PCR-based differential display protocol that uses selected primers. A detailed description of each step is given in the text. An autoradiogram is depicted schematically to demonstrate the band selection approach. Lane A: control tissue; lane B: test tissue; lane C: the same as A except that the 5′ primer is omitted in the PCR step; and lane D: the same as lane A except that reverse transcriptase is omitted in the RT reaction. Band 1 is dependent on RT but appears in lane C, suggesting that it contains sequences that are amplified by the 3′ primer alone. Band 2 has a differential appearance in test and control tissue lanes and is dependent on the 2 primers for amplification. Such bands are selected for reamplification, cloning, heterogeneity analysis and testing for reproducible amplification and confirmation of differential expression of the individual cDNAs [adapted from Figure 1 of Ikonomov and Jacob (12)].

oped for arbitrarily primed differential display. In particular, it differs from RNA differential display and RAP-PCR by the use of: *(i)* experimentally selected primer pairs that avoid the amplification of highly abundant transcripts, *(ii)* a higher PCR annealing temperature (50°C instead of 40°–42°C) in all cycles, *(iii)* band selection from the sequencing gel based on their dependence on the two primers (in addition to a differential appearance in test and control tissue lanes and dependence on the presence of reverse transcriptase), *(iv)* elimination of false positives by amplification using the modified procedure with independent sets of RNA extractions and by Southern blot analysis of the products with an already isolated radiolabeled cDNA clone and *(v)* highly sensitive quantitative RT-PCR instead of Northern blot analysis to confirm the differential expression of individual cDNAs. We present the methodology of the modified differential display protocol that uses selected primers and include relevant tips from the recent literature. Our studies demonstrate that this method provides the sensitivity and selectivity required to detect the differential expression of mRNAs present at relatively low levels, ranging from 50 to 1000 transcript copies per cell (12,14).

PROTOCOLS

A flow-chart summary of the protocol and the principle of band selection is depicted in Figure 1. The major steps are detailed below.

Extraction of Total RNA

The modified differential display procedure works well with small amounts of total RNA. We have developed the method using 0.1–0.2 µg of total RNA, equivalent to the amount present in a single embryonic day 8 chick ciliary ganglion (12). Ganglia are rapidly dissected and immediately frozen on dry ice. Total RNA is extracted using any of the available guanidinium isothiocyanate protocols, modified by the scaling down of volumes and the addition of glycogen (1–5 µg) as carrier (5,9,14). DNase treatment is not necessary. This differential display procedure provides a widely applicable, attractive way to identify differentially expressed genes in a microscopic biological system, even when the starting amount of total RNA is not visible on a formaldehyde–agarose gel stained with ethidium bromide.

Primer Sequences

Primer selection is the key variable that increases the bias toward amplifying moderate- to low-abundance mRNAs. Three distinct types of primer pairs have worked successfully in this procedure, including arbitrarily designed primers, specific primers that target a known mRNA and arbitrary combinations of specific primers (all 15–21-mers that have a 50% GC content). Surprisingly, specific primers are effective for amplifying unexpected transcripts. Although the targeted sequence is the predominant, and often the only visible, band on the agarose gel, several additional weaker bands are detected on a polyacrylamide gel after overnight or longer exposure. Thus, specific primers already known to avoid the amplification of highly abundant sequences can be readily used in target-predicting or arbitrary combinations to identify differentially expressed mRNAs using our procedure.

Reverse Transcription

To improve priming specificity, first-strand cDNA synthesis is performed at 42°C.

Materials and Reagents

- RT master mixture: 6.9 µL sterile distilled water, 4.0 µL 5× RT buffer (250 mM Tris-HCl, pH 8.3, 375 mM KCl, 15 mM $MgCl_2$) (Life Technologies, Gaithersburg, MD, USA), 1.6 µL dNTPs (250 µM each; Amersham Pharmacia Biotech, Piscataway, NJ, USA), 0.5 µL RNase inhibitor (1 U; Promega, Madison, WI, USA), 2.0 µL dithiothreitol (DTT) (0.1 M; Life Technologies), 2.0 µL antisense primer (10 µM)
- Diethyl pyrocarbonate (DEPC)-treated water
- 1 µL Moloney murine leukemia virus (MMLV) Reverse Transcriptase (Life Technologies)
- PTC-100™ Programmable Thermal Cycler (MJ Research, Watertown, MA, USA)

Procedure

1. Heat denature the total RNA samples for 5 min at 65°C to eliminate secondary structures. Centrifuge for 2 min at the highest speed in a minicentrifuge and keep on ice before use.
2. Label the required number of 0.5-mL tubes, including a control without reverse transcriptase.
3. Prepare an RT master mixture. The components are listed in the recommended order to be added and at the required volume per tube. Multiply the volume by

the total number of tubes plus one. Vortex mix briefly, spin down the samples and then place 17 µL of the mixture into each tube.
4. Add 2 µL (0.1 µg) total RNA in DEPC-treated water to the corresponding tube. Set the tubes in the PTC-100 thermal cycler at 65°C and use the following program: denature the 3′ primer at 65°C for 5 min and anneal at 42°C for 10 min. Add 1 µL (200 U) MMLV reverse transcriptase. Reverse transcribe at 42°C for 50 min and denature at 94°C for 5 min. Store at 4°C until tubes are removed. At the end of the cycle, spin the tubes briefly to collect condensation. The total reaction volume is 20 µL per tube.

PCR Amplification

To improve selectivity of priming and reproducibility, single-stranded cDNAs are amplified using a PCR annealing temperature of 50°C instead of the 40°–42°C used in other differential display protocols. To avoid saturation (the plateau phase of amplification) and thereby increase the detection of differentially expressed mRNAs, the number of PCR cycles is set at 30. Two-microliter aliquots of the 20-µL RT mixture are used for each PCR, making it possible to use up to 10 different sense primers to amplify cDNAs.

Materials and Reagents

- PCR master mixture: 6.2 µL sterile distilled water, 2.0 µL 10× PCR buffer (Promega) (500 mM KCl, 100 mM Tris-HCl, pH 9.0, 0.1% Triton® X-100), 1.5 µL MgCl$_2$ (25 mM; Promega), 4.0 µL dNTP mixture (250 µM each; Amersham Pharmacia Biotech), 0.1 µL [α-^{32}P]dCTP (1.25 µCi; Amersham Pharmacia Biotech), 0.2 µL *Taq* DNA polymerase (1 U; Promega), 2.0 µL antisense primer (10 µM), 2.0 µL sense primer (10 µM)
- Mineral oil (Sigmal Chemical, St. Louis, MO, USA)

Procedure

1. Label the required number of tubes and include controls that lack sense primer and cDNA template (RT mixture).
2. Prepare a PCR master mixture without the sense primer using the recipe per tube above. Multiply the volume by the number of tubes plus one. It is important to take a 16.0-µL aliquot as a control to identify products that are amplified with the antisense primer alone (see Results and Discussion). Subsequently, add the necessary amount of sense primer to the remaining master mixture. Separate 18.0-µL aliquots into the rest of the tubes.
3. Overlay the reaction mixture with 25 µL mineral oil to prevent evaporation.
4. Add 2 µL of the cDNAs (the RT mixture) under the oil. Mix gently and spin for 10 s. Adjust the volume of the control samples with sterile distilled water. The final PCR mixture volume is 20 µL. Heat denature the samples in the PCR thermal cycler at 94°C for 5 min, followed by 30 amplification cycles. The PCR cycling parameters are: denaturation at 94°C for 30 s, annealing at 50°C for 30 s and extension at 72°C for 1.5 min. To complete extension, the PCR cycles are followed by 5 min at 72°C. To obtain larger products (up to 2 kb), two

thermostable DNA polymerases can be used (7) (also see References 10 and 11 for discussion on DNA polymerases in differential display).

Band Selection and Reamplification

RNA fingerprints from test and control cells are compared by separating radiolabeled RT-PCR products on a 6% polyacrylamide sequencing gel (17). The dried gel is exposed to X-ray film. Multiple bands appear after overnight exposure at room temperature. An example of differential display using selected primers is shown in Figure 2.

Bands are selected on the basis of three criteria: *(i)* differential appearance in test and control lanes, *(ii)* dependence on both sense and antisense primers for amplification (see Results and Discussion) and *(iii)* dependence on the addition of reverse transcriptase to eliminate products amplified from genomic DNA contamination of extracted RNA (Figure 1). Bands larger than 250 bp that did not appear to be differentially expressed are also sampled occasionally because of the frequent presence of multiple co-migrating species in the bigger bands, which might mask differences in the levels of a subset of these products. Selected bands are excised after careful alignment of the autoradiogram on the gel. DNA is recovered and reamplified with the same primers and PCR protocol (16) (also see Reference 13 for alignment tips).

Cloning, Sequencing and Testing for Band Heterogeneity

Reamplified PCR products are cloned using the TA Cloning® Kit (Invitrogen, Carlsbad, CA, USA). To speed up this step, a new product (the TOPO™ TA Cloning Kit; Invitrogen) can be used for 5-min cloning of *Taq* DNA polymerase-generated PCR fragments. Plasmid DNA is isolated using the Wizard™ minipreps protocol (Promega). Sequencing is done by the dideoxy chain termination method (Sequenase® Version 2.0; Amersham Pharmacia Biotech) (24) with T7 and M13 reverse primers. Nucleotide and predicted amino acid sequences are tested for homology to known sequences in the DNA and protein databases using the Basic Local Alignment Search Tool (BLAST) (**http://www.ncbi.nlm.nih.gov/BLAST/**). When the nonredundant search (using the blastn or blastx program in the nr database) shows no significant homology, it is frequently informative to search the Database of Expressed Sequence Tags (dbEST) (using the blastn program in

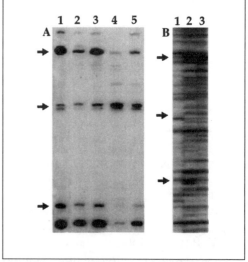

Figure 2. **Autoradiograms demonstrating the differential appearance of bands in adjacent test and control tissue lanes after RT-PCR with the selected primer procedure and PAGE of the radiolabeled products.** Shown are RNA fingerprints of ciliary ganglia that have developed in the absence as compared to the presence of synaptic interactions (A) and in ganglia at selected stages of normal development (B). Multiple bands are detected after overnight exposure of the dried gel to X-ray film. Arrows indicate some differentially appearing bands.

the dbest database), which contains 5' and 3' ends of cDNAs.

Because single bands bigger than 250 bp often contain co-migrating sequences, 5–6 plasmid DNA minipreps are routinely prepared from the cloning of each reamplified band. After sequencing the DNA from one miniprep, inserts of the same size from the other minipreps are tested for sequence heterogeneity by restriction endonuclease mapping or Southern blotting. Inserts that are not cleaved by a selected enzyme, those producing unexpected fragment sizes or those that do not hybridize to the already identified cDNA clone are subsequently sequenced.

Reproducible Amplification and Elimination of False Positives

Selected sequences are analyzed for reproducible amplification using Southern blotting. Radiolabeled cDNAs are hybridized to Southern filters of RT-PCR products that are amplified from novel, independent RNA extractions with the original primer pair used to amplify the selected cDNA. Specifically, the PCR products are separated by agarose gel electrophoresis and transferred to a Zeta-Probe® nylon membrane (Bio-Rad Laboratories, Hercules, CA, USA) by alkaline blotting. The Southern blot filter is hybridized and washed under high-stringency conditions using standard procedures. For a probe, the cloned cDNA is labeled with [α-^{32}P]dCTP to high specific activity by random priming using a commercially available kit and purified using the CHROMA SPIN™-100 Column protocol (CLONTECH Laboratories, Palo Alto, CA, USA).

An internal control must be included to establish the quality and the presence of equivalent quantities of total RNA from test and control cells. An aliquot of the same RNA samples is amplified by RT-PCR with specific primers for the control transcript. Sequences that show a differential appearance in test and control lanes relative to the control mRNA are subjected to further analysis. For our studies, cβ4-tubulin (25) is used because the levels of this transcript are not altered in ciliary ganglion neurons developing in the absence as compared to the presence of synaptic interactions in vivo (14).

Confirmation of Differential Expression

Differential expression of selected cDNAs is confirmed by Northern blot analysis or quantitative RT-PCR. Most differential display protocols use Northern blot analysis. Total RNA is separated on a 1% agarose–formaldehyde-containing gel (23) and transferred to a nylon membrane by alkaline blotting. Probes are radiolabeled as for Southern blotting.

To achieve greater sensitivity for detecting mRNAs of moderate to low abundance that are identified by differential display using selected primers, relative RT-PCR is used to confirm differential expression. Primers and RT-PCR conditions are optimized to amplify only the targeted sequence, as determined by agarose gel electrophoresis, restriction endonuclease mapping and sequence analysis. The specificity of RT-PCR is improved by: *(i)* extending the 5' primer inward or replacing it with an internal sequence for the isolated cDNA, *(ii)* increasing the PCR annealing temperature to 63° from 50°C and *(iii)* using SUPERSCRIPT™ (RNase H-) Reverse Transcriptase (Life Technologies), enabling RT to be performed at 50°C instead of 42°C.

Aliquots of the same total RNA sample are simultaneously amplified with the optimized primer pair for the sequence of interest and the control transcript, which was, in our case, cβ4-tubulin-specific primers. Care must be taken to remain in the exponential phase of amplification for each sequence. Typically, 20 cycles are used for cβ4-tubulin cDNA amplification, whereas 30 cycles (sometimes with a larger amount of RT mixture) are necessary for the cDNA of interest.

For quantitation, PCR is carried out in the presence of [α-^{32}P]dCTP, and the radio-labeled products are separated on a 2% agarose gel. Visible bands are dissected, transferred into a liquid scintillator and their radioactivity counted. Corresponding areas of gel from control lanes (samples minus reverse transcriptase) are also counted and used to subtract out background. The ratio between the counts in the test versus control sample is normalized to the ratio for cβ4-tubulin mRNA.

RESULTS AND DISCUSSION

Our modified differential display procedure that uses selected primers is a highly efficient method for identifying differentially expressed transcripts of moderate to low abundance in small amounts of RNA extracted from a microscopic biological system. This procedure differs from other differential display protocols in the use of: *(i)* both primer and band selection to avoid the amplification and isolation of highly abundant sequences, respectively, *(ii)* a higher PCR annealing temperature, *(iii)* 30 instead of 40 PCR cycles to remain in the linear phase of amplification for most transcripts and *(iv)* RT-PCR-based techniques to establish reproducible amplification and confirm differential expression of specific transcripts. When all of these modifications are applied, none of the cDNAs identified by this procedure show significant homology to highly abundant sequences. In contrast, a large proportion of sequences identified by mRNA differential display and RAP-PCR are characterized as highly abundant.

Primer Selection

Primer selection is crucial for the predominant amplification of moderate- to low-abundance transcripts in our procedure. Our studies demonstrate that primers avoiding the amplification of highly abundant ribosomal and mitochondrial RNA must be selected experimentally. This is necessary because only a 6-base perfect match between the 3′ end of the primer and the template sequence is sufficient for priming, even at the higher PCR annealing temperature (50°C) used in the selected primer differential display procedure. If we take 7 bases as a prerequisite for priming, all possible combinations of primer ends will be present in the large 16 775-bp intronless sequence of the chicken mitochondrial genome (6). Primers with high (50%–65%) GC content at their 3′ ends tend to be sticky and frequently amplify ribosomal and mitochondrial RNA. These primers should be avoided. An easy way to implement the selected primer differential display procedure is to use specific primer pairs already shown to amplify a targeted sequence from total RNA and avoid unwanted abundant sequences. Such primers could be used in sequence targeting and arbitrary combinations.

Band Pattern

The higher PCR annealing temperature used in this procedure improves priming specificity and band pattern reproducibility but decreases the total number of amplification products per PCR. The total number of bands ranges from 10 to 80, with a mean of 34 ± 7 (standard error of the mean) for both specific and arbitrary primer pairs (Figure 2; Reference 12). Two general band patterns are distinguished. Specific primer pairs that target an identified gene produce a predominant band of the expected size accompanied by several weaker bands. In contrast, arbitrarily designed primers and specific primers in arbitrary combinations result in multiple bands of similar strength. Importantly, for each primer pair, the band pattern is specific and reproducible with independent sets of RNA extractions, the same RNA in separate experiments and RNA from the same source extracted using different protocols.

Band Selection and Sequence Analysis

Sequence analysis of reaction products (total of 61 cDNAs) reveals that the selection of products that are dependent on the two primers significantly increases the bias towards identifying transcripts of moderate to low abundance. Only 5% of products dependent on both primers are highly abundant sequences. In contrast, 60% of the cDNAs that are amplified by the 3′ primer alone are highly abundant mitochondrial, ribosomal and glyceraldehyde-3-phosphate dehydrogenase (GAPD) transcripts (12). Moreover, these primers often have a high (50%–65%) GC content at their 3′ ends. These results demonstrate that band selection dramatically increases the efficiency of identifying differentially expressed genes of interest. This selection is accomplished by comparing the RNA fingerprint produced with both primers versus that with the antisense primer alone (Figure 1). Amplification products that are identical in size between the two fingerprints are avoided, thus eliminating cDNAs that have the same 3′ primer sequence at both ends. Alternatively, such band selection can be carried out by labeling the 5′ primer using published procedures (11,26).

Reproducibility and Differential Expression

To eliminate false positives and establish reproducibility, we use Southern blot analysis of RT-PCR products amplified from independent RNA extractions with the original primer pair. To control for differences in the relative amount of total RNA between test and control tissue samples, RT-PCR with specific primers for cβ4-tubulin is included. Twelve of the cDNA clones identified by the selected primer differential display procedure were tested and show a single, differentially appearing band of the expected size, with one exception. Alternative Southern blot and reverse Northern blot approaches have been used for eliminating false positives (4,15,20,22,28,32).

Confirmation of Differential Regulation

Northern blot analysis is routinely used in differential display protocols to confirm differential expression. However, it is not sufficiently sensitive to detect moderate- to low-abundance sequences identified by the selected primer differential display procedure, with few exceptions. The factors limiting our Northern blot sensitivity include: *(i)* the small size of the probes (usually 200–350 bp), *(ii)* the small amount of total RNA (because of the need for embryonic surgery, we use 2–3 μg of total RNA extracted from 20–30 ciliary ganglia dissected at E8) and *(iii)* the low abundance of the isolated transcripts. No signal is detectable with most of the isolated cDNAs, even after exposing the blots to X-ray film for 7 days with an intensifying screen at -80°C. In contrast, when the same filters are hybridized to a 340-bp 18S ribosomal probe, strong signals are detected after a short exposure (2 h) at room temperature. Importantly, nicotinic acetylcholine receptor α5 subunit mRNA, included as a control because it is known to be present at 200 transcript copies per E8 ciliary ganglion neuron (14), is not detected on these Northern blot filters. These results suggest that the majority of the sequences identified by the selected primer differential display procedure are expressed at moderate to low abundance.

To confirm the differential expression of reproducibly amplified cDNAs, we use relative RT-PCR. Six selected cDNAs were studied with optimized primers, a higher RT temperature using SUPERSCRIPT reverse transcriptase and a higher PCR annealing temperature of 63°C. All of these cDNAs show differential regulation, indicating the efficient elimination of false positives by Southern blot analysis. In the case of transcript levels that could be measured by both relative RT-PCR and Northern blot analysis, we found a good correlation between the results.

CONCLUSION

An ideal RT-PCR protocol for identifying differentially expressed mRNAs should provide selectivity of priming and reproducibility, while still enabling the simultaneous amplification of multiple mRNAs from a total RNA sample. Moreover, it is important to increase the bias toward identifying moderate- to low-abundance mRNAs because of their high proportion in the total cellular mRNA pool (1–3). The complexity of this task stems from 2 limitations: *(i)* mRNA constitutes only 2%–5% of the total RNA with relatively rare mRNAs representing 50% of mRNA mass, and *(ii)* a relatively short perfect match at the 3′ primer end is sufficient for priming. Diminishing the selectivity of priming increases the number of products but also increases the bias towards sequences of high abundance, whereas an increase in selectivity decreases the number of amplification products. At the expense of lower throughput (i.e., smaller number of bands per PCR), the selected primer differential display procedure is tailored to increase the selectivity of priming and improve reproducibility. The procedure has two key advantages in comparison to other differential display protocols: *(i)* it dramatically increases the efficiency of identifying relatively rare transcripts, and *(ii)* it is widely applicable to microscopic biological systems.

ACKNOWLEDGMENTS

This research was supported by NIH Grant No. NS 21725 (to M.H.J.).

REFERENCES

1. **Axel, R., P. Fiegelson and G. Schutz.** 1976. Analysis of the complexity and diversity of mRNA from chicken liver and oviduct. Cell 7:247-254.
2. **Bertioli, D.J., U.H.A. Schlichter, M.J. Adams, P.R. Burrows, H.H. Steibniss and J. Antoviw.** 1995. An analysis of differential display shows a strong bias towards high copy number mRNAs. Nucleic Acids Res. 23:4520-4523.
3. **Bishop, J.O., J.G. Morton, M. Rosbash and M. Richardson.** 1974. Three abundance classes in HeLa messenger RNA. Nature 250:199-204.
4. **Callard, D., B. Lescure and L. Mazzolini.** 1994. A method for the elimination of false positives generated by the mRNA differential display technique. BioTechniques 16:1096-1103.
5. **Chomczynski, P.** 1993. A reagent for the single-step simultaneous isolation of RNA, DNA and proteins

from cell and tissue samples. BioTechniques *15*:532-537.
6. **Desjardins, P. and R. Morais.** 1990. Sequence and gene organization of the chicken mitochondrial genome: a novel gene order in higher vertebrates. J. Mol. Biol. *212*:599-634.
7. **Diachenco, L.B., J. Ledesma, A.A. Chenchik and P.D. Siebert.** 1996. Combining the technique of RNA fingerprinting and differential display to obtain differentially expressed mRNA. Biochem. Biophys. Res. Commun. *219*:824-828.
8. **Eberwine, J., P. Crino and M. Dichter.** 1995. Single-cell mRNA amplification: implications for basic and clinical neuroscience. The Neuroscientist *1*:200-211.
9. **Feramisco, J.R., J.E. Smart, K. Burridge, D.M. Helfman and J.P. Thomas.** 1982. Co-existence of vinculin-like protein of higher molecular weight in smooth muscle. J. Biol. Chem. *257*:11024-11031.
10. **Haag, E. and V. Raman.** 1994. Effects of primer choice and source of *Taq* DNA polymerase on the banding patterns of differential display RT-PCR. BioTechniques *17*:226-228.
11. **Hadman, M., B.L. Adam, G.L. Wright and T.J. Bos.** 1995. Modifications to the differential display technique reduce background and increase sensitivity. Anal. Biochem. *226*:383-386.
12. **Ikonomov, O.C. and M.H. Jacob.** 1996. Differential display protocol with selected primers that preferentially isolates mRNAs of moderate- to low-abundance in a microscopic system. BioTechniques *20*:1030-1042.
13. **Kim, A., S. Roffler-Tarlov and C.S. Lin.** 1995. New technique for precise alignment of an RNA differential display gel with its image. BioTechniques *19*:346.
14. **Levey, M.S., C.L. Brumwell, S.E. Dryer and M.H. Jacob.** 1995. Innervation and target tissue interactions differentially regulate acetylcholine receptor subunit mRNA levels in developing neurons *in situ*. Neuron *14*:153-162.
15. **Li, F., E.S. Barnathan and K. Kariko.** 1994. Rapid method for screening and cloning cDNAs generated in differential mRNA display: application of northern blot for affinity capturing of cDNAs. Nucleic Acids Res. *22*:1764-1765.
16. **Liang, P., L. Averboukh and A.B. Pardee.** 1993. Distribution and cloning of eukaryotic mRNAs by means of differential display: refinements and optimizations. Nucleic Acids Res. *21*:3269-3275.
17. **Liang, P. and A.B. Pardee.** 1992. Differential display of eukaryotic messenger RNA by means of the polymerase chain reaction. Science *257*:967-971.
18. **Liang, P. and A.B. Pardee.** 1995. Recent advances in differential display. Curr. Opin. Immunol. 7:274-280.
19. **Livesey, F.J. and S.P. Hunt.** 1996. Identifying changes in gene expression in the nervous system: mRNA differential display. Trends Neurosci. *19*:84-88.
20. **Mathieu-Daude, F., R. Cheng, J. Welsh and M. McClelland.** 1996. Screening of differentially amplified cDNA products from arbitrarily primed PCR fingerprints using single strand conformation polymorphism (SSCP) gels. Nucleic Acids Res. *24*:1504-1507.
21. **McClelland, M., F. Mathieu-Daude and J. Welsh.** 1995. RNA fingerprinting and differential display using arbitrarily primed PCR. Trends Genet. *11*:242-246.
22. **Mou, L., H. Miller, J. Li, E. Wang and L. Chalifour.** 1994. Improvements to the differential display method for gene analysis. Biochem. Biophys. Res. Commun. *199*:564-569.
23. **Sambrook, J., E.F. Fritsch and T. Maniatis.** 1989. Molecular Cloning: A Laboratory Manual, 2nd ed. CSHL Press, Cold Spring Harbor, NY.
24. **Sanger, F., S. Nicklen and A.R. Coulson.** 1977. DNA sequencing with chain-terminating inhibitors. Proc. Natl. Acad. Sci. USA *74*:5463-5467.
25. **Sullivan, K.F., J.C. Havercroft, P.S. Machlin and D.W. Cleveland.** 1986. Sequence and expression of the chicken β5- and β4-tubulin genes define a pair of divergent β-tubulins with complementary patterns of expression. Mol. Cell Biol. *6*:4409-4418.
26. **Tokuyama, Y. and J. Takeda.** 1995. Use of ^{33}P-labeled primer increases the sensitivity and specificity of mRNA differential display. BioTechniques *18*:424-425.
27. **Velculescu, V.E., L. Zhang, W. Zhou, J. Vogelstein, M.A. Basrai, D.E. Basset Jr., P. Hieter, B. Vogelstein and K.W. Kinzler.** 1997. Characterization of the yeast transcriptome. Cell *88*:243-251.
28. **Voegeli-Lange, R., N. Brueckert, T. Boller and A. Wiemken.** 1996. Rapid selection and classification of positive clones generated by mRNA differential display. Nucleic Acids Res. *24*:1385-1386.
29. **Wan, J.S., S.J. Sharp, G.M.-C. Poirier, P.C. Wagman, J. Chambers, J. Pyati, Y.-L. Hom, J.E. Galindo et al.** 1996. Cloning differentially expressed mRNAs. Nature Biotechnol. *14*:1685-1691.
30. **Wang, X., R.R. Ruffolo and G.Z. Feuerstein.** 1996. mRNA differential display: application in the discovery of novel pharmacological targets. Trends Pharmacol. Sci. *17*:276-279.
31. **Welsh, L., K. Chada, S.S. Dalal, R. Cheng, D. Ralph and M. McClelland.** 1992. Arbitrarily primed PCR fingerprinting of RNA. Nucleic Acids Res. *20*:4965-4970.
32. **Zhang, H., R. Zhang and P. Liang.** 1996. Differential screening of gene expression difference enriched by differential display. Nucleic Acids Res. *24*:2454-2455.

14 Solid-Phase Differential Display

Joakim Lundeberg[1], Øystein Røsok[2], Anne Hansen Ree[3] and Jacob Odeberg[1]
[1]*Department of Biochemistry and Biotechnology, KTH, Royal Institute of Technology, Stockholm, Sweden and Departments of* [2]*Immunology and* [3]*Tumor Biology, Institute of Cancer Research, The Norwegian Radium Hospital, Oslo, Norway*

OVERVIEW

The mRNA differential display method has been adopted for use in a solid-phase format. This system offers an attractive alternative to solution-based differential display because minute amounts of mRNA per cell can be analyzed in considerably less time than previously. In addition, the suggested protocol simplifies the procedure substantially because no precipitations or centrifugations are required. The solid support used, here exemplified by monodisperse paramagnetic beads, has also allowed for optimization of the enzymatic and preparative steps in the differential display procedure. Automated systems can also be envisioned that would further reduce the variations described with the differential display procedure.

BACKGROUND

Magnetic particles have frequently been used in many fields of biochemistry, molecular biology and medicine. Several manufacturers now offer a variety of ligands (antibodies, nucleic acids, streptavidin etc.) coupled to a solid support. The main reason for using solid-phase approaches lies in the improved reproducibility and efficiency, as clearly demonstrated in peptide and nucleic acid synthesis. Furthermore, in development of closed systems, the use of coupled magnetic particles has circumvented the difficult steps in automation such as centrifugation or precipitation because the particles allow for a rapid change of reaction buffers and reagents using a simple magnetic field. Introduction of magnetic particles into molecular biology techniques (DNA sequencing and mRNA purification) has proven to be an attractive and robust alternative compared to traditional procedures.

The differential display technique has rapidly become a frequently used technique in analysis of differentially expressed genes (6–9). Although the method is very attractive conceptually, reports have shown problems with contaminating rRNA, cDNA or DNA (1,5,9) and with reproducibility (2,9). We describe a protocol to perform solid-phase differential display that minimizes these problems in samples containing small amounts of target mRNA. The approach is based on using differential display oligo(dT) probes to capture the mRNA. These oligo(dT) primers are also used in the subsequent primer extension of first-strand synthesis and polymerase chain reaction (PCR). Capture of mRNA to a solid phase during the procedure simplified the purification steps, limited sample loss and enabled rapid handling of mRNA. Interference of genomic DNA was also minimized by the washing procedure. The protocol has also facilitated optimization of the enzymatic conditions as well as the determination of required amounts of mRNA per cell, permitting analysis of genes expressed in the most immature and rare hematopoietic progenitor cells (11).

PROTOCOLS

Materials and Reagents

- Phosphate-buffered saline (PBS): 137 mM NaCl, 2.7 mM KCl, 4.3 mM $Na_2HPO_4 \times 7H_2O$, 1.4 mM KH_2PO_4, pH 7.5
- Lysis buffer: 100 mM Tris-HCl, pH 8.0, 500 mM LiCl, 10 mM EDTA, pH 8.0, 1% sodium dodecyl sulfate (SDS), 5 mM dithiothreitol (DTT)
- Binding buffer: 10 mM Tris-HCl, pH 8.0, 2 M LiCl, 1 mM EDTA, pH 8.0
- Washing buffer: 10 mM Tris-HCl, pH 7.5, 0.15 M LiCl, 1 mM EDTA, 0.1% SDS
- 5× first-strand buffer: 250 mM Tris-HCl, pH 8.3, 375 mM KCl, 15 mM $MgCl_2$
- 10× PCR buffer: 100 mM Tris-HCl, pH 8.3, 20 mM $MgCl_2$, 500 mM KCl
- Reverse transcription (RT) washing buffer: 10 mM Tris-HCl, pH 8.3, 50 mM KCl, 5 mM $MgCl_2$
- PCR washing buffer: 20 mM Tris-HCl, pH 8.3, 50 mM KCl, 0.1% Tween® 20
- AmpliTaq® DNA Polymerase and reaction buffer (Perkin-Elmer, Norwalk, CT, USA)
- Moloney murine leukemia virus (MMLV) (or SUPERSCRIPT™) Reverse Transcriptase and reaction buffers (Life Technologies, Gaithersburg, MD, USA)
- Nucleotide solutions (10 and 20 mM each dNTP) (Amersham Pharmacia Biotech, Uppsala, Sweden)
- ^{35}S-labeled dATP (1200 Ci/mmol; Amersham Pharmacia Biotech, Bucks, England, UK)
- Oligo(dT) differential display primers, each having a concentration of 25 µM. D427: 5′-biotin-$(T)_{11}$VG-3′; D434: 5′-biotin-$(T)_{11}$VT-3′; D435: 5′-biotin-$(T)_{11}$VC-3′; D505: 5′-biotin-$(T)_{11}$VA-3′ (V denotes a degenerate position consisting of nucleotides A, G or C)
- Arbitrary differential display primers, each having a concentration of 5 µM. D388: 5′-GCAGATGATG-3′; D389: 5′- GATCTCCTCA-3′; D395: 5′-CTTGATTGCC-3′; D492: 5′-CAGTGTAGTC-3′; D493: 5′-GTTTTCGCAG-3′; D495: 5′-TCGATACAGG-3′; AP2: 5′-CTGATCCATG-3′; AP3: 5′-CTGATCCATC-3′

Note: Biotinylated oligonucleotides can be obtained from several commercial sources offering oligonucleotide synthesis services. It is of great importance that the biotinylated oligonucleotide be purified from unbound biotin, preferably by reverse-phase

FPLC or HPLC, because free biotin will occupy binding sites on the beads and reduce the binding capacity of biotinylated oligo(dT) probe.

- M-280 streptavidin-coated Dynabeads® (10 mg/mL) (Dynal AS, Oslo, Norway)
- Neodymium-iron-boron magnet (Magnetic Particle Concentrator [MPC®]; Dynal AS)
- Thermal cycler
- Rotator
- Stop solution (shake 100 mL formamide with 5 g Amberlite MB-1 resin and 300 mg dextran blue for 30 min. Filter through 0.45-μm pore size filter)
- A polyacrylamide sequencing gel and electrophoresis equipment

Preparation of the Magnetic Beads

Standard precautions for RNase degradation should be taken throughout the protocol (11).

Procedure

1. Resuspend the beads by pipetting. Use 20 μL (200 mg) of resuspended beads per template, pipet the suspended beads into an RNase-free 1.5-mL microcentrifuge tube. The beads can be washed in bulk for the total number of templates.
2. Place the tube in the magnetic tube holder and allow the beads to adhere to the magnet at the side of the wall. After approximately 20 s, remove the supernatant using a pipet (do not remove from the magnetic holder).
3. Add an equal volume of binding buffer and gently pipet to resuspend.
4. Repeat using the magnetic holder, allow the beads to adhere to the side of the tube and remove the supernatant.
5. Resuspend the beads in 100 μL binding buffer and add 4 μL of the biotinylated oligo(dT) probe.
6. Incubate at room temperture for 15 min. Mix during the immobilization reaction once or twice by gentle pipetting or tapping.
7. Collect the beads by moving the vials into the magnetic holder and remove the supernatant.
8. Wash the beads once with 100 μL binding buffer and twice with 100 μL lysis buffer.
9. Remove the supernatant.
10. Resuspend the beads in 100 μL lysis buffer and then put the tube on ice.

Preparation of Bone Marrow Cells (Optional)

An example for preparation of bone marrow cells is given below, although other protocols for solid tissues can easily be adopted.

Procedure

1. Wash the isolated cells with ice-cold PBS.
2. Centrifuge the cells at 200–500× g for 5 min.

3. Resuspend the cell pellet in PBS and separate into aliquots into tubes to contain the desired number of starting cells in the following experiments.
4. Centrifuge the cells at 200–500× g for 5 min.
5. Resuspend the cell pellet in 100 µL lysis buffer and vortex mix immediately.
6. Store the lysed cells at -130°C until analyzed.

Preparation of the Master Mixture for cDNA Synthesis and PCR

Prepare master mixtures for first-strand synthesis, PCR and primer hot start in a microcentrifuge tube.

Procedure

1. Components per sample (20 µL total volume): 4 µL 5× first-strand buffer, 2 µL 0.1 M DTT, 1 µL dNTPs (10 mM), 200 U MMLV reverse transcriptase and sterile water to 20 µL.
2. Prewarm the cDNA mixture at 37°C for 1–2 min before adding to the beads.
3. PCR master mixture components per sample (17.5 µL total volume): 2.5 µL 10× PCR buffer, 2.5 µL dNTPs (20 mM), 1.25 µL [^{35}S]dATP, 1 U *Taq* DNA polymerase and sterile water to 17.5 µL.
4. Primer hot-start mixture components per sample (5 µL total volume): 2.5 µL biotinylated oligo(dT) primer (25 µM), 2.5 µL arbitrary primer (5 µM).

Capture of mRNA, cDNA Synthesis and PCR

Procedure

1. Thaw the cell lysate on ice (ca. 3 min).
2. Add the cell lysate (100 µL) to the prepared streptavidin-coated magnetic beads with coupled oligo(dT) probe (100 µL).
3. Incubate on ice for 5 min. Mix gently during the hybridization.
4. Collect the beads with the magnetic holder and remove the supernatant.
5. Wash the beads twice with 100 µL washing buffer and once with 200 µL RT washing buffer.
6. Collect the beads with the magnetic holder and remove the supernatant.
7. Resuspend the beads in 100 µL RT washing buffer and transfer the solution to a new tube.
8. Wash once with 200 µL RT washing buffer.
9. Collect the beads with the magnetic holder and remove the supernatant.
10. Resuspend the beads in 20 µL prewarmed cDNA mixture.
11. Incubate 1 h at 37°C using a rotator to keep the beads in continuous suspension.
12. Collect the beads with the magnetic holder and remove the supernatant. (Note that the first-strand products remain immobilized onto the magnetic particles.)
13. Wash the beads twice with 100 µL PCR washing buffer.

14. Collect the beads with the magnetic holder and remove the supernatant.
15. Resuspend the beads with 25 µL PCR washing buffer.
16. Transfer 2.5 µL to a PCR tube. (Store the rest at -70°C.)
17. Add 17.5 µL of the PCR master mixture into the PCR tube.
18. Overlay with 2 drops of mineral oil.
19. Warm the tubes to 75°C in the PCR block and perform PCR hot start by adding 5 µL of the primer hot-start mixture. Continue directly to the first denaturation step in the PCR program.
20. Run the PCR as follows: denaturation at 94°C for 30 s, annealing at 40°C for 1 min and extension at 72°C for 2 min for 35 cycles. A final extension step at 72°C for 10 min is strongly suggested.

Gel Electrophoresis and Detection

The samples are run on a 6% denaturating polyacrylamide gel prepared in a standard fashion (11). Run the gel until the xylene cyanol marker (the slowest-migrating blue color, corresponding to approximately 110 nucleotides) is about 5 cm from the bottom of the gel. Dry the gel and expose it to film (e.g., Kodak® X-Omat S; Scientific Imaging Systems [Eastman Kodak], New Haven, CT, USA). Before developing the film, orient the gel and film by piercing them with a needle.

Isolation of PCR Fragments

Procedure

1. Orient the gel and film according to the piercings.
2. Mark the bands of interest with a needle.
3. Cut out the fragments with a scalpel. Sterilize the scalpel in a burning flame between each fragment to avoid cross-contamination.
4. Place the fragments into separate 1.5-mL microcentrifuge tubes and add 200 µL PCR washing buffer.
5. Vortex mix and incubate at room temperature for at least 15 min.
6. Incubate at 100°C for 5 min and then allow to cool to room temperature.
   Note: Extracted DNA fragments can either be precipitated (steps 6–8), purified using the DNA DIRECT™ Kit (Dynal AS) or, in the case of biotinylated PCR products, re-immobilized to streptavidin-coated magnetic beads (as described below, steps 7–14).
7. For each sample, use 20 µL (200 mg) resuspended streptavidin-coated beads and wash the beads as described above in washing buffer.
8. Resuspend the beads in 200 µL washing buffer.
9. Add the sample (200 µL) to the resuspended streptavidin-coated beads (total volume 400 µL).
10. Incubate for 15 min at room temperature. Mix during immobilization.
11. Collect the beads with the magnetic holder and remove the supernatant.
12. Wash the beads twice with 100 µL PCR washing buffer.

13. Resuspend the beads with 20 µL PCR washing buffer.
14. Take 2.5–5 µL of the template (resuspended beads) to the PCR.
15. Reamplify the purified products essentially as described above but in a 50-µL PCR volume and with 0.5 µL biotinylated oligo(dT) probe and 2.5 µL decamer primer used in the previous amplification.
16. Clone the generated fragments by either using pGEM®-T cloning vectors (Promega, Madison, WI, USA) (10) for subsequent PCR sequencing using general vector primers or performing direct solid-phase sequencing (3,4).
17. Perform database searches and adequate expression confirmations.

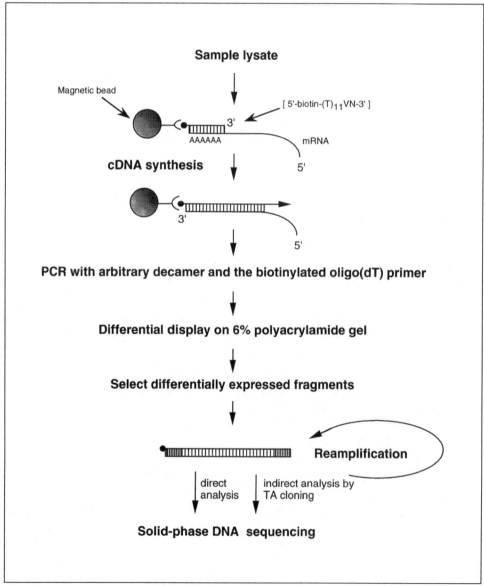

Figure 1. Schematic outline of solid-phase differential display.

RESULTS AND DISCUSSION

The purpose of the developed procedure has been to simplify sample preparation and to increase the throughput in the differential display methodology (6–8) by taking advantage of magnetic bead technology, schematically depicted in Figure 1. Briefly, the cell populations to be analyzed are directly lysed in a LiCl–SDS-based buffer and incubated with streptavidin-coated magnetic beads with a coupled biotinylated probe. Selective mRNA capture is achieved by the differential display oligo(dT) probe having a poly(dT) stretch and two additional partly degenerated positions at the 3' end. Uncaptured, potentially interfering DNA and rRNA are removed from the immobilized mRNA by magnetic separation and subsequent washing steps. The beads (with mRNA template) can then be conditioned into suitable and optimal buffers for solid-phase cDNA synthesis. After RT is completed, the supernatant can be removed, leaving the beads with immobilized first-strand product

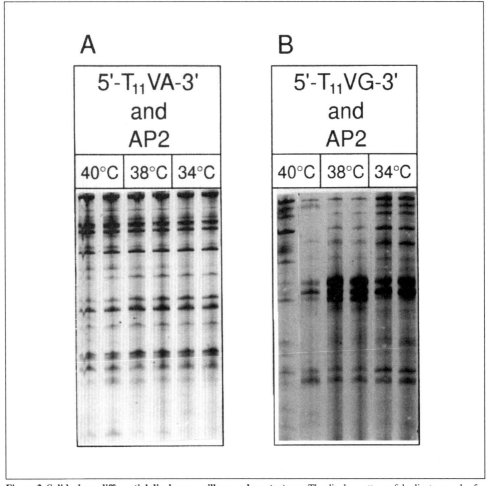

Figure 2. Solid-phase differential display on axillary node metastases. The display pattern of duplicate samples for each set of conditions demonstrates the reproducibility and specificity of the protocol for two (of the four) differential display oligo(dT) primers used at three alternating PCR annealing temperatures, as indicated.

for subsequent buffer conditioning prior to the next enzymatic step, PCR. The PCR was performed essentially as described previously (6,7) using a primer pair consisting of an arbitrary decamer and a biotinylated oligo(dT)$_{11}$VN primer (V = degenerate position corresponding to A, G or C; N = ultimate position corresponding to A, G, C or T). The difference in this protocol is that the template consists of beads with immobilized first-strand synthesis. The first heating cycle of PCR releases the single-stranded template into solution, allowing for efficient amplification. Differentially expressed fragments labeled with [^{35}S]dATP are identified by conventional comparative gel electrophoresis and can be isolated from the polyacrylamide gel using either magnetic beads or ethanol precipitation. The reamplified material can be either directly sequenced using solid-phase sequencing (3,4) with dye-labeled terminators or cloned into a pGEM-T vector to allow amplification with vector-specific primers followed by solid-phase sequencing.

The solid-phase cDNA-coupled synthesis was investigated in terms of yield and selectivity by variation of probe length and different reverse transcriptases using streptavidin-coated magnetic beads (10). For example, by extending the oligo(dT) probe with up to 25 thymine nucleotides, mRNA hybridization and cDNA synthesis can be performed at higher and more stringent temperatures. This modification is possible without affecting the subsequent PCR because the cDNA primer will remain coupled to the support after cDNA synthesis. However, it was demonstrated that only the oligo(dT)$_{11}$ primers resulted in the specificity required in differential display, and it also showed that the reproducibility generally was higher using mesophilic reverse transcriptases compared to the thermostable alternatives. Likewise, the reproducibilty was markedly lower for decreasing amounts of target mRNA, and for the analysis of bone marrow cells, a minimum limit was established to correspond to approximately 10 ng total RNA.

The solid-phase differential display protocol has been developed primarily for gene expression analysis in samples with a very limited amount of target mRNA, such as stem cells and lymph node micrometastases (Figure 2). The optimized solid-phase differential display protocol uses oligo(dT)$_{11}$VN-coated beads for capture and cDNA priming combined with MMLV (or a derivative) in first-strand synthesis for the highest possible reproducibility. In Figure 2, two different cDNA probes are compared, showing the reproducibility at different PCR annealing temperatures. The figure also shows the expected differences in display patterns obtained when alternating the oligo(dT) primers. The average length of the solid-phase differential display fragments is approximately 450 bp. An interesting feature using this approach is that the majority (ca. 65%) of the obtained fragments are primed only by the arbitrary primer during amplification, which will benefit the subsequent homology searches because these fragments are more likely to be within the coding region.

In conclusion, the use of magnetic particles in differential display has proven to be an attractive alternative to identify differentially expressed genes. Although a focus in our studies has been to optimize the protocol for small amounts of mRNA, it is likely that the method can easily be used in other situations when target amounts are not limiting.

ACKNOWLEDGMENTS

We thank Frank Larsen and Erik Hornes (Dynal AS) for support.

REFERENCES

1. **Callard, D., B. Lescure and L. Mazzolini.** 1994. A method for the elimination of false positives generated by the mRNA differential display technique. BioTechniques *16*:1096-1103.
2. **Chen, Z., K. Swisshelm and R. Sager.** 1994. A cautionary note on reaction tubes for differential display and cDNA amplification in thermal cycling. BioTechniques *16*:1002-1006.
3. **Hultman, T., S. Bergh, T. Moks and M. Uhlén.** 1991. Bidirectional solid-phase sequencing of *in vitro*-amplified plasmid DNA. BioTechniques *10*:84-93.
4. **Hultman, T., S. Ståhl, E. Hornes and M. Uhlén.** 1989. Direct solid phase sequencing of genomic and plasmid DNA using magnetic beads as solid support. Nucleic Acids Res. *17*:4937-4946.
5. **Li, F., E.S. Barnathan and K. Kariko.** 1994. Rapid method for screening and cloning cDNAs generated in differential mRNA display: application of Nothern blot for affinity capturing of cDNAs. Nucleic Acids Res. *22*:1764-1765.
6. **Liang, P., L. Averboukh, K. Keyomarsi, R. Sager and A.B. Pardee.** 1992. Differential display and cloning of mRNAs from human breast cancer versus mammmary epithelial cells. Cancer Res. *52*:6966-6968.
7. **Liang, P., L. Averboukh and A.B. Pardee.** 1993. Distribution and cloning of eukaryotic mRNAs by means of differential display: refinements and optimisations. Nucleic Acids Res. *21*:3269-3275.
8. **Liang, P. and A.B. Pardee.** 1992. Differential display of eukaryotic messenger RNA by means of the polymerase chain reaction. Science *257*:967-971.
9. **Liang, P. and A.B. Pardee.** 1995. Recent advances in differential display. Curr. Opin. Immunol. 7:274-280.
10. **Røsok, Ø., J. Odeberg, M. Rode, T. Stokke, S. Funderud, E. Smeland and J. Lundeberg.** 1996. Solid-phase method for differential display of genes expressed in hematopoietic stem cells. BioTechniques *21*:114-121.
11. **Sambrook, J., E.F. Fritsch and T. Maniatis.** 1987. Molecular Cloning: A Laboratory Manual, 2nd ed. CSHL Press, Cold Spring Harbor, NY.

15 Isolating Differentially Expressed Genes by Representational Difference Analysis

David Chang[1,3] and Christopher Denny[2,3]
Departments of [1]Medicine, [1]Microbiology & Immunology and [2]Pediatrics, [2]Gwynne-Hazen Memorial Laboratories and [3]Jonsson Comprehensive Cancer Center, UCLA School of Medicine, University of California, Los Angeles, CA, USA

OVERVIEW

Representational difference analysis (RDA) was first described by Lisitsyn et al. (11) to identify unique fragments between two populations of genomic fragments. Hubank and Schatz (8) were the first to show that this procedure could be adapted to cDNA populations. This acheivement was more than just recapitulation of the RDA protocol using a different DNA source. Though genomic RDA and cDNA RDA are similar, different determinants need to be taken into consideration for these procedures to be successful. In genomic RDA, the populations of fragments being compared are enormously complex, but most single-copy sequences are present in stoichiometric amounts. By comparison, in cDNA RDA, the cDNA pools are much less complex, but there is large variability in expression levels of different genes. In spite of these theoretical concerns, cDNA RDA has been successfully used to identify differentially expressed genes in a number of different cell systems (1–10).

BACKGROUND

Strategy

RDA was developed as a method to detect DNA fragments present in one pool (tester) but not in another (driver). Representative fragments from each population are first generated by restriction endonuclease digestion of DNAs followed by polymerase chain reaction (PCR) amplification. The amplified DNA fragments are termed "amplicons". The amplicons are then subject to successive rounds of subtractive cross-hybridization followed by differential PCR amplification. This leads to progressive enrichment of fragments that are present in the tester pool but not in the driver.

To use cDNA RDA to detect differentially expressed genes, cDNAs generated from the two cell populations being compared are used instead of genomic DNA (Figure 1). Basically, this protocol breaks down into three steps: *(i)* cDNA synthesis, *(ii)* amplicon preparation and *(iii)* subtractive hybridization/differential PCR amplification. Repeating step three 3–4 times usually results in cDNA fragments that derive

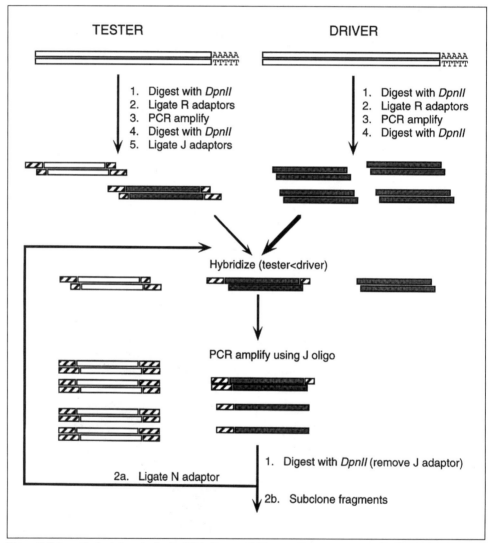

Figure 1. Schematic depicting cloning by RDA. Tester and driver amplicons are generated by ligating R adaptors to *Dpn*II-digested oligo(dT)-primed cDNAs. After PCR amplification, R adaptors are removed by *Dpn*II digestion, and J adaptors are ligated only to tester fragments, which are then hybridized to an excess of driver. Fragments present predominantly in the tester will form homodimers with an adaptor at each end (clear box) and amplify efficiently. Fragments present in both driver and tester will form heterodimers with an adaptor only at one end (gray box) and amplify inefficiently. J adaptors are then removed by *Dpn*II digestion, and fragments are either subcloned into plasmids or ligated to N adaptors and subjected to another round of RDA subtraction.

exclusively from differentially expressed transcripts. These fragments can then be subcloned and analyzed further.

Critical Determinants

Not all differentially expressed genes are equally enriched by cDNA RDA. The ability for any single gene fragment to be enriched depends in part on the relative level of expression of its parent gene. The absolute level of expression (whether a gene is present at many or only a few copies per cell) is much less important. All things being equal, those cDNA fragments that derive from genes 100× differentially expressed will be enriched by cDNA RDA more efficiently than those 5× differentially expressed.

This effect of favoring highly differentially expressed genes is most pronounced if 3–4 cycles of cDNA RDA are performed rather than 1–2 cycles. In the end, there is a trade-off between the relative enrichment for differentially expressed genes and the complexity of cDNA RDA output. If only a few (i.e., 6–12) highly differentially expressed genes are sought with very low nondifferentially expressed gene contamination, 3–4 cycles of cDNA RDA should be performed. On the other hand, if larger populations of differentially expressed genes are desired, the cDNA RDA should be limited to 1–2 cycles. Under these conditions, significant numbers of nondifferentially expressed genes will be present, necessitating screening of the cDNA RDA output.

The ability of some cDNA fragments to be PCR-amplified better than others also has an effect on cDNA RDA output. For a gene to be represented in cDNA RDA amplicon pools, it must have appropriate restriction sites positioned within it that yield a fragment upon digestion that amplifies well. Considering this, it seems unlikely that a single restriction enzyme would suffice for all differentially expressed genes. In fact, it appears that any single cDNA RDA detects only a subset of the genes that are differentially expressed and that changing the restriction enzyme used to make amplicons alters the subset being sampled.

General Practical Considerations

RDA is a pairwise comparison between two cDNA populations (e.g., A and B). It is generally a good idea that both possible comparisons be made at the same time (i.e., A-B and B-A subtractions are performed in parallel). One serves as a good control for the other. If bands appear in one comparison that are not in the reciprocal comparison, there is a very high likelihood that they derive from differentially expressed genes.

Volumes and temperatures are important. A micropipettor that is accurate to 0.1 µL is very helpful.

Spin column technologies have been used frequently to try to circumvent phenol–chloroform extractions. While spin columns from Qiagen (Chatsworth, CA, USA) have been adequate, other similar devices could probably be used as well.

Protocol 1: cDNA Synthesis

There are many protocols to synthesize cDNA, and in theory, any one of them is sufficient. The following method is quick and easy. What is important though is to be consistent. Usually, RNA harvesting and cDNA preparation are performed for both cell populations in lock step. What is done to one is done to the other. At the end, the gel profiles of the two cDNAs should be nearly identical. A convenient amount of starting material is 100 µg of total RNA from each cell population, but as few as 30 µg have been successfully used.

Materials and Reagents

- RNasin® (10 U/µL; Promega, Madison, WI, USA)
- SUPERSCRIPT™ II Reverse Transcriptase (200 U/µL; Life Technologies, Gaithersburg, MD, USA)
- *E. coli* polymerase I (10 U/µL; Life Technologies)
- RNase H (3 U/µL; Life Technologies)
- *E. coli* ligase (10 U/µL; Life Technologies)
- 5× first-strand buffer (SUPERSCRIPT; Life Technologies)
- 0.1 M dithiothreitol (DTT)
- 5 mM dNTP mixture (final concentration 5 mM each)
- Oligo(dT)$_{12-18}$ (0.5 µg/µL; Life Technologies)
- 15 mM β-nicotinamide adenine dinucleotide (βNAD) (Sigma Chemical, St. Louis, MO, USA)
- 5× double-strand (DS) buffer: 100 µL 1 M Tris-HCl (pH 7.5), 200 µL 2.5 M KCl, 30 µL 1 M MgCl$_2$, 50 µL bovine serum albumin (BSA) (10 µg/µL), 595 µL distilled H$_2$O
- PolyATtract® Paramagnetic Beads (Promega) or Oligotex™ Latex Beads (Qiagen)

Procedure

1. Harvest total RNA from cells by guanidinium isothiocyanate lysis followed by CsCl gradient centrifugation (2) or acid phenol extraction (Stat-60™; Tel-Test, Friendswood, TX, USA).
2. Purify poly(A+) RNA from total RNA using oligo(dT) paramagnetic beads (PolyATtract) or oligo(dT) latex beads (Oligotex).
3. First-strand cDNA synthesis:

poly(A+) RNA	2.5 µg
distilled H$_2$O	final volume 29 µL

    Heat to 70°C for 10 min, spin briefly and put on ice. Add:

5× first-strand buffer	10.0 µL
RNasin (10 U/µL)	0.5 µL
0.1 M DTT	0.5 µL
5 mM dNTP mixture	5.0 µL
oligo(dT)$_{12-18}$ (0.5 µg/µL)	3.0 µL
BSA (10 µg/µL)	0.5 µL
SUPERSCRIPT II reverse transcriptase (200 U/µL)	1.5 µL

    Leave at 37°C for 1–2 h.
4. Second-strand cDNA synthesis. Place the above mixture on ice and add the following:

distilled H$_2$O	25 µL
5× DS buffer	20 µL
15 mM βNAD	1 µL
0.1 M DTT	0.5 µL
E. coli polymerase I (10 U/µL)	2.5 µL
RNase H (3 U/µL)	0.4 µL
E. coli ligase (10 U/µL)	0.5 µL

Cool to 16°C for 3 h.

5. Clean up. Perform phenol–chloroform extraction. Ethanol precipitate using 25 µL 10 M NH$_4$OAc and 250 µL ethanol. Resuspend in 20 µL distilled H$_2$O.

6. Check cDNAs by running 2–4 µL of each on a 1% agarose minigel run in 1× TAE buffer (40 mM Tris, pH 7.2, 20 mM NaOAc, 1 mM EDTA). Profiles should be identical (most important), and highest species should extend to 7–10 kb (less important).

Protocol 2: Amplicon Preparation

The following protocol details the creation of amplicons from cDNA that will eventually be used as both driver and tester. Though only one set of reactions is listed, it is assumed that it will be repeated for both cDNA populations being compared by RDA. Both pairwise comparisons should be made (i.e., A-B and B-A) so that tester and driver amplicons need to be generated from both cDNA populations.

Materials and Reagents

- DpnII (10 U/µL; New England Biolabs, Beverly, MA, USA)
- T4 DNA ligase (New England Biolabs)
- Taq DNA polymerase (5 U/µL; Perkin-Elmer, Norwalk, CT, USA)
- 10× DpnII buffer (New England Biolabs)
- 10× T4 DNA ligase buffer (New England Biolabs)
- GeneAmp® 10× PCR Buffer (no Mg^{2+}; Perkin-Elmer)
- 25 mM MgCl$_2$
- 2.5 mM dNTP mixture
- Buffer PB (Qiagen)
- Buffer PE (Qiagen)
- 10 mM Tris-HCl (pH 8.6)

Oligonucleotides

The following oligonucleotides are used for cDNA RDA. They frequently can be used without further purification from several vendors.

- RBgl24:   5′-AGCACTCTCCAGCCTCTCACCGCA-3′
- RBgl12:   5′-GATCTGCGGTGA-3′
- JBgl24:   5′-ACCGACGTCGACTATCCATGAACA-3′
- JBgl12:   5′-GATCTGTTCATG-3′
- NBgl24:   5′-AGGCAACTGTGCTATCCGAGGGAA-3′
- NBgl12:   5′-GATCTTCCCTCG-3′

Procedure

1. Digest about half of each cDNA with *Dpn*II.

cDNA	8 µL
10× *Dpn*II buffer	10 µL
distilled H$_2$O	80 µL
*Dpn*II (10 U/µL)	2 µL

   Leave at 37°C for 1–2 h. Perform phenol–chloroform extraction, ethanol precipitate using sodium acetate and resuspend in 43 µL distilled H$_2$O.

2. Ligate RBgl adaptors to cDNA fragments. Adaptors consist of two oligonucleotides: a 24-mer matched with a 12-mer. The 3′ 8 bp of the 24-mer are complementary to the 3′ 8 bp of the 12-mer. The 5′ 4 bp of the 12-mer complement the 5′ overhang of a *Dpn*II site (GATC).

cDNA/*Dpn*II fragments	43 µL
10× ligase buffer	6 µL
RBgl24 (2 µg/µL)	4 µL
RBgl12 (1 µg/µL)	4 µL

   Heat at 55°C for 1–2 min in a heating block and cool heating block to <15°C in a cold room for 1 h. Add 3 µL T4 DNA ligase and leave at 16°C overnight. Bring to 100 µL final volume with distilled H$_2$O.

3. Generate amplicons by PCR. The following recipe is for a single 100-µL PCR. Depending on PCR yield, 8–16 such PCRs are run for each cDNA. This protocol assumes that a PCR machine with a heated lid is being used, obviating the need for mineral oil.

cDNA/RBgl ligation mixture	1 µL
10× PCR buffer (no Mg^{2+})	10 µL
25 mM MgCl$_2$	16 µL
2.5 mM dNTP mixture	10 µL
RBgl24 (1 µg/µL)	1 µL
distilled H$_2$O	61 µL
Taq DNA polymerase (5 U/µL)	0.5 µL

   Add solution to preheated block and heat at 72°C for 5 min. Perform 20 cycles of denaturation at 95°C for 1 min and extension at 72°C for 3 min, with a final extension at 72°C for 10 min. Pool reaction mixtures into 1.5-mL microcentrifuge tubes (400 µL/tube), perform phenol–chloroform extraction, ethanol precipitate using sodium acetate, resuspend in 75 µL distilled H$_2$O and pool tubes.

4. Check amplicons on a 1.2% 0.5× agarose minigel run in 1× TBE buffer (5 mM Tris base, 5 mM boric acid, 1 mM EDTA). They should appear as a smear ranging from 150 to 1600 bp.

5. Digest amplicons with *Dpn*II to cleave RBgl adaptors. The following reaction corresponds to all of the amplicons from 8 PCRs. After digestion, these reactions are cleaned up by running them over QIAquick® Spin Columns (Qiagen). This removes all of the free nucleotides left over from PCR and some of the cleaved adaptors, making amplicon quantitation easier. Because their capacity is 10 µg, three columns are needed for each of the following digestions.

RBgl amplicons	150 µL

10× *Dpn*II buffer	30 µL
distilled H$_2$O	110 µL
*Dpn*II (10 U/µL)	10 µL

Leave at 37°C for 2–3 h. For each 100 µL of reaction mixture, add 500 µL Buffer PB. Mix thoroughly, load the spin column and spin for 1 min. Add 750 µL Buffer PE. Wash column and spin dry for 3 min. Add 50 µL 10 mM Tris-HCl (pH 8.6). Collect eluate.

6. Quantitate concentration by optical density (OD)$_{260}$. Check the amplicon profiles on a 1.2% 0.5× agarose minigel run in TBE buffer. The majority of species should be greater than 200 bp.

Protocol 3: Subtractive Hybridization/Differential PCR Amplification

This section describes an iterative procedure that consists of: *(i)* ligation of a new adaptor to the tester, *(ii)* hybridization of tester with driver amplicons and *(iii)* differential PCR amplification of hybridization mixtures. This section can be repeated 3–4 times to progressively enrich for fragments present in greater quantity in the tester than in the driver. With each repetition, the tester–driver ratio decreases: 1st round, 1:40–100; 2nd round, 1:400–800; 3rd round, 1:4000–8000; 4th round, 1:20 000. The amount of driver stays constant (about 5–10 µg). The tester gets titrated progressively downward after each successive ligation step.

Materials and Reagents

- T4 DNA ligase (New England Biolabs)
- *Taq* DNA polymerase (5 U/µL; Perkin-Elmer)
- 10× T4 DNA ligase buffer (New England Biolabs)
- 10 M NH$_4$OAc
- 3× EEP buffer: 30 µL 1 M EPPS (pH 8.0) (Sigma), 6 µL 0.5 M EDTA, 250 µL 50% polyethylene glycol (PEG) (Sigma), 7 µL distilled H$_2$O
- GeneAmp 10× PCR Buffer (no Mg^{2+}; Perkin-Elmer)
- 25 mM MgCl$_2$
- 2.5 mM dNTP mixture

Procedure

1. Create tester fragments by ligating a portion of the amplicons to JBgl.

*Dpn*II-digested amplicons (0.1 µg/µL)	5 µL
10× ligase buffer	6 µL
JBgl24 (2 µg/µL)	4 µL
JBgl12 (1 µg/µL)	4 µL
distilled H$_2$O	38 µL

   Heat at 55°C for 1–2 min in a heating block. Cool heating block to <15°C in a cold room for 1 h. Add 3 µL T4 DNA ligase and leave at 16°C overnight. Dilute with distilled H$_2$O to 100 µL.

2. Hybridize tester with driver amplicons. Driver amplicons are the same as the *Dpn*II-digested amplicons from step 6 of Protocol 2. This recipe is written for a subtractive hybridization with a tester–driver ratio of 1:100. Future hybridizations will be identical to this, except that less tester will be used. A good posi-

tive control to strongly consider is setting up a mock hybridization using the same amount of tester but using tRNA instead of driver amplicons.

JBgl tester amplicons (0.005 µg/µL)	20 µL
driver amplicons (0.1 µg/µL)	100 µL
phenol–chloroform	60:60

Vortex mix, spin and remove aqueous layer. Add:

chloroform	120 µL

Vortex mix, spin and remove aqueous layer. Add:

10 M NH_4OAc	30 µL
ethanol	300 µL

Ice, spin for 20 min, wash twice with 70% ethanol and dry completely. Add:

3× EEP buffer	4 µL

Resuspend thoroughly. Add:

mineral oil	40 µL

Heat at 95°–96°C for 4 min. Add:

5 M NaCl	1 µL

Heat at 67°C for >21 h (up to 2 days), remove oil and dilute by serially mixing well with 10 µL TE buffer, 25 µL TE buffer, then 60 µL distilled H_2O.

3. Differentially amplify tester–driver hybridization mixture. This is accomplished by an initial PCR amplification followed by a dilution step and a final PCR amplification. The number of cycles for the latter PCR can vary and generally increases with each successive round of cDNA RDA (see step 7). The cycle number provided is an estimate to allow for adequate amplification but not overamplification. This can be explicitly determined for each cDNA RDA by setting up a single PCR 2, analyzing by gel electrophoresis and removing aliquots at varying PCR cycles.

PCR 1:

tester–driver hybridization mixture	10 µL
10× PCR buffer (no Mg^{2+})	10 µL
25 mM $MgCl_2$	16 µL
2.5 mM dNTP mixture	10 µL
distilled H_2O	53 µL
JBgl24 (2 µg/µL)	0.5 µL
Taq DNA polymerase	0.5 µL

Add the solution to a preheated block. Heat at 72°C for 5 min, perform 10 cycles at 95°C for 1 min and 70°C for 3 min, then perform a final extension at 72°C for 10 min.

PCR 2:

Note: Reactions are done in quadruplicate to yield sufficient amount of product.

PCR 1 reaction mixture	10 µL
10× PCR buffer (no Mg^{2+})	10 µL
25 mM $MgCl_2$	16 µL
2.5 mM dNTP mixture	10 µL
distilled H_2O	53 µL
JBgl24 (2 µg/µL)	0.5 µL
Taq DNA polymerase	0.5 µL

Add solution to a preheated block. Perform 18 cycles at 95°C for 1 min and 70°C for 3 min, with a final extension at 72°C for 10 min.

4. Run 10 µL of each RDA reaction (there should be two: A-B and B-A) on a 1.2% 0.5× agarose minigel run in TBE buffer. There should be a smear 200–1200 bp long with some predominant bands.
5. To start round 2, purify RDA reaction products using spin columns and digest with *Dpn*II to remove JBgl adaptors using step 5 of Protocol 2.
6. Quantitate *Dpn*II-digested RDA reaction products by OD_{260}. Using 0.5 µg of each, set up ligation to NBgl adaptors using step 1 of Protocol 3.
7. The next rounds of RDA are essentially repeats of steps 1–6. The output of the previous tester–driver subtraction becomes the tester for the next round. The changes between rounds can be summarized:

Round	Adaptor	PCR annealing temperature	PCR 2 cycles (approximately)
1	JBgl	70	18
2	NBgl	72	18
3	JBgl	70	20–21
4	NBgl	72	20–21

Screening RDA Output

As mentioned earlier, RDA stringency and diversity of RDA output are inversely proportional. If many rounds of RDA are performed, most if not all of the resultant cDNA fragments will be from differentially expressed transcripts. To confirm this, these bands can be individually cloned into plasmid vectors and used as probes in Northern blot experiments using starting RNAs.

If only a few rounds of RDA are performed, the diversity of the output will be high, but so will the contamination by nondifferentially expressed species. In this case, a differential screening of RDA output has proved very helpful in identifying fragments from differentially expressed genes. *Dpn*II-digested RDA fragments are first shotgun-subcloned into plasmids. Colony replica filters or PCR-amplified plasmid DNAs are then gridded onto filters and probed with radiolabeled starting amplicon preparations. Those DNAs that hybridize more strongly to one amplicon population than the other are very likely to originate from differentially expressed genes.

REFERENCES

1. **Braun, B.S., R. Freiden, S.L. Lessnick, W.A. May and C.T. Denny.** 1995. Identification of target genes to the Ewing's sarcoma EWS/FLI fusion protein by representational difference analysis. Mol. Cell. Biol. *15*:4623-4630.
2. **Chirgwin, J.M., A.E. Przybyla, R.J. MacDonald and W.J. Rutter.** 1979. Isolation of biologically active ribonucleic acid from sources enriched in ribonuclease. Biochemistry *18*:5294-5299.
3. **Chu, C.C. and W.E. Paul.** 1997. Fig1, an interleukin 4-induced mouse B cell gene isolated by cDNA representational difference analysis. Proc. Natl. Acad. Sci. USA *94*:2507-2512.
4. **Dron, M. and L. Manuelidis.** 1996. Visualization of viral candidate cDNAs in infectious brain fractions from Creutzfeldt-Jakob disease by representational difference analysis. J. Neurovirol. *2*:240-248.
5. **Edman, C.F., S.A. Prigent, A. Schipper and J.R. Feramisco.** 1997. Identification of ErbB3-stimulated genes using modified representational difference analysis. Biochem. J. *323*:113-118.
6. **Fu, X. and M.P. Kamps.** 1997. E2a-Pbx1 induces aberrant expression of tissue-specific and developmentally regulated genes when expressed in NIH 3T3 fibroblasts. Mol. Cell. Biol. *17*:1503-1512.
7. **Gress, T.M., C. Wallrapp, M. Frohme, F. Muller-Pillasch, U. Lacher, H. Friess, M. Buchler, G. Adler**

and J.D. Hoheisel. 1997. Identification of genes with specific expression in pancreatic cancer by cDNA representational difference analysis. Genes, Chromosomes and Cancer *19*:97-103.
8. **Hubank, M. and D.G. Schatz.** 1994. Identifying differences in mRNA expression by representational difference analysis of cDNA. Nucleic Acids Res. *22*:5640-5648.
9. **Lerner, A, L.K. Clayton, E. Mizoguchi, Y. Ghendler, W. van Ewijk, S. Koyasu, A.K. Bhan and E.L. Reinherz.** 1996. Cross-linking of T-cell receptors on double-positive thymocytes induces a cytokine-mediated stromal activation process linked to cell death. EMBO J. *15*:5876-5887.
10. **Lewis, B.C., H. Shim, Q. Li, C.S. Wu, L.A. Lee, A. Maity and C.V. Dang.** 1997. Identification of putative c-Myc-responsive genes: characterization of *rcl*, a novel growth-related gene. Mol. Cell. Biol. *17*:4967-4978.
11. **Lisitsyn, N., N. Lisitsyn and M. Wigler.** 1993. Cloning the differences between two complex genomes. Science *259*:946-951.

16 Linker-Capture Subtraction

Meiheng Yang and Arthur J. Sytkowski
Laboratory for Cell and Molecular Biology, Division of Hematology and Oncology, Beth Israel Deaconess Medical Center, Harvard Medical School, Boston, MA, USA

OVERVIEW

Linker-capture subtraction (LCS) is a simple and efficient method applicable to the identification and isolation of genes expressed differentially between similar cell types (17). LCS achieves enrichment of target gene sequences by specifically preserving polymerase chain reaction (PCR) priming sites (linkers) of target sequences. Figure 1 represents the procedure of LCS. In the first step, both tester and driver DNA are prepared. This is accomplished by digesting the double-stranded cDNA with restriction enzymes of choice, ligating the fragments to linkers and carrying out a PCR with linker sequence as primer. The driver DNA is digested with restriction enzymes to remove the linker sequence. In the second step, the linkered tester DNA is hybridized to an excess of driver DNA (with linkers removed) followed by incubation with mung bean nuclease, which digests single-stranded DNA specifically. This leaves only linkered tester–tester homohybrids and unlinkered homo- and heterohybrids. In the following step, the linkered tester–tester homohybrids are amplified by PCR with linker sequence as primer to fulfill the first round of enrichment. The amplified PCR products are then used as tester for another round of subtraction. The process of subtractive hybridization, mung bean nuclease digestion and PCR amplification is carried out three times. Finally, the PCR products of the third round of subtraction are used to prepare a subtraction library by inserting them into a vector.

BACKGROUND

The isolation and identification of differentially expressed genes are of great importance in the study of embryogenesis, cell growth and differentiation and neoplastic transformation. A variety of methods have been used to achieve this end. They include differential screening of cDNA libraries with selective probes, subtractive hybridization using DNA–DNA hybrids or DNA–RNA hybrids, RNA fingerprinting and differential display (4,5,7,9,13). Recently, PCR-coupled subtractive processes have been reported (10–12,14,18). Each of these methods has achieved some success, and each has some inherent limitations. Differential display (7) has the problems of false positives, redundancy and underrepresentation of certain mRNA species. cDNA representational difference analysis (RDA) (6) is a labor-intensive process, and its efficiency remains to be evaluated.

We have sought to overcome some of these problems and have developed a method designated LCS (17). This method does not rely on a kinetic mechanism of enrichment as does RDA (8). Rather, it achieves enrichment by specifically preserving PCR priming sites of target sequences using a single-stranded nuclease (e.g., mung bean nuclease) as the mediator. Also, it is a much less labor-intensive process. We have applied LCS to the human prostate cancer cell lines LNCaP and PC-3, which have different tumorigenic and metastatic potentials. It has resulted in the rapid and effective isolation of genes expressed differentially between the two cell lines.

PROTOCOLS

Materials and Reagents

- Oligo(dT)$_{25}$ Dynabeads® (Dynal, Lake Success, NY, USA)
- SUPERSCRIPT™ Choice System (Life Technologies, Gaithersburg, MD, USA)
- GENECLEAN® Kit (Bio 101, Vista, CA, USA)
- *Alu*I, *Rsa*I and *Sac*I (New England Biolabs, Beverly, MA, USA)
- 4 Weiss U/μL T4 DNA ligase, 10× ligation buffer (Stratagene, La Jolla, CA, USA)
- 5 U/μL *Taq* DNA polymerase (Promega, Madison, WI, USA)
- 50–100 U/μL mung bean nuclease (Promega)
- *N*-(2-hydroxyethyl)piperazine-*N'*-(3-propanesulfonic acid) (EPPS; Sigma Chemical, St. Louis, MO, USA)
- NuSieve® Agarose (FMC BioProducts, Rockland, ME, USA)
- pGEM®-7Zf(+) (Promega)
- GeneScreen™ Plus membrane (NEN Life Science Products, Boston, MA, USA)
- 10× PCR buffer: 100 mM Tris-HCl, pH 8.9, 500 mM KCl, 20 mM MgCl$_2$, 1% Triton® X-100
- Enzyme solution: 10 mM Tris-HCl, pH 8.9, 50 mM KCl, 0.1% Triton X-100, 1 mM dNTPs, 5 μM amplification primer (AP), 10 mM MgCl$_2$, 5 U *Taq* DNA polymerase
- pH-shift buffer A: 1 mM ZnCl$_2$, 10 mM sodium acetate, pH 5.0
- pH-shift buffer B: 10 mM Tris-HCl, pH 8.9, 50 mM KCl, 0.1% Triton X-100

Procedures

Protocol 1: Preparation of cDNA

1. Isolate total RNA with a guanidinium thiocyanate–phenol method (16).
2. Select poly(A$^+$) RNA through oligo(dT)$_{25}$ Dynabeads.

3. Synthesize cDNA from 2 µg of poly(A+) RNA using the SUPERSCRIPT Choice System according to the manufacturer's instructions. Oligo(dT)$_{12-18}$ is used to prime the first strand of cDNA synthesis.

Protocol 2: Preparation of the Tester DNA and the Driver DNA

1. Digest the double-stranded cDNA with *Alu*I and *Rsa*I. Purify the cDNA fragments using the GENECLEAN Kit. Be sure to set up parallel reactions of the tester and driver preparations.

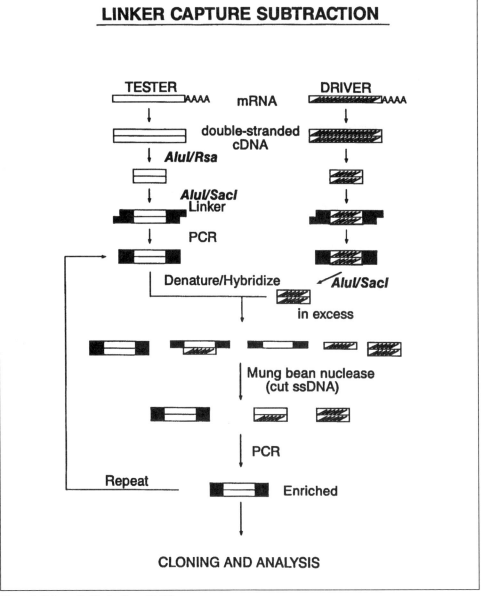

Figure 1. Schematic illustration of LCS.

2. Ligate the DNA with a double-stranded oligodeoxyribonucleotide linker, which has a blunt end and a 2-base 3' protruding end: ACTCTTGCTTGGAC-GAGCTCTACTGAGAACGAACCTGCTCGAGA-p.

   The linker contains an *AluI/SacI* site near the blunt end, as indicated. The top strand is designated AP. The bottom strand is phosphorylated at the 5' end.

   Prepare the linker by annealing the two strands. Mix an equal mass of each of the two oligodeoxyribonucleotides. Heat the mixture to 90°C for 2 min and allow to cool to room temperature.

   Carry out the ligation by mixing the following components: 1 µg cut cDNA, 5 µg linker, 1 µL 10× ligation buffer, 1 µL T4 DNA ligase (4 Weiss U/µL) in a total volume of 10 µL. Set reaction to 8°C for 20 h.

   Electrophorese the reaction mixture through a 2% low-melting agarose gel to remove the unligated linkers. Collect the linker-ligated cDNA fragments in the size range of 0.1–1.0 kb.

3. Amplify the linker-ligated cDNA fragments by PCR using AP as primer. Place the following components into each PCR tube: 1 µL melted agarose containing cDNA fragments, 79 µL sterile H_2O, 10 µL 10× PCR buffer, 8 µL dNTP mixture (2.5 mM each), 1 µL AP (100 µM) and 1 µL *Taq* DNA polymerase (5 U/µL) in a total volume of 100 µL. Run for 30 cycles (94°C for 1 min, 55°C for 1 min and 72°C for 1 min).

4. Purify the amplified cDNA fragments using the GENECLEAN Kit. To prepare the ready-for-use driver DNA, digest 20 µg of PCR-amplified driver DNA with *Alu*I (50 U) at 37°C for 2 h, followed by *Sac*I (50 U) at 37°C for 1 h. After digestion, purify the products using GENECLEAN.

Protocol 3: Subtractive Hybridization

1. Mix the digested driver DNA (2.5 µg) and nondigested tester DNA (0.1 µg) and vacuum dry the mixture.

2. Redissolve the mixture in 4 µL of a buffer containing 15 mM EPPS (pH 8.0) and 1.5 mM EDTA, overlay the sample with mineral oil and denature the DNA by heating for 5 min at 100°C. Then add 1 µL of 5 M NaCl and hybridize the DNA for 20 h at 67°C.

3. After hybridization, add 20 µL of pH-shift buffer A (1 mM $ZnCl_2$, 10 mM sodium acetate, pH 5.0) and divide the solution into 5 aliquots. Incubate each of the samples with 0, 0.85, 1.75, 3.5 or 7 U of mung bean nuclease at 37°C for 30 min.

4. Add 80 µL of pH-shift buffer B (10 mM Tris-HCl, pH 8.9, 50 mM KCl and 0.1% Triton X-100) to each sample and heat them at 95°C for 5 min. Then add 20 µL of enzyme solution to each sample. Run PCR under same conditions as above.

5. Electrophorese each sample on a 2% agarose gel. Select the sample with the most abundant products of 0.1–1.0 kb as tester for another round of subtraction.

6. Repeat steps 1–5 twice with 2.5 µg of driver DNA and 0.025 µg of tester DNA.

7. Finally, to test for enrichment of target sequences, electrophorese the PCR products derived from subtraction cycles 0–3 on 4% NuSieve agarose, transfer the DNA to the GeneScreen Plus membrane and probe the membrane with the

randomly labeled PCR products (with linkers removed) of the third round of subtraction (see Results).

Protocol 4: Construction of Subtractive Library and Clonal Analysis

1. After three rounds of subtraction, purify the PCR-amplified products using the GENECLEAN Kit and digest the DNA with *Sac*I.
2. Insert the digested DNA into dephosphorylated pGEM-7Zf(+) at the *Sac*I site and transform the plasmid into competent *E. coli* JM109 cells.
3. Randomly pick white colonies from each library and inoculate them into LB medium with ampicillin in individual wells of a 96-well plate.
4. Prepare two replica DNA dot blots on GeneScreen Plus filters using 25 µL of bacterial cells per well. Process the replica dot blots according to the procedure of Brown and Knudson (3) and probe the membrane with randomly labeled driver DNAs.
5. Boil candidate positive colonies for 5 min in 20 µL H$_2$O and centrifuge the solutions. Amplify DNA in the supernatant by PCR using universal vector primers T7 and SP6 for 20 cycles at 94°C for 1 min, 55°C for 1 min and 72°C for 1 min.
6. Electrophorese the PCR products on 2% agarose. Excise the desired bands and purify them (GENECLEAN). Do direct DNA sequencing (15) on those products.
7. Also, use those products as probes in Northern blot analyses.

RESULTS

We have used the method to clone and identify genes expressed differentially between the human prostate cancer cell lines LNCaP and PC-3, which have different tumorigenic and metastatic potentials. After three cycles of subtraction, the PCR products were cleaved with *Sac*I, inserted into pGEM-7Zf(+) and transformed into *E. coli* JM109 cells. Figure 2A shows the electrophoretic analysis of the PCR-amplified DNA derived from subtraction cycles 0–3. The original unsubtracted DNAs from LNCaP (section L-P, lane 0) and PC-3 (section P-L, lane 0) moved as a smear between 0.1 and 1.0 kb. As subtraction rounds were performed, distinct bands were seen (section L-P, lanes 1–3; section P-L, lanes 1–3). The intensity and resolution of these bands increased progressively with successive subtraction. When labeled PCR products of the third round of subtraction were electrophoresed on a 6% sequencing gel, 50–60 bands could be seen (not shown). DNA of the agarose gel of Figure 2A was transferred to the GeneScreen Plus membrane and probed with the labeled PCR products of the third round of subtraction (Figure 2B, section L-P, lane 3 or Figure 2C, section P-L, lane 3). The results indicate strong enrichment of differentially expressed sequences.

After three rounds of subtraction, the PCR-amplified products were inserted into pGEM-7Zf(+) and transformed into *E. coli* JM109 cells. We randomly picked 48 white colonies from each of the libraries and grew them in LB medium in individual wells of a 96-well plate. Two replica DNA dot blots were prepared and probed with the labeled driver DNAs from LNCaP (section P-L, lane 0) and PC-3 (section L-P, lane 0). A comparison of the hybridization intensity of a clone in two replica membranes revealed the relative abundance of the transcript in the two cell types. Over two thirds of the selected clones demonstrated significant differences in abundance. We tested clones further with Northern blot and sequence analyses. From 78 colonies, 15 distinct clones were identified that correspond to mRNAs expressed differentially between LNCaP and PC-3 cell lines. The extent of differential expression ranged from several-fold to >100-fold. Besides five novel genes, the identified genes included some very interesting known genes that are or might be involved in signal transduction, tumor growth, tumor invasion and metastasis. A Northern blot exemplifying differential expression of two genes is shown in Figure 3. DNA sequence analyses demonstrated that the LNCaP-specific gene is prostate-specific antigen (PSA), which is known to be expressed in LNCaP but not in PC-3 (2). The PC-3-specific gene was found to be vimentin, the differential expression of which has not been reported previously in these prostate cancer cells.

DISCUSSION

LCS offers several important features. First, it is highly effective. We achieved a strong stepwise enrichment of target sequences as shown in Figure 2. When the PCR-

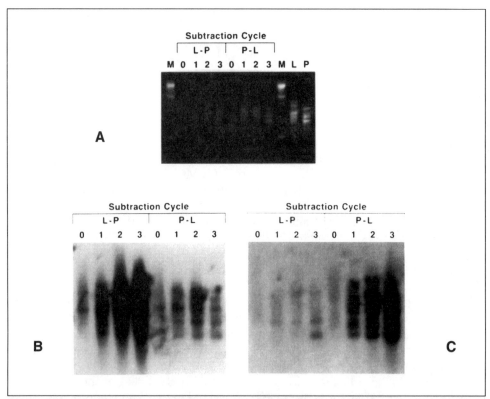

Figure 2. Enrichment of specific sequences from LNCaP (L-P) and PC-3 (P-L) cell lines. (A) 10 µL of PCR mixture (lanes 0–3) were electrophoresed on a 4% NuSieve agarose gel. Lane M: 100-bp markers. Lane L: 20 µL of final PCR product L-P. Lane P: 20 µL of final PCR product P-L. (B) Effect of subtraction cycles on enrichment of LNCaP-specific sequences. DNA shown in panel A was blotted onto a GeneScreen Plus filter and was probed with radiolabeled PCR product of L-P. (C) Effect of subtraction cycles on enrichment of PC-3-specific sequences. Filter shown in panel B was stripped and reprobed with radiolabeled PCR product of P-L.

amplified products of the third round of subtraction were cloned, 81% of randomly picked clones (78 out of 96 colonies) corresponded to mRNAs expressed differentially between LNCaP and PC-3 cell lines. Such a high efficiency has not been reported by others, and it obviates the need to screen the subtraction library by differential hybridization, an essential step in some other methods. Second, LCS is simple to carry out and contains fewer steps than other methods. In particular, a variety of labor-intensive and potentially error-prone physical partitioning steps, such as biotinylation or repeated phenol extraction/ethanol precipitation, were eliminated. In LCS, all these steps of subtractive hybridization, mung bean nuclease digestion and PCR amplification can be performed in one PCR tube, which makes the process very easy for operation and feasible for automation. Third, LCS is a fast and economical process. The materials required are kept to a minimum. The procedure, from isolation of mRNA to construction of the subtraction library, can be completed within one week.

Unlike RDA (8), which uses a kinetic mechanism of enrichment, LCS achieves enrichment by specifically preserving PCR priming sites (linkers) of target sequences. Mung bean nuclease plays a central role in the process. It removes linkers of all other linkered sequences except for tester–tester homohybrids. The nuclease also digests single-stranded (ss)DNA in the hybridization solution, which is an abundant species that might otherwise cause high background or even failure of enrichment. The use of the enzyme appears to be more reliable and efficient than other physical partitioning methods, as indicated by the high enrichment of target sequences and efficient isolation of numerous differentially expressed genes in our experiments. Exonuclease VII, also specific for ssDNA, could be used in the LCS protocol. It has the added advantage of a pH optimum near to those of the subtractive hybridization and PCR, thus eliminating the need for pH-shift buffers in the process.

The digestion of double-stranded cDNA with different restriction enzymes gives a representation of the mRNA population. In the example of LCS presented here, we used *Alu*I and *Rsa*I. Obviously, the use of other enzymes would give different representations and might result in the isolation of genes different from those achievable using *Alu*I and *Rsa*I. Also, the use of different PCR conditions such as additional concentrations of magnesium, different annealing temperatures and the addition of reagents such as dimethyl sulfoxide (DMSO), formamide or glycerol would achieve other representations of the mRNA population of the cells under study. Moreover, thermostable DNA polymerases from different vendors might also give different representations because these enzymes appear to have different efficiencies in amplifying large fragments.

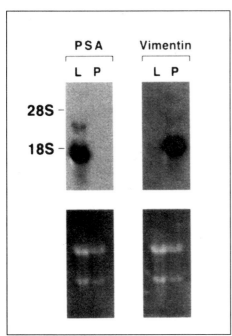

Figure 3. **LCS reveals differential expression of PSA and vimentin in LNCaP ("L") and PC-3 ("P") cells**. Northern analysis. Total RNA from either cell was electrophoresed and probed with radiolabeled cDNA of two differentially expressed clones isolated by LCS. Top panels: autoradiograms. Bottom panels: ethidium bromide-stained gels.

We added the same linker to the tester and driver in our experiment. Previously, Balzer and Baumlein (1) reported the addition of different linkers to the tester and driver to avoid the necessity of restriction enzyme digestion of driver and to eliminate contamination of residual linkered driver. We found that the addition of different linkers gave an unequal representation of starting mRNAs for tester and driver, probably because of the tendency to amplify some sequences preferentially, also known as PCR "bias". Therefore, we found it advantageous to use the same linker for both tester and driver rather than different linkers. While it is true that the driver linkers cannot be completely removed by a restriction enzyme, we believe that our protocol itself has a mechanism to eliminate the contamination problem of the residual linkered driver. Because the unlinkered driver is present in high excess in the reaction, the residual linkered driver should be driven out by the unlinkered driver. Of course, too high a level of linkered driver would still pose a problem for efficiency of enrichment. Therefore, we designed a linker with both *Alu*I and *Sac*I sites included to ensure maximum removal of linker sequences from driver. Incubation with *Alu*I first and then *Sac*I was used to avoid incorporation of driver DNA into the library, because the *Sac*I site was used for library construction.

How to achieve an enrichment for both abundant and rare target genes is always an issue for the methods of cloning differentially expressed genes. For LCS, we think that the hybridization time might be the determining factor. Kinetically, short hybridization times favor enrichment of more abundant sequences, while longer times allow rare sequences to bind. As long as enough time is given for hybridization, rare target sequences can remain in the reaction. Indeed, moderate to rare genes are included in the genes we identified (data not shown). Moreover, as noted by Hubank and Schatz (6), unwanted dominant sequences (i.e., already identified sequences in the reaction) can be driven out by supplementing driver with unlinkered corresponding sequences. This would allow less dominant species to be isolated. One can also try to normalize tester and driver before subtraction. We suggest a procedure based on reassociation kinetics. First, both tester and driver are denatured and hybridized for a short time (e.g., one hour). Then, the linkers of hybrids (presumably more abundant sequences) are removed by restriction enzymes. Finally, the remaining single-stranded fraction of DNA is amplified by PCR. Proper sampling of target genes that differ in abundance by only a few fold is another issue. By adjustment of the tester–driver ratios and by more cycles of subtraction, these genes should be isolated.

We suggest that LCS will be generally applicable to experiments such as those reported here as well as to studies of differential gene expression in cells incubated in the absence or presence of cytokines, growth factors or other biologically active molecules. Moreover, the method should also prove useful in finding differences among genomic DNAs. Obviously, this method will not detect important genes critical for biological events whose mRNAs are not expressed differentially. One should also not expect that in one experiment, this method will provide the entire array of differentially expressed genes between cell types, although by modification of the conditions described above, it might be possible to achieve this goal.

ACKNOWLEDGMENTS

This work was supported by NIH Grant No. NRSA 1 F32 DK 09364 to M.Y., and US Navy Grant No. N00014-93-1-0776 and NIH Grant No. DK38841 to A.J.S.

REFERENCES

1. **Balzer, H.J. and H. Baumlein.** 1994. An improved gene expression screen. Nucleic Acids Res. *14*:2853-2854.
2. **Blok, L.J., M.V. Kumar and D.J. Tindall.** 1995. Isolation of cDNAs that are differentially expressed between androgen-dependent and androgen-independent prostate carcinoma cells using differential display PCR. Prostate *26*:213-224.
3. **Brown, S.E. and D.L. Knudson.** 1991. 96-well plates for recombinant library maintenance and dot-blot production. BioTechniques *10*:719-722.
4. **Davis, R.L., H. Weintraub and A. Lassar.** 1987. Expression of a single transfected cDNA converts fibroblasts to myoblasts. Cell *51*:987-1000.
5. **Hedrick, S.M., D.I. Cohen, E.A. Nielsen and M.M. Davis.** 1984. Isolation of cDNA clones encoding T cell-specific membrane-associated proteins. Nature *308*:149-153.
6. **Hubank, M. and D.G. Schatz.** 1994. Identifying differences in mRNA expression by representational difference analysis of cDNA. Nucleic Acids Res. *22*:5640-5648.
7. **Liang, P. and A.B. Pardee.** 1992. Differential display of eukaryotic messenger RNA by means of the polymerase chain reaction. Science *257*:967-971.
8. **Lisitsyn, N., N. Lisitsyn and M. Wigler.** 1993. Cloning the differences between two complex genomes. Science *259*:946-951.
9. **Mather, E.L., F.W. Alt, A.L.M. Bothwell, D. Baltimore and M.E. Koshland.** 1981. Expression of J chain RNA in cell lines representing different stages of B lymphocyte differentiation. Cell *23*:369-378.
10. **Sive, H.L. and T.S. John.** 1988. A simple subtractive hybridization technique employing photoactivatable biotin and phenol extraction. Nucleic Acids Res. *16*:10937.
11. **Straus, D. and F.M. Ausubel.** 1990. Genomic subtraction for cloning DNA corresponding to deletion mutations. Proc. Natl. Acad. Sci. USA *87*:1889-1893.
12. **Wang, Z. and D.D. Brown.** 1991. A gene expression screen. Proc. Natl. Acad. Sci. USA *88*:11505-11509.
13. **Welsh, J., K. Chada, S.S. Dalal, R. Cheng, D. Ralph and M. McColelland.** 1992. Arbitrarily primed PCR fingerprinting of RNA. Nucleic Acids Res. *20*:4965-4970.
14. **Wieland, I., G. Bolger, G. Asouline and M. Wigler.** 1990. A method for difference cloning: gene amplification following subtractive hybridization. Proc. Natl. Acad. Sci. USA *87*:2720-2724.
15. **Winship, P.R.** 1989. An improved method for directly sequencing PCR amplified material using dimethyl sulphoxide. Nucleic Acids Res. *17*:1266.
16. **Xie, W.Q. and L.I. Rothblum.** 1991. Rapid, small-scale RNA isolation from tissue culture cells. BioTechniques *11*:325-327.
17. **Yang, M. and A.J. Sytkowski.** 1996. Cloning differentially expressed genes by linker capture subtraction. Anal. Biochem. *237*:109-114.
18. **Zeng, J., R.A. Gorski and D. Hamer.** 1994. Differential cDNA cloning by enzymatic degrading subtraction (EDS). Nucleic Acids Res. *22*:4381-4385.

17. Suppression Subtractive Hybridization: A Method for Generating Subtracted cDNA Libraries Starting from Poly(A+) or Total RNA

Luda Diatchenko, Alex Chenchik and Paul D. Siebert
Gene Cloning and Analysis Group, CLONTECH Laboratories, Palo Alto, CA, USA

OVERVIEW

Suppression subtractive hybridization (SSH) is a new and highly effective method for generating subtracted cDNA libraries. Based primarily on a recently described technique called suppression polymerase chain reaction (PCR), SSH combines normalization and subtraction in a single procedure. The normalization step equalizes the abundance of cDNAs within the target population, and the subtraction step excludes sequences that are common to the target and driver populations. Consequently, only one round of subtractive hybridization is needed, and the subtracted library is normalized in the abundance of different cDNAs. The SSH method dramatically increases the probability of obtaining low-abundance differentially expressed cDNAs and simplifies analysis of the subtracted library. The SSH technique has many applications in molecular genetic and positional cloning studies, including the identification of disease, developmental, tissue-specific or other differentially expressed genes. This chapter provides a detailed protocol for generating subtracted cDNA from total and poly(A+) RNA and includes instructions for differential screening of the subtracted cDNA library.

BACKGROUND

Subtractive hybridization methods are valuable molecular tools for identifying differentially regulated genes important for cellular growth and differentiation. In the past decade, numerous subtractive hybridization techniques have been developed and used to isolate significant genes in many systems (7–9,15,18). However, many of these techniques require either tedious and complicated procedures or large amounts of starting material, thereby reducing their overall efficiencies. Hence, these tech-

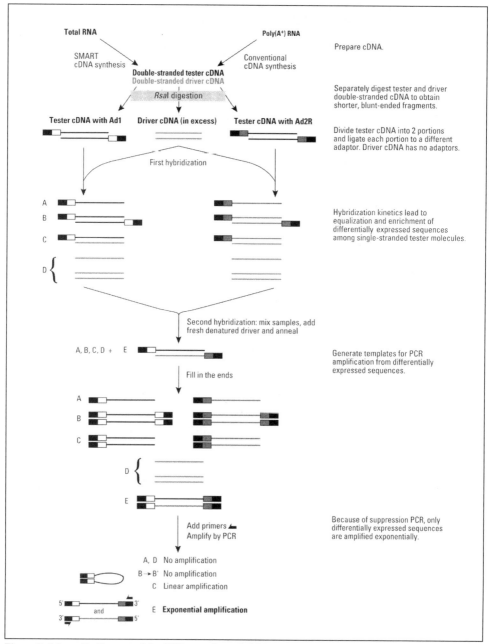

Figure 1. Schematic diagram of the SSH procedure. Solid boxes represent the outer part of the Ad1 and Ad2R longer strands and corresponding P1 sequence. Clear boxes represent the inner part of the Ad1 longer strand and corresponding NP1. Shaded boxes represent the inner part of the Ad2R longer strand and corresponding NP2R. Note that after filling in the recessed 3′ ends with DNA polymerase, type A, B and C molecules having Ad2R are also present but are not shown.

niques can be greatly improved if the procedures can be simplified and if they require less starting material. We have developed a new PCR-based cDNA subtraction method, SSH, and have demonstrated its effectiveness in isolating tissue-specific and differentially regulated cDNAs (4,6,10).

The central feature of the SSH technique is the suppression PCR effect (12,16). By incorporating this suppression effect in a PCR amplification scheme, the SSH method normalizes sequence abundance within the target cDNA population, eliminates any intermediate steps for physical separation of single-stranded and double-stranded cDNAs, requires only one round of subtractive hybridization and can achieve greater than 1000-fold enrichment for differentially expressed cDNAs.

Usually, 2–4 μg of poly(A+) RNA are needed for a comprehensive SSH experiment. However, such amounts of poly(A+) RNAs can sometimes be difficult to obtain. To circumvent this problem, a preliminary amplification step can be incorporated to generate sufficient quantities of cDNA before initiating the SSH procedure (6). Moreover, total RNA can be used successfully as starting material for preamplification of the cDNA.

In this chapter, we describe an improved version of the SSH technique for generating subtracted cDNA from either poly(A+) or total RNA. A detailed protocol for preamplification of cDNA is provided in Chenchik et al. (in this volume). We also describe a differential screening procedure for rapid isolation of truly differentially expressed cDNA clones from the subtracted library, and we discuss the use of "virtual" Northern blots for the verification of differential gene expression.

THE PRINCIPLE OF SSH

An overview of the SSH procedure is shown in Figure 1. First, cDNA is synthesized from the two types of tissues or cell types being compared. The cDNA in which specific transcripts are to be found is called tester cDNA, and the reference cDNA is called driver cDNA. If the researcher has 2 μg of poly(A+) RNA, the conventional method of Gubler and Hoffman (5) can be used for cDNA synthesis. If less RNA is available, the switch mechanism at the 5′ end of RNA templates (SMART™) amplification technology (CLONTECH Laboratories, Palo Alto, CA, USA) can be used to preamplify high-quality cDNA from total RNA (see Chenchik et al. in this volume). In the second step, the tester and driver cDNAs are digested with a four-base-cutting restriction enzyme that yields blunt ends. The tester cDNA is then subdivided into two portions, and each is ligated to a different double-stranded adaptor (adaptors 1 and 2R [Ad1 and Ad2R]). The ends of the adaptors lack a phosphate group, so only one strand of each adaptor covalently attaches to the 5′ ends of the cDNAs.

In the first hybridization, an excess of driver cDNA is added to each sample of tester cDNA. The samples are then heat-denatured and allowed to anneal. Figure 1 shows the type A, B, C and D molecules generated in each sample. Type A molecules include equal concentrations of high- and low-abundance sequences because reannealing is faster for the more abundant molecules because of the second-order kinetics of hybridization (1). At the same time, type A molecules are significantly enriched for differentially expressed sequences because common nontarget cDNAs form type C molecules with the driver. During the second hybridization, the two primary hybridization samples are mixed together. The type A cDNAs from each tester sample are now able to associate and form type B, type C and new type E hybrids.

Type E hybrids are double-stranded tester molecules with different single-stranded ends that correspond to Ad1 and Ad2R. Freshly denatured driver cDNA is added to further enrich fraction E for differentially expressed sequences.

The entire population of molecules is then subjected to two rounds of PCR to selectively amplify the differentially expressed sequences. During the first cycle of primary PCR, the adaptor ends are filled in, creating the complementary primer binding sites needed for amplification. Type A and D molecules are missing primer annealing sites and cannot be amplified. Type B molecules form a pan-like structure that prevents their exponential amplification (12,16). Type C molecules have one primer annealing site and can only be amplified linearly. Only type E molecules, which have two different primer annealing sites, can be amplified exponentially, and these represent the differentially expressed sequences.

The mathematical model for the formation of fraction E and the rate of enrichment has been described elsewhere (6).

PROTOCOLS

Materials and Reagents

Some modifications in the procedure and adaptor sequences have been incorporated since the original SSH technique was published (4). These modifications improve the overall efficiency of the subtractive hybridization and are described here.

Oligonucleotides

The following oligonucleotides are used at a concentration of 10 µM. Whenever possible, oligonucleotides should be gel-purified.
- cDNA synthesis primer: 5′-TTTTGTACAAGCTT$_{30}$-3′
- Ad1: 5′-CTAATACGACTCACTATAGGGCTCGAGCGGCCGCCCGGGCAGGT-3′
  3′-GGCCCGTCCA-5′
- Ad2R: 5′-CTAATACGACTCACTATAGGGCAGCGTGGTCGCGGCCGAGGT-3′
  3′-GCCGGCTCCA-5′
- Primer 1 (P1): 5′-CTAATACGACTCACTATAGGGC-3′
- Nested primer 1 (NP1): 5′-TCGAGCGGCCGCCCGGGCAGGT-3′
- Nested primer 2R (NP2R): 5′-AGCGTGGTCGCGGCCGAGGT-3′
- *GAPD* 5′ primer: 5′-ACCACAGTCCATGCCATCAC-3′
- *GAPD* 3′ primer: 5′-TCCACCACCCTGTTGCTGTA-3′
- Blocking solution: a mixture of the cDNA synthesis primer, nested primers (NP1 and NP2R) and their respective complementary oligonucleotides (2 mg/mL each)

Buffers and Enzymes

All chemical reagents were obtained from Sigma Chemical (St. Louis, MO, USA).

First-strand synthesis

- SUPERSCRIPT™ II Reverse Transcriptase (200 U/µL; Life Technologies, Gaithersburg, MD, USA)

- 5× first-strand buffer: 250 mM Tris-HCl, pH 8.3, 30 mM MgCl$_2$, 375 mM KCl

Second-strand synthesis

- 20× second-strand enzyme cocktail: DNA polymerase I (6 U/mL; New England Biolabs, Beverly, MA, USA), RNase H (0.2 U/mL; Epicentre Technologies, Madison, WI, USA), *E. coli* DNA ligase (1.2 U/µL; New England Biolabs)
- 5× second-strand buffer: 500 mM KCl, 50 mM ammonium sulfate, 25 mM MgCl$_2$, 0.75 mM β nicotinamide-adenine dinucleotide (β-NAD), 100 mM Tris-HCl, pH 7.5, 0.25 mg/mL bovine serum albumin (BSA)
- T4 DNA polymerase (3 U/µL; New England Biolabs)

Endonuclease digestion

- 10× *Rsa*I restriction buffer: 100 mM Tris-HCl, pH 7.0, 100 mM MgCl$_2$, 10 mM dithiothreitol (DTT)
- *Rsa*I (10 U/mL; New England Biolabs)

Ligation

- T4 DNA ligase (400 U/µL; New England Biolabs)
- 5× DNA ligation buffer: 250 mM Tris-HCl, pH 7.8, 50 mM MgCl$_2$, 10 mM DTT, 0.25 mg/mL BSA
- 3 mM ATP

Hybridization

- 4× hybridization buffer: 4 M NaCl, 200 mM HEPES, pH 8.3, 4 mM cetyltrimethylammonium bromide (CTAB)
- Dilution buffer: 20 mM HEPES-HCl, pH 8.3, 50 mM NaCl, 0.2 mM EDTA

PCR amplification

- Advantage® cDNA PCR Mix (CLONTECH): KlenTaq-1 DNA Polymerase, Deep Vent™ DNA Polymerase (New England Biolabs), TaqStart™ Antibody
- 10× PCR buffer: 40 mM Tricine-KOH (pH 9.2 at 22°C), 3.5 mM Mg(OAc)$_2$, 10 mM KOAc, 75 mg/mL BSA

Note: *Taq* DNA polymerase alone can be used, but five additional thermal cycles will be needed in both the primary and secondary PCR, and the additional cycles can cause higher background. The TaqStart antibody provides automatic hot-start PCR (11). If the Advantage cDNA PCR Mix is not used, manual hot start or hot start with wax beads is strongly recommended to reduce nonspecific DNA synthesis.

General reagents

- dNTP mixture (10 mM each dNTP) (Amersham Pharmacia Biotech, Piscataway, NJ, USA)

- TNE buffer: 10 mM Tris-HCl, pH 8.0, 10 mM NaCl, 0.1 mM EDTA
- ExpressHyb™ Hybridization Solution (CLONTECH)

Purification reagents

- CHROMA SPIN™-1000 Columns (CLONTECH)
- Advantage PCR-Pure Kit (CLONTECH)
- Microfiltration columns (0.45 µm; PGC Scientific, Gaithersburg, MD, USA)

Note: All cycling parameters were optimized using a DNA Thermal Cycler 480 (Perkin-Elmer, Norwalk, CT, USA). Perkin-Elmer GeneAmp® PCR Systems 2400 and 9600 have also been tested. For the latter, the denaturation time needs to be reduced from 30 to 10 s. For other types of thermal cycler, the cycling parameters must be optimized.

Preparation of the Subtracted cDNA Library from Poly(A+) RNA

The following protocol is recommended for generating a subtracted library from 2 µg of poly(A+) RNA. If not enough poly(A+) RNA is available, the cDNA can be preamplified using the SMART amplification technology. After amplification of cDNA, follow the purification procedure starting from the "Preparation of the Subtracted cDNA Library from Total RNA" section.

First-Strand cDNA Synthesis Procedure

Perform this procedure individually with each tester and driver poly(A+) RNA sample.

1. For each tester and driver sample, combine the following components in a sterile 0.5-mL microcentrifuge tube (do not use a polystyrene tube):
   poly(A+) RNA (2 µg) to 2–4 µL
   cDNA synthesis primer (10 µM) to 1 µL
   if needed, add sterile H$_2$O to a final volume of 5 µL.
2. Incubate the tubes at 70°C in a thermal cycler for 2 min.
3. Cool at room temperature for 2 min and briefly centrifuge using a PicoFuge® microcentrifuge (Stratagene, La Jolla, CA, USA) at maximum rotation speed (6400 rpm).
4. Add the following to each reaction tube:
   5× first-strand buffer to 2 µL
   dNTP mixture (10 mM each) to 1 µL
   sterile H$_2$O* to 0.5 µL
   0.1 M DDT to 0.5 µL
   SUPERSCRIPT II Reverse Transcriptase (200 U/µL) to 1 µL.

   *Optional: To monitor the progress of cDNA synthesis, dilute 0.5 µL of [α-^{32}P]dCTP (10 mCi/mL, 3000 Ci/mmol) with 9 µL of H$_2$O and replace the H$_2$O above with 1 µL of the diluted label.
5. Incubate at 42°C for 1.5 h in an air incubator.
6. Place the tubes on ice to terminate first-strand cDNA synthesis and immediately proceed to second-strand cDNA synthesis.

Second-Strand cDNA Synthesis Procedure

1. Add the following components, previously cooled on ice, to the first-strand synthesis reaction tubes:
   sterile H$_2$O to 48.4 µL
   5× second-strand buffer to 16.0 µL
   dNTP mixture (10 mM each) to 1.6 µL
   20× second-strand enzyme cocktail to 4.0 µL.
2. Mix contents and briefly spin the tubes. The final volume should be 80 µL.
3. Incubate the tubes at 16°C (water bath or thermal cycler) for 2 h.
4. Add 2 µL (6 U) of T4 DNA polymerase. Mix contents well.
5. Incubate at 16°C for 30 min in a water bath or a thermal cycler.
6. Add 4 µL of 0.2 M EDTA to terminate second-strand synthesis.
7. Perform phenol–chloroform extraction and ethanol precipitation. We recommend using 0.5 vol of 4 M NH$_4$OAc rather then NaOAc for ethanol precipitation.
8. Dissolve each pellet in 50 µL of H$_2$O.
9. Transfer 6 µL of each reaction mixture to a fresh microcentrifuge tube. Store these samples at -20°C until after *Rsa*I digestion for agarose gel electrophoresis to estimate yield and size range of the double-stranded cDNA products synthesized.

*Rsa*I Digestion Procedure

Perform the following procedure with each experimental double-stranded tester and driver cDNA. This step generates shorter, blunt-ended double-stranded cDNA fragments optimal for subtractive hybridization.

1. Add the following reagents into each tube from step 8 above:
   double-stranded cDNA to 43.5 µL
   10× *Rsa*I restriction buffer to 5.0 µL
   *Rsa*I (10 U/µL) to 1.5 µL.
2. Mix and incubate at 37°C for 2 h.
3. Analyze 5 µL of each digestion mixture on a 2% agarose gel along with the undigested cDNA from step 9 of the second-strand synthesis to analyze the efficiency of *Rsa*I digestion.
4. Add 2.5 µL of 0.2 M EDTA to terminate each reaction.
5. Perform phenol–chloroform extraction and ethanol precipitation.
6. Dissolve each pellet in 5.5 µL of H$_2$O and store at -20°C.

Adaptor Ligation Procedure

We strongly recommend performing subtractions in both directions for each tester–driver cDNA pair. Forward subtraction is designed to enrich for differentially expressed transcripts present in poly(A$^+$) RNA sample 1 but not poly(A$^+$) RNA sample 2. Reverse subtraction is designed to enrich for differentially expressed transcripts present in poly(A+) RNA sample 2 but not in poly(A+) RNA sample 1. The availability of such forward- and reverse-subtracted cDNA will be useful for differential screening of the resulting subtracted tester cDNA library. In forward subtraction, RNA sample 1 functions as a tester (tester 1), and RNA sample 2 functions as a driver (driver 1). In the reverse subtraction, RNA sample 1 functions a driver (driver 2), and RNA sample 2 functions a tester (tester 2).

Three adaptor ligations must be performed for each experimental cDNA that will function as a tester in the subtractive hybridization procedure. Each cDNA is separated into two tubes: one aliquot is ligated with Ad1, and the second is ligated with Ad2R. Tester cDNA with ligated Ad1 will be called tester 1-1, tester 2-1 etc.; tester cDNA with ligated Ad2R will be called tester 1-2, tester 2-2 etc. After the ligation reactions are set up, a portion of each corresponding tester ligation reaction (e.g., tester 1-1 and tester 1-2) is combined so that the cDNA is ligated with both adaptors. This reaction serves as a negative control for subtraction (unsubtracted tester controls 1-c and 2-c).

1. Dilute 1 µL of each *Rsa*I-digested tester cDNA from the above section with 5 µL of sterile H_2O.
2. Prepare a master ligation mixture according to the following quantities for each reaction:
   sterile H_2O to 2 µL
   5× ligation buffer to 2 µL
   3 mM ATP to 1 µL
   T4 DNA ligase (400 U/µL) to 1 µL.
3. For each tester cDNA mixture, combine the reagents in a 0.5-mL microcentrifuge tube in the order shown. Pipet the solution up and down to mix thoroughly.

	Tube 1 Tester 1-1	Tube 2 Tester 1-2
Diluted tester cDNA	2 µL	2 µL
Ad1 (10 µM)	2 µL	0 µL
Ad2R (10 µM)	0 µL	2 µL
Master ligation mixture	6 µL	6 µL
Final volume	10 µL	10 µL

4. In a fresh microcentrifuge tube, mix 2 µL of tester 1-1 and 2 µL of tester 1-2. This is the unsubtracted tester control 1-c. Do the same for each tester cDNA sample. After ligation, approximately 1/3 of the cDNA molecules in each unsubtracted tester control tube will have two different adaptors on their ends and are therefore suitable for exponential PCR amplification using adaptor-derived primers.
5. Centrifuge the tubes briefly and incubate at 16°C overnight.
6. Stop ligation reactions by adding 1 µL of 0.2 M EDTA.
7. Heat samples at 72°C for 5 min to inactivate the ligase.
8. Briefly centrifuge the tubes. Remove 1 µL from each unsubtracted tester control (1-c, 2-c etc.) and dilute into 1 mL of H_2O. These samples will be used for PCR amplification. Preparation of the experimental adaptor-ligated tester cDNAs is now complete.

Ligation Efficiency Test

The following PCR experiment is recommended to verify that at least 25% of the cDNAs have adaptors on both ends. This experiment is designed to amplify fragments that span the adaptor–cDNA junctions of testers 1-1 and 1-2 using the adaptor-specific P1 primer and a gene-specific primer. On a gel, PCR products generated using one gene-specific primer and adaptor-specific primer should be about the same intensity as the PCR products amplified using two gene-specific primers, as shown in Figure 2 for the glyceraldehyde-3-phosphate dehydrogenase (*GAPD*) gene. It is important that the amplified gene-specific fragment has no *Rsa*I restriction site. The selected *GAPD* primers listed in the Materials and Reagents section work well for human, mouse and rat cDNA samples.

1. Dilute 1 µL of each ligated cDNA (e.g., testers 1-1 and 1-2) into 200 µL of H$_2$O.
2. Combine the reagents in four separate tubes as follows.

Tube:	1	2	3	4
Tester 1-1 (ligated to Ad1)	1 µL	1 µL	0 µL	0 µL
Tester 1-2 (ligated to Ad2R)	0 µL	0 µL	1 µL	1 µL
GAPD 3′ primer (10 µM)	1 µL	1 µL	1 µL	1 µL
GAPD 5′ primer (10 µM)	0 µL	1 µL	0 µL	1 µL
P1 (10 µM)*	1 µL	0 µL	1 µL	0 µL
Total volume	3 µL	3 µL	3 µL	3 µL

*P1 contains 22 nucleotides corresponding to the 5′-end sequence of both Ad1 and Ad2R.

3. Prepare a master mixture for all of the reaction tubes plus one additional tube. For each reaction, combine the reagents in the following order.

	Amount per reaction tube (µL)	Amount for 4 reactions (µL)
Sterile H$_2$O	18.5	92.5
10× PCR buffer	2.5	12.5
dNTP mixture (10 mM)	0.5	2.5
50× Advantage cDNA Polymerase Mix	0.5	2.5
Total volume	22.0	110.0

4. Mix thoroughly and briefly centrifuge the tubes.
5. Separate 22-µL aliquots of master mixture into each reaction tube from step 2.
6. Overlay with 50 µL of mineral oil. Skip this step if an oil-free thermal cycler is used.
7. Commence 20 cycles of denaturation at 94°C for 30 s, annealing at 65°C for 30 s and extension at 68°C for 2.5 min.
8. Examine the products by electrophoresis on a 2% agarose gel.

Typical results are shown in Figure 2. If no products are visible after 20 cycles, perform 5 more cycles of amplification and again analyze the product by gel electrophoresis. The number of cycles will depend on the abundance of the specific gene. The efficiency of ligation is estimated to be the ratio of the intensities of the bands corresponding to the PCR products from tube 2 to tube 1 for Ad1 and from tube 4 to tube 3 for Ad2R. Any ligation efficiency of 25% or less will substantially reduce the subsequent subtraction efficiency. In this case, the ligation reaction should be repeated with fresh samples before proceeding to the next step.

For mouse or rat cDNAs, the PCR products amplified with the *GAPD* 3′ primer and P1 will be approximately 1.2 kb instead of the 0.75-kb band observed for human cDNA (because rat and mouse *GAPD* cDNAs lack the *Rsa*I restriction site in the nucleotide [nt] 340 position). However, for the human cDNA (which contains the *Rsa*I site), the presence of a 1.2-kb band suggests that the cDNAs are not completely digested by *Rsa*I. If a significant amount of this longer PCR product persists, the *Rsa*I digestion should be repeated.

First Hybridization

1. For each tester sample, combine the reagents in the following order.

	Hybridization tube 1-1	Hybridization tube 1-2
*Rsa*I-digested driver (after *Rsa*I digestion step)	1.5 µL	1.5 µL
Ad1-ligated tester 1-1 (after ligation step)	1.5 µL	0 µL
Ad2R-ligated tester 1-2 (after ligation step)	0 µL	1.5 µL
4× hybridization buffer	1.0 µL	1.0 µL
Final volume	4.0 µL	4.0 µL

2. Overlay samples with one drop of mineral oil and centrifuge briefly.
3. Incubate samples in a thermal cycler at 98°C for 1.5 min.
4. Incubate samples at 68°C for 7–12 h and then proceed immediately to the second hybridization.

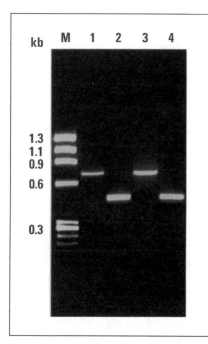

Figure 2. Typical results of the ligation efficiency analysis. The results shown here are for human samples; if mouse or rat samples are used, the PCR product amplified using the *GAPD* 3′ primer and P1 is ca. 1.2 instead of 0.75 kb. Lane 1: PCR products using tester 1-1 (Ad1-ligated) as the template and the *GAPD* 3′ primer and P1. Lane 2: PCR products using tester 1-1 (Ad1-ligated) as the template and the *GAPD* 3′ and 5′ primers. Lane 3: PCR products using tester 1-2 (Ad2R-ligated) as the template and the *GAPD* 3′ primer and P1. Lane 4: PCR products using tester 1-2 (Ad2R-ligated) as the template and the *GAPD* 3′ and 5′ primers. Shown is an EtdBr-stained 2% agarose gel. Lane M: fX174 DNA/*Hae*III digest size markers.

Second Hybridization

Repeat the following steps for each experimental driver cDNA.
1. Add the following reagents into a sterile 0.5-mL microcentrifuge tube:
   driver cDNA to 1 µL
   4× hybridization buffer to 1 µL.
2. Overlay it with one drop of mineral oil.
3. Incubate in a thermal cycler at 98°C for 1.5 min.
4. Remove the tube of freshly denatured driver from the thermal cycler.
5. To the tube of freshly denatured driver cDNA, add hybridized material from hybridization tubes 1-1 and 1-2 consecutively. This ensures that the two hybridization samples are mixed only in the presence of excess driver cDNA.
6. Incubate the hybridization reaction at 68°C overnight.
7. Add 200 µL of dilution buffer to the tube and mix well by pipetting.
8. Incubate in a thermal cycler at 68°C for 7 min.

PCR Amplification Procedure

Each experiment should include at least four reactions: *(i)* subtracted tester cDNAs, *(ii)* unsubtracted tester control (1-c), *(iii)* reverse-subtracted tester cDNAs and *(iv)* unsubtracted driver control for the reverse subtraction (2-c).
1. Place 1 µL of each diluted cDNA (i.e., each subtracted sample from the second hybridization and the corresponding diluted unsubtracted tester control from the adaptor ligation) into an appropriately labeled tube.
2. Prepare a master mixture for all of the primary PCR tubes plus one additional tube. For each reaction, combine the reagents in the order shown.

	Amount per reaction (µL)
Sterile H$_2$O	19.5
10× PCR buffer	2.5
dNTP mixture (10 mM)	0.5
P1 (10 µM)	1.0
50× Advantage cDNA Polymerase Mix	0.5
Total volume	24.0

3. Separate into aliquots 24 µL of master mixture into each reaction tube prepared in step 1 above.
4. Overlay with 50 µL of mineral oil. Skip this step if an oil-free thermal cycler is used.
5. Incubate the reaction mixture in a thermal cycler at 75°C for 5 min to extend the adaptors. Do not remove the samples from the thermal cycler.
6. Immediately commence 27 cycles of denaturation at 94°C for 30 s, annealing at 66°C for 30 s and extension at 72°C for 1.5 min.
7. Analyze 8 µL from each tube on a 2% agarose gel.

Note: The sequence for P1 is present at the 5′ ends of both Ad1 and Ad2R and hence can be used in a single-primer PCR amplification. If no PCR product is observed after 27 cycles, amplify for three more cycles and analyze the products again by gel electrophoresis. If no PCR product is observed in the subtracted or unsubtracted (unsubtracted tester control 1-c) samples, the activity of the *Taq* DNA polymerase needs to be examined. If the problem is not with the polymerase mixture, optimize the PCR cycling parameters by decreasing the annealing and extension temperatures in small increments. Lowering the temperature by only 1°C can dramatically increase the background. Initially, try reducing the annealing temperature from 66° to 65°C and the extension temperature from 72° to 71°C. If PCR products are observed in the unsubtracted (unsubtracted tester control 1-c) samples but not in the subtracted sample, proceed to step 8 and perform more cycles of secondary PCR in step 13.

8. Dilute 3 µL of each primary PCR mixture in 27 µL of H_2O.
9. Place 1 µL of each diluted primary PCR product mixture from step 8 into an appropriately labeled tube.
10. Prepare a master mixture for the secondary PCR samples plus one additional reaction by combining the reagents in the following order.

	Amount per reaction (µL)
Sterile H_2O	18.5
10× PCR buffer	2.5
NP1 (10 µM)	1.0
NP2R (10 µM)	1.0
dNTP mixture (10 mM)	0.5
50× Advantage cDNA Polymerase Mix	0.5
Total volume	24.0

11. Separate into aliquots 24 µL of master mixture into each reaction tube from step 9.
12. Overlay with one drop of mineral oil. Skip this step if an oil-free thermal cycler is used.
13. Immediately commence 10–12 cycles of denaturation at 94°C for 30 s, annealing at 68°C for 30 s and extension at 72°C for 1.5 min.
14. Analyze 8 µL from each reaction mixture on a 2% agarose gel.

Note: The patterns of secondary PCR products of the subtracted samples usually look like smears with or without a number of distinct bands. If no product is observed after 12 cycles, perform 3 more cycles of amplification and again check the products by gel electrophoresis. Add cycles sparingly—too many cycles will increase background.

PCR Analysis of Subtraction Efficiency

At this point, it is important to determine the efficiency of the SSH procedure by comparing the abundance of known cDNAs before and after subtraction. Ideally, one should use a nondifferentially expressed gene (e.g., a housekeeping gene) and a gene previously demonstrated to be differentially expressed between the two RNA sources. These comparisons can be performed using either PCR or hybridization techniques.

The test described below uses the *GAPD* primers to confirm the reduced relative abundance of *GAPD* following the SSH procedure on subtracted samples.

1. Dilute the subtracted and unsubtracted (unsubtracted tester controls 1-c and 2-c) secondary PCR products 10-fold in H_2O.
2. For each sample, combine the following reagents in a 0.5-mL microcentrifuge tube in the order shown.

	Tube 1	Tube 2
Diluted subtracted cDNA (2nd PCR product)	1.0 µL	–
Diluted unsubtracted tester control 1-c (2nd PCR product)	–	1.0 µL
GAPD 5′ primer (10 µM)	1.2 µL	1.2 µL
GAPD 3′ primer (10 µM)	1.2 µL	1.2 µL
Sterile H_2O	22.4 µL	22.4 µL
10× PCR buffer	3.0 µL	3.0 µL
dNTP mixture (10 mM)	0.6 µL	0.6 µL
50× Advantage cDNA Polymerase Mix	0.6 µL	0.6 µL
Total volume	30.0 µL	30.0 µL

3. Mix and briefly centrifuge the tubes.
4. Overlay with one drop of mineral oil. Skip this step if an oil-free thermal cycler is used.
5. Immediately commence 18 cycles of denaturation at 94°C for 30 s, annealing at 60°C for 30 s and extension at 68°C for 2 min.
6. Remove 5 µL from each reaction mixture into a clean tube and store on ice. Return the rest of the reaction mixture to the thermal cycler for 5 more cycles.
7. Repeat step 6 twice (i.e., remove 5 µL after 28 and 33 cycles).
8. Examine the 5-µL samples (i.e., the aliquots that were removed from each reaction after 18, 23, 28 and 33 cycles) on a 2% agarose gel.

Figure 3 shows an example of *GAPD* reduction in a testis-specific subtracted

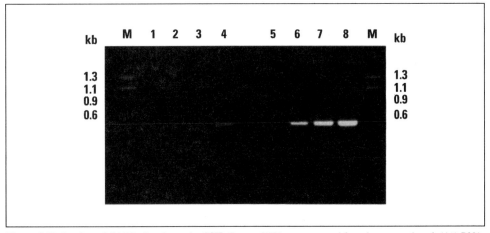

Figure 3. Reduction of *GAPD* abundance by SSH. Tester cDNA was prepared from human testis poly(A+) RNA, and driver cDNA was prepared from a mixture of poly(A+) RNA samples prepared from 10 different human tissues. PCR was performed on subtracted (lanes 1–4) and unsubtracted (lanes 5–8) secondary PCR products using the *GAPD* 5′ and 3′ primers. Lanes 1 and 5: 18 cycles; lanes 2 and 6: 23 cycles; lanes 3 and 7: 28 cycles; lanes 4 and 8: 33 cycles. Lane M: fX174 DNA/*Hae*III digest size markers.

mixture. In general, for the unsubtracted cDNA, a GAPD product is observed after 18–23 cycles, depending on the abundance of *GAPD* in the particular cDNA. For example, in skeletal muscle and heart poly(A+) RNA, *GAPD* is extremely abundant. However, in the subtracted samples, a product should be observed about 5–15 cycles later than it is seen in the unsubtracted samples.

As a positive control for the enrichment of differentially expressed genes, repeat the PCR procedure above using PCR primers for a gene known to be expressed in the tester poly(A+) RNA but not in the driver poly(A+) RNA. This cDNA should be enriched by the subtraction operation. Do not use PCR primers that amplify a cDNA fragment that contains an *Rsa*I restriction site between the PCR priming sites because it will not be amplified (because of *Rsa*I digestion prior to the subtraction procedure).

Generation of a Subtracted cDNA Library

Once a subtracted sample has been confirmed to be enriched for cDNAs derived from differentially expressed genes, the PCR products can be subcloned using several conventional cloning techniques. The following describes two such methods currently used in our laboratories.

The first is TA cloning. Use 3 µL of the secondary PCR product for a TA-based cloning system, such as the AdvanTAge™ PCR Cloning Kit (CLONTECH) according to the manufacturer's protocol. Typically, 10^5 independent clones from 1 µL of secondary PCR product can be obtained using the above cloning kit and electroporation. It is important to optimize the cloning efficiency because a low cloning efficiency will result in a high background of colonies that are not derived from subtracted cDNA.

The second is site-specific or blunt-end cloning. For site-specific cloning, cleave at the *Eag*I, *Not*I and *Xma*I (*Sma*I, *Srf*I) sites embedded in the adaptor sequences and then ligate the products to an appropriate plasmid vector. Keep in mind that all of these sites might be present in the cDNA fragments. For blunt-end cloning, cleave at the *Rsa*I site at the adaptor–cDNA junction. For either method, use a commercially available cloning kit or consult Reference 14.

Once it has been generated, the subtracted cDNA library can be arrayed into 96- or 384-well microplates for storage according to established protocols (14). The number of independent colonies obtained for each library depends on the estimated number of differentially expressed genes as well as the subtraction and subcloning efficiencies. In general, 2000–5000 colonies can be initially arrayed and studied. The complexity of the library can be increased by additional subcloning of the secondary PCR products.

Preparation of the Subtracted cDNA Library from Total RNA

If 2 µg of poly(A+) RNA cannot be obtained for both tester and driver, cDNA can be preamplified from total RNA using the SMART method (see Chenchik et al., in this volume). In our experience, each 100-µL SMART reaction typically yields 1–3 µg of double-stranded cDNA after PCR and the subsequent purification. Subtraction usually requires 2 µg of driver cDNA, so two tubes of SMART cDNA should be sufficient.

Column Chromatography Procedure

This section describes postamplification purification steps that should be performed for SMART PCR-amplified cDNA before starting the SSH subtraction. Follow this procedure for both the tester and driver samples.

1. Transfer 7 µL of SMART PCR product to a clean microcentrifuge tube and store at -20°C. This sample will be used for analysis of column chromatography (step 10).
2. Combine two reaction tubes of each PCR product into a 1.5-mL microcentrifuge tube and add an equal volume of phenol–choloroform–isoamyl alcohol (25:24:1). Vortex mix thoroughly.
3. Centrifuge the tubes at 12 000× g for 10 min to separate the phases.
4. Remove the top (aqueous) layer and place it in a clean 1.5-mL tube.
5. Add 700 µL of n-butanol and vortex mix the solution thoroughly. Butanol extraction allows one to concentrate the PCR product to a volume of 40–70 µL.

   Note: Addition of too much n-butanol might remove all the water and precipitate the nucleic acid. If this happens, add water to the tube and vortex mix until an aqueous phase reappears.
6. Centrifuge the solution at room temperature at 12 000× g for 1 min.
7. Remove and discard the upper (n-butanol organic) phase.
8. If the volume does not equal 40–70 µL, repeat steps 5–7 with the same volume of n-butanol.

   Note: If the volume is <40 µL, add H_2O to the aqueous phase to adjust the volume to 40–70 µL.
9. Purify the cDNA by gel-filtration chromatography with an exclusion size of 1000 nt. As an example, we included a protocol for CHROMA SPIN-1000 columns with a matrix volume of 0.75 mL.

   a. Equilibrate the column with 1.5 mL of 1× TNE buffer.
   b. Carefully and slowly apply the sample to the center of the gel bed's flat surface.
   c. Apply 25 µL of 1× TNE buffer and allow the buffer to completely drain out of the column.
   d. Apply 150 µL of 1× TNE buffer and allow the buffer to completely drain out of the column.
   e. Transfer column to a clean 1.5-mL microcentrifuge tube. Apply 320 µL of TNE buffer and collect this as the purified cDNA fraction. Transfer 10 µL of this fraction to a clean microcentrifuge tube and store at -20°C. Use this aliquot for agarose gel analysis (step 10).
   f. Apply 75 µL of 1× TNE buffer and collect the eluate in a clean microcentrifuge tube. Store at -20°C. Save this fraction until after agarose–ethidium bromide (EtdBr) gel analysis (step 10).
10. To confirm that the PCR product is present in the purified cDNA fraction, perform agarose gel electophoresis. Electrophorese 3 µL of the unpurified PCR product (step 1) alongside 10 µL of the PCR product purified by column chromatography (step 9e) and 10 µL of the second fraction (step 9f) on a 1.2% agarose gel. Compare the band intensities and estimate what percentage of the PCR product remains after column chromatography. The yield of cDNA after

column chromatography is typically 50%. If the yield is <30%, check to see if the cDNA is present in the second fraction. If this second fraction has a higher yield of cDNA than the first, combine the fractions and proceed with *Rsa*I digestion. Otherwise, repeat the PCR and column chromatography steps.

*Rsa*I Digestion Procedure

This step is equivalent to the *Rsa*I digestion procedure in the "Preparation of the Subtracted cDNA Library from Poly(A+) RNA Section" and generates shorter, blunt-ended double-stranded cDNA fragments that are necessary for adaptor ligation and optimal for subtraction. Before proceeding with *Rsa*I digestion, set aside 10 µL of purified double-stranded cDNA for agarose gel analysis to estimate the size range of the double-stranded cDNA products (step 4).

1. Add 36 µL 10× *Rsa*I restriction buffer and 1.5 µL *Rsa*I (10 U/µL) to 320 µL of the purified cDNA fraction collected from the CHROMA SPIN column.
2. Vortex mix and spin briefly in a microcentrifuge.
3. Incubate at 37°C for 3 h.
4. To confirm that the *Rsa*I digestion was successful, electrophorese 10 µL of uncut double-stranded cDNA and 10 µL of *Rsa*I-digested cDNA on a 1.2% agarose gel. Compare the profiles of both samples. Before *Rsa*I digestion, double-stranded cDNA should appear as a smear from 0.5 to 7 kb with bright bands corresponding to abundant mRNAs. After *Rsa*I digestion, the smear should range from 0.1 to 2 kb.
5. Add 8 µL of 0.5 M EDTA to terminate the reaction.
6. Transfer 10 µL of the digested cDNA to a clean microcentrifuge tube and store at -20°C. You will compare this sample to the digested cDNA after final purification.

Purification of Digested cDNA

Digested cDNA should be purified using silica matrix-based PCR purification systems. We routinely use the Advantage PCR-Pure Kit, but other systems such as those from QIAGEN (Chatsworth, CA, USA) or BIO 101 (Vista, CA, USA) can be used. Alternatively, phenol–chloroform extraction can be performed; however, this might decrease the efficiency of SSH subtraction.

1. Follow the manufacturer's protocol. Measure the concentration and estimate the yield of cDNA. For each reaction, we usually obtain 1–3 µg of PCR-amplified cDNA after purification; for two tubes, obtain a total of 2–6 µg of cDNA. If the yield is lower than this, troubleshoot the purification step.
2. Adjust the concentration of the cDNA up to >300 ng/µL. Now the cDNAs are ready for subtraction. Follow the subtraction protocol starting from the adaptor ligation step.

Differential Screening of the Subtracted cDNA Library

Two approaches can be used for differential screening of the arrayed subtracted cDNA clones—cDNA dot blots and colony dot blots. For colony dot blots, the arrayed cDNA colonies are spotted on nylon filters, grown on antibiotic plates and processed for colony hybridization. This method is usually convenient and cheaper, but it is less sensitive and gives a higher background than cDNA dot blots. The cDNA array approach is highly recommended and the following protocol is provided.

Amplification of cDNA Inserts by PCR

For high-throughput screening, a 96-well-format PCR from one of several thermal cycler manufacturers is recommended. Alternatively, single tubes can be used.

1. Replicate arrayed subtracted cDNA clones in 96-well plates. Alternatively, random colonies can be picked and processed directly without initially arraying the clones.
2. Grow each colony in 100 µL of LB medium with ampicillin in a 96-well plate at 37°C for at least 2 h (up to overnight) with gentle shaking.
3. Prepare a master mixture for 100 PCRs.

	Amount per reaction (µL)
10× PCR buffer	2.0
NP1*	0.6
NP2R*	0.6
dNTP mixture (10 mM)	0.4
H_2O	15.0
50× Advantage cDNA Polymerase Mix	0.4
Total volume	19.0

*Alternatively, primers flanking the insertion site of the vector can be used in PCR amplification of the inserts.

4. Separate into aliquots 19 µL of the master mixture into each tube or well of the reaction plate.
5. Transfer 1 µL of each bacterial culture (from step 2) to each tube or well containing master mixture.
6. Perform PCR in an oil-free thermal cycler with the following conditions: denaturation at 94°C for 2 min, then 22 cycles of denaturation at 94°C for 30 s and extension at 68°C for 3 min.
7. Analyze 5 µL from each reaction on a 2% agarose gel.

Preparation of cDNA Dot Blots of the PCR Products

1. For each PCR, combine 5 µL of the PCR product and 5 µL of 0.6 *N* NaOH (freshly made or at least freshly diluted from concentrated stock).
2. Transfer 1–2 µL of each mixture to a nylon membrane. This can be accomplished by dipping a 96-well replicator in the corresponding wells of a microplate used for PCR amplification and spotting it onto a dry nylon filter. Make at least two identical blots for hybridization with forward- and reverse-subtracted probes.
3. Neutralize the blots for 2–4 min in 0.5 M Tris-HCl (pH 7.5) and wash in 2× standard saline citrate (SSC).
4. Immobilize the DNA on the membrane using a UV linking device (Stratalinker®; Stratagene) or bake the blots for 4 h at 68°C.

Preparation of Subtracted cDNA Probes

Before the forward- and reverse-subtracted cDNA can be used as probes for differential screening, the adaptor sequences must be removed. Despite their small size, these sequences cause a very high background when hybridized to the arrayed subtracted library.

Note: If the adaptors for blunt-end or site-specific cloning were removed, skip this section.

1. Combine two tubes (ca. 40 µL total) of each forward- and reverse-subtracted secondary PCR product.
2. Purify the PCR products using a silica matrix-based purification system such as the Advantage PCR-Pure Kit. The average concentration of each subtracted PCR product is about 10–40 ng/µL. Precipitate the purified cDNA mixture with ethanol.
3. Adjust the volume after purification up to 28 µL with H_2O. Set aside 3 µL for subsequent gel analysis.
4. Digest cDNA mixtures with a combination of *Rsa*I, *Eag*I and *Sma*I restriction enzymes to remove the adaptors as follows.
    a. Add 3 µL of 10× Restriction Enzyme Buffer 4 (New England Biolabs) and 1 µL of *Rsa*I restriction enzyme. Incubate for 1 h at 37°C.
    b. Add 1 µL of *Sma*I restriction enzyme and incubate for 1 h at room temperature. Add 59 µL of H_2O, 10 µL of 10× Restriction Enzyme Buffer 3 (New England Biolabs) and 1 µL of *Eag*I restriction enzyme. Incubate for 1 h at 37°C.
5. Analyze 3 µL of digested and undigested cDNA on a 2% agarose gel. The digested adaptor should appear on the gel as a low-molecular-weight band.
6. Separate the adaptors from the cDNA using a silica matrix-based purification resin as before. Adjust the volume of each sample up to 50 µL with H_2O.
7. Label subtracted probes by random-primer labeling using commercially available kits.

Hybridization with the Forward- and Reverse-Subtracted cDNA Probes

Hybridization conditions have been optimized for ExpressHyb Hybridzation Solution. Determine the optimal hybridization conditions for other systems.

1. Prepare a prehybridization solution for each membrane.
    a. Combine 50 µL of 20× SSC, 50 µL of sheared salmon sperm DNA (10 mg/mL stock solution) and 10 µL of blocking solution (containing 2 mg/mL of unpurified NP1, NP2R, cDNA synthesis primers and their complementary oligonucleotides).
    b. Boil the blocking solution for 5 min, then chill on ice.
    c. Combine the blocking solution with 5 mL of ExpressHyb Hybridization Solution.
2. Place each membrane in the prehybridization solution prepared in step 1. Prehybridize for 40–60 min with continuous agitation at 72°C.
3. Prepare hybridization probes.
    a. Mix 50 µL of 20× SSC, 50 µL of sheared salmon sperm DNA (10 mg/mL),

10 µL blocking solution and purified probe (at least 10^7 counts per minute [cpm]/100 ng of subtracted cDNA). Make sure the specific activity of each probe is approximately equal.

  b. Boil the probe for 5 min, then chill on ice.

  c. Add the probe to the prehybridization solution.

4. Hybridize overnight with continuous agitation at 72°C.
5. Prepare low-stringency (2× SSC, 0.5% sodium dodecyl sulfate [SDS]) and high-stringency (0.2× SSC, 0.5% SDS) washing buffers and warm them to 68°C.
6. Wash membranes with low-stringency buffer (4×, 20 min at 68°C), then wash with high-stringency buffer (2×, 20 min at 68°C).
7. Analyze by autoradiography or phosphor imaging.
8. If desired, remove probes from the membranes by boiling for 7 min in 0.5% SDS. Blots can typically be reused at least 3 times.

Note: To minimize hybridization background, we recommend storing the membranes in plastic wrap at -20°C when they are not in use.

RESULTS AND DISCUSSION

Using Total RNA as Starting Material for SSH

The SSH technique has been used to identify many differentially expressed genes in our laboratory (4,6,10) and in other laboratories (3,17,19,20). The basic procedure requires 2–4 µg of poly(A+) RNA for both the tester and driver. However, it is often impossible to obtain more than several hundred nanograms of total RNA, e.g., when working with small amounts of tissue, or unstable or slow-growing cell lines. In these cases, a preamplification step using the SMART technology can be used to generate high-quality cDNA from total RNA. Although amplification of total cDNA can lead to changes in the relative abundance of each single cDNA species (because of their size, GC content and efficiency of amplification), when tester and driver samples are amplified in parallel, these changes take place for all RNA species in both tester and driver. In general, the use of preamplification should be minimized whenever possible to avoid any artificial alteration of individual cDNA concentrations. However, when only limited amounts of total RNA are available, we have found that preamplified cDNAs can be successfully used in the SSH procedure.

To illustrate the utility of preamplification, we will describe our efforts to isolate genes that are differentially expressed during globin gene switching. In this study, we used heterospecific cell hybrids between human fetal erythroblasts and mouse erythroleukemia cells. While in continuous culture, these hybrids will spontaneously switch their genetic program from expression of predominantly human fetal to γ-globin (γ line) to almost exclusively adult β-globin (β line) (13). Total RNA from these cell line hybrids was purified using the procedure described by Chomczynski and Sacchi (2) and was kindly provided to us by J.Y. Chan and Y.W. Kan (University of California and Howard Hughes Medical Institute). The SSH method was used to isolate and identify genes that differentially expressed during globin gene switching. Because only a few micrograms of total RNA from each hybrid line were available as starting material, we used SMART PCR cDNA synthesis to generate cDNA.

Tester cDNA was synthesized from 1 µg of total RNA isolated from a γ line and driver cDNA from 1 μg of total RNA from a β line in accordance with the protocol described in Chenchik et al. of this book. These cDNAs were then used for the SSH procedure. Both forward and reverse subtractions were done. Forward subtraction was performed using γ-line cDNA as tester and β-line cDNA as driver, and the forward-subtracted cDNA was enriched for messages specific to the γ-line hybridoma. Reverse-subtracted cDNA (obtained by reversing tester and driver cDNAs) was enriched for messages specific to the β-line hybridoma. PCR analysis confirmed that *GAPD* and α-tubulin sequences, expected to be present in both populations, were reduced by a factor of over 10^3 (data not shown), indicating that the subtraction was successful.

Differential Screening

Although the SSH method greatly enriches for differentially expressed genes, the

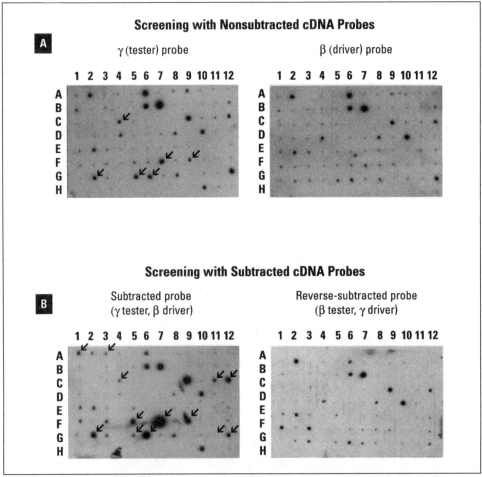

Figure 4. Differential screening approach. (A) Dot blots hybridized with nonsubtracted cDNA probes made from tester (γ-line) and driver (β-line) RNA. (B) Dot blots hybridized with cDNA probes made from forward-subtracted cDNA (γ-line tester, β-line driver) and reverse-subtracted cDNA (β-line tester, γ-line driver). On each membrane, the following clones correspond to controls. H10: γ-globin cDNA fragment (positive control); H11: prostate-specific cDNA (negative control); H12: prostate-specific cDNA (negative control).

subtracted sample will still contain some cDNAs that correspond to mRNAs common to both the tester and driver samples. Although this background depends on the quality of the RNA and the performance of the particular subtraction, it mainly depends on the complexity of the mRNAs that are differentially expressed in tester and driver. Even if one obtains a good enrichment of differentially expressed cDNAs, a higher background occurs when the difference between the tester and driver pool is small. When the background is high, identification of differentially expressed cDNAs can be achieved by choosing random clones from the subtracted library and performing Northern blot analysis on each clone. As an alternative to this time-consuming approach, we have incorporated a differential screening step to minimize the false positives prior to embarking on Northern blot analysis.

In our globin gene-switching study, we anticipated that the differences between tester and driver poly(A$^+$) RNA populations would be small. Thus, we proceeded with a differential screening procedure (Figure 4). Eighty-four randomly selected cDNA clones isolated from the γ-enriched cDNA pool were arrayed (DNA dot blot) onto nylon membranes for quadruplicate screening. When the DNA dot blots were hybridized with the nonsubtracted cDNA probes, six clones showed a differential signal (Figure 4A). The same blots were hybridized to probes prepared from the γ-enriched pool and from a β-enriched pool generated from the reverse subtraction (Figure 4B). Figure 4 shows a typical result of a differential screen of a subtracted cDNA library obtained using the new SSH method. These results reveal the following types of clones.

1. **Clones hybridizing to the forward-subtracted probe but not to the reverse-subtracted probe (e.g., clones F5 and F7).** These clones correspond to the differentially expressed mRNA with very high probability of being differentially expressed. The efficiency of hybridization can vary significantly (compare F7 and G5). This means that the concentration of the F7 cDNA fragment in subtracted cDNA is much higher than G5, and that is why the F7 cDNA fragment was enriched during the subtraction more efficiently. The enrichment efficiency for each particular fragment mainly depends on the difference in abundance of the corresponding mRNA in the tester and driver pools. In general, one would expect that genes with greater difference in abundance would show better enrichment. However, this is not always true because some cDNA sequences might not be rehybridized or amplified very efficiently, and such sequences will be enriched less efficiently. Furthermore, the intensity of the signal depends on the length of the cDNA insert—the longer the size, the more efficient the hybridization.

2. **Clones hybridizing to both subtracted probes with the same efficiency (e.g., A2, B6).** These clones do not correspond to the differentially expressed mRNA, and they represent background.

3. **Clones hybridizing to both subtracted probes but with different efficiencies of hybridization.** In our experience, if the difference in signal intensity is 2–3-fold, the clone does not correspond to a differentially expressed mRNA (C9). This random difference is a result of random fluctuation in efficiency of forward and reverse subtractions.

4. **Clones that do not hybridize noticeably to either hybridization probe (e.g., A5, F12).** Such clones do not represent differentially expressed mRNAs (18) and usually cause most of the background in an SSH-subtracted cDNA library. These clones happen to be in the subtracted library because they randomly escaped sub-

traction, and thus each is present as a single copy in the subtracted probe. Because of very low concentration of each of this kind of molecule in the subtracted probe, they will not hybridize to the corresponding clone.

It is useful to compare the results from the differential screening using subtracted and nonsubtracted probes (Figure 4, A and B). Use of nonsubtracted tester and driver cDNA as probes has some limitations. Only abundant mRNAs in the probe [abundance >0.1% of the entire poly(A^+) RNA population] will hybridize to the corresponding clone (1). Nevertheless, in this case, hybridization of a particular clone with the tester cDNA probe but not with the driver cDNA probe (e.g., C4) or more efficient hybridization to the tester cDNA probe will confirm the results of differential screening obtained using subtracted probes (e.g., G6). But if the RNA is not abundant, nonsubtracted cDNA probes will not generate visible hybridization signals with the cDNA clone (e.g., A1, A3).

Occasionally, the differential signal obtained by hybridization with subtracted probes cannot be confirmed by hybridization with nonsubtracted cDNA probes (e.g., C12). In such cases, Northern blot analysis shows that the cDNA fragment does not correspond to a differentially expressed mRNA or that it is hybridizing to the tester and driver poly(A^+) RNA, but: *(i)* transcripts have different sizes and probably present alternatively spliced forms, or *(ii)* fragments are hybridizing to the transcripts of several sizes but one or more bands specific for the tester poly(A^+) RNA. The latter case usually occurs when a differentially expressed cDNA fragment possesses some sequence homology to different members of a multigene family but is specific for tester poly(A^+) RNA (e.g., C12).

After analysis of the differential screening results shown in Figure 4, 13 clones have been judged to be candidates for differentially expressed sequences.

Verification of Differential Expression

Confirmation of differential expression of putative clones obtained by differential screening can be done by Northern blot analysis, virtual Northern blot analysis (see Chenchik et al., in this volume) or RT-PCR. In our hands, the differential screening procedure results in about 2% of false positives. In other words, if 100 clones are screened, 1–2 clones that show differential signals will not show differential expression by Northern blot analysis. In our study, differential expression was confirmed by virtual Northern blot analysis for 13 out of 13 candidates; three blots are shown in Figure 5. All 13 clones represented genes that are expressed primarily or only in the γ-line hybridoma but are not expressed in the β-line hybridoma. Eleven clones were sequenced, and nucleic acid homology searches were performed using the Basic Local Alignment Search Tool (BLAST) program (**http://www.ncbi.nih.gov/BLAST/**) (Table 1). All 11 clones represented unique sequences, and four represented novel genes. No clones corresponded to the γ-globin gene.

Complexity of the Subtracted Library

Massive sequencing of random clones from SSH-subtracted cDNA libraries has revealed that they are highly complex (4,17). This high complexity results from the inclusion of a normalization step, which enriches for differentially expressed genes of both high and low abundance, many of which are likely to be novel. We would like to stress that the complexity of any subtracted cDNA library mainly depends on the

Table 1. Identification of the Isolated cDNA Clones

Name of Clone	Identity
A1	No significant homology
A3	Human heat-shock protein (M27024)
C4	Significant homology to mouse pim-1 protooncogene encoding protein kinase (M13945)
C11	Human proteosome subunit HC8 (D00762)
F5	Human heat-shock protein (M27024)
F7	Human proteosome subunit MB1 (S74378)
F9	Mouse dihydrofolate reductase (V00733)
G2	Significant homology to rat nucleoporin Nup84 mRNA (U93692) and human mRNA for Nup88 protein (Y08612)
G6	Significant homology to mouse adult brain mRNA for phospholipase A2 (D78647)
G11	Human proteosome subunit HC3 (D00760)
C12	Human phospholipase A2 mRNA (M86400)

Numbers in parentheses are GenBank® accession numbers.

nature of the tester and driver mRNA samples. For tissue-specific subtracted libraries (4), when the differences between tester and driver poly(A+) RNA population are expected to be very high, the complexity of the subtracted library is significantly higher than in the case of β- and γ-producing cell lines. In the latter case, even when all random clones represent the unique sequences, some correspond to different restriction fragments of the same protein (Table 1, A3 and F5) or different subunits of the same protein complex (Table 1, C11, G11 and F7).

The normalization step of the SSH procedure also provides the high level of enrichment for rare transcripts. Although the mathematical model of our subtraction procedure (6) and our model experiments (4) have demonstrated over 1000-fold

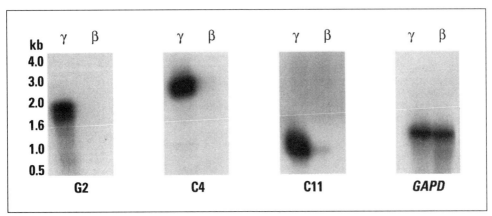

Figure 5. Confirmation of putative differentially expressed cDNAs by virtual Northern blot analysis. Each lane contains 0.5 µg of SMART PCR-amplified cDNA from either γ- or β-globin-producing cell lines (as indicated). To produce each blot, SMART cDNA was electrophoresed on a 1.2% agarose gel and transferred to a nylon membrane. Each blot was hybridized to the indicated radioactively labeled cDNA clone. Exposure times were G2 and C4: overnight; C11: 1 week; and *GAPD*: 2 h.

enrichment for the rare specific transcript, the practical level of enrichment for a particular gene depends greatly on its original abundance, the ratio of its concentrations in driver and tester and the number of other differentially expressed genes. With the incorporation of a normalization step in our subtraction procedure, the highest enrichment level can be obtained for differentially expressed mRNAs exhibiting low abundance and/or large differences in expression levels in the tester and driver RNA populations. However, as with other subtractive hybridization techniques, the efficiency of SSH is lower in experiments designed to detect mRNAs that show only moderate (i.e., 2- to 4-fold) differences between the tester and driver populations. Nevertheless, these types of differentially expressed genes can still be identified in our subtracted libraries (6).

Conclusions

We have presented a versatile subtractive hybridization technique based on the suppression PCR effect. Our results demonstrate that SMART cDNA synthesis and SSH cDNA subtraction can be used sequentially for efficient cDNA subtraction from small samples in which only a few micrograms of total RNA are available, e.g., biopsy specimens, embryos, flow-sorted cells or cell lines. In this case, virtual Northern blots can be used as a general tool for confirmation of differential gene expression. With the incorporation of various modifications, such as differential screening with forward- and reverse-subtracted cDNAs, the SSH technique should be applicable to many other studies requiring the enrichment of cDNAs derived from differentially expressed genes of particular tissue or cell types.

ACKNOWLEDGMENTS

We thank Sharada Vijaychander for excellent technical assistance, Stephanie Trelogan for critically reading the manuscript, Theresa Provost for helping in figure preparation and Jenifer Fishel for helping with the manuscript preparation.

REFERENCES

1. **Britten, R.J. and E.H. Davidson.** 1985. Hybridization strategy, p. 3-15. *In* B.D. Hames and S.J. Higgins (Eds.), Nucleic Acid Hybridization—A Practical Approach. IRL Press, Oxford.
2. **Chomczynski, P. and N. Sacchi.** 1987. Single-step method of RNA isolation by acid guanidinium thiocyanate-phenol-chloroform extraction. Anal. Biochem. *162*:156-159.
3. **Chu, A.-L., T.A. McKinsy, L. Liu, J.J. Gentry, M.H. Malim and D.W. Ballard.** 1997. Suppression of tumor necrosis factor-induced cell death by inhibitor of apoptosis c-IAP2 is under NF-κB control. Proc. Natl. Acad. Sci. USA *94*:10057-10062.
4. **Diatchenko, L., Y.-F.C Lau, A. Campbell, A. Chenchik, F. Moqadam, B. Huang, S. Lukyanov, K. Lukyanov et al.** 1996. Suppression subtractive hybridization: a method for generating differentially regulated or tissue-specific cDNA probes and libraries. Proc. Natl. Acad. Sci. USA *93*:6025-6030.
5. **Gubler, U. and B.J. Hoffman.** 1983. A simple and very efficient method for generating complementary DNA libraries. Gene *25*:263-269.
6. **Gurskaya, N.G., L. Diatchenko, A. Chenchik, P.D. Siebert, G.L. Khaspekov, K.A. Lukyanov, L.L. Vagner, O.D. Ermolaeva, S.A. Lukyanov and E.D. Sverdlov.** 1996. Equalizing cDNA subtraction based on selective suppression of polymerase chain reaction: cloning of Jurkat cell transcripts induced by phytohemagglutinin and phorbol 12-myristate 13-acetate. Anal. Biochem. *240*:90-97.
7. **Hara, E., T. Kato, S. Nakada, S. Sekiya and K. Oda.** 1991. Subtractive cDNA cloning using oligo(dT)$_{30}$-latex and PCR: isolation of cDNA clones specific to undifferentiated human embryonal carcinoma cells.

Nucleic Acids Res. *19*:7097-7104.
8. **Hedrick, S.M., D.I. Cohen, E.A. Nielsen and M.M. Davis.** 1984. Isolation of cDNA clones encoding T cell-specific membrane-associated proteins. Nature *308*:149-153.
9. **Hubank, M. and D.G. Schatz.** 1994. Identifying differences in mRNA expression by representational difference analysis of cDNA. Nucleic Acids Res. *22*:5640-5648.
10. **Jin, H., X. Cheng, L. Diatchenko, P.D. Siebert and C.-C. Huang.** 1997. Differential screening of a subtracted cDNA library: a method to search for genes preferentially expressed in multiple tissues. BioTechniques *23*:1084-1086.
11. **Kellogg, D.E., I. Rybalkin, S. Chen, N. Mukhamedova, T. Vlasik, P.D. Siebert and A. Chenchik.** 1994. TaqStart Antibody™: "hot start" PCR facilitated by a neutralizing monoclonal antibody directed against *Taq* DNA polymerase. BioTechniques *16*:1134-1137.
12. **Lukyanov, K.A., G.A. Launer, V.S. Tarabykin, A.G. Zaraisky and S.A. Lukyanov.** 1995. Inverted terminal repeats permit the average length of amplified DNA fragments to be regulated during preparation of cDNA libraries by polymerase chain reaction. Anal. Biochem. *229*:198-202.
13. **Papayannopoulou, T., M. Brice and G. Stamatoyannopoulos.** 1986. Analysis of human hemoglobin switching in MEL x human fetal erythroid cell hybrids. Cell *46*:469-476.
14. **Sambrook, J., E.F. Fritsch and T. Maniatis.** 1989. Molecular Cloning: A Laboratory Manual, 2nd ed. CSHL Press, Cold Spring Harbor, NY.
15. **Sargent, T.D. and I.B. Dawid.** 1983. Differential gene expression in the gastrula of *Xenopus laevis*. Science *222*:135-139.
16. **Siebert, P.D., A. Chenchik, D.E. Kellogg, K.A. Lukyanov and S.A. Lukyanov.** 1995. An improved method for walking in uncloned genomic DNA. Nucleic Acids Res. *23*:1087-1088.
17. **von Stain, O.D., W.-G. Thies and M. Hofmann.** 1997. A high throughput screening for rarely transcribed differentially expressed genes. Nucleic Acids Res. *25*:2598-2602.
18. **Wang, Z. and D.D. Brown.** 1991. A gene expression screen. Proc. Natl. Acad. Sci. USA *88*:11505-11509.
19. **Wong, B.R., J. Rho, J. Arron, E. Robinson, J. Orlinick, M. Chao, S. Kalachikov, E. Cayani et al.** 1997. TRANCE is a novel ligand of the tumor necrosis factor receptor family that activates c-Jun N-terminal kinase in T cells. J. Biol. Chem. *272*:25190-25194.
20. **Yokomizo, T., T. Izumi, C. Chang, Y. Takuwa and T. Shimizu.** 1997. A G-protein-coupled receptor for leukotriene B4 that mediates chemotaxis. Nature *387*:620-624.

18 Optimized Subtraction-Enhanced Display Technique

Marylene Denijn, Theodorus B.M. Hakvoort and Wouter H. Lamers
Department of Anatomy and Embryology, Academic Medical Center, University of Amsterdam, Amsterdam, The Netherlands

OVERVIEW

Each cell type is characterized by its own pattern of gene expression (its "transcriptome"). Many genes are expressed in most cells (housekeeping genes), whereas other genes confer a specific phenotype to the cell. The identification of this latter group of gene products can lead to the elucidation of biological or pathological processes in which specific cells are involved. The first attempts to evaluate cell-specific gene expression were based on the differential colony screening of total cDNA libraries of tissues or cells of interest. This approach is very cumbersome, and success is often limited to abundant gene products.

Several techniques have been developed to increase the sensitivity and the efficiency of differential screening. Serial analysis of gene expression (SAGE) (16,19) aims at the identification and quantification of all gene products present in a specified cell type by making an inventory of the gene products based on sequence analysis of small fragments (10-bp tags). When the expression pattern of two different tissues is catalogued, both inventories are screened for differential gene expression. Approximately 300 000 tags per tissue sample have to be determined to get a representative sample of the transcriptome (19).

Differential display (9–11) and RNA fingerprinting by arbitrarily primed PCR (RAP-PCR) (13,18) are based on the creation of subclasses of cDNA fragments to be analyzed using reverse transcription polymerase chain reaction (RT-PCR) technology. The PCR products, representing the cDNA subclasses derived from two tissue or cell sources, are analyzed in pairs on a display gel. Because up-regulated gene products in one cDNA population will appear as bands with increased staining intensity when compared to their counterpart in the other cDNA population, the techniques are critically dependent in a quantitative amplification of fragments to maintain differences in expression of the same mRNA in two cell types.

Subtractive hybridization aims at the reduction of the complexity of the cDNA libraries by the removal of common gene products from the cDNA populations prior to analysis. The population of fragments analyzed is enriched for the differentially expressed gene products, so that the probability of finding rare, differentially expressed gene products is increased. Basically, gene products derived from two different cDNA populations are mixed and denatured, forming single-stranded fragments. During the following hybridization step, fragments common to both populations will form hybrids between strands derived from tracer and driver. These hybrids are eliminated in a subsequent extraction step. Several variations on the elimination of the common gene products have been reported (1,3,4,6,8,17), one of which is described in a chapter of this book (Diatchenko et al.). All the approaches use double-stranded cDNA fragments that are modified to allow selective PCR amplification. PCR allows for the preparation of sufficient amounts of cDNA from small quantities of tissues or cells, permits repeated subtraction steps (6,17) or can be used as a tool for the enrichment of the up-regulated and unique sequences in the tracer population (3,6). After the subtractions, the fragments are cloned to obtain a cDNA fragment library enriched for differentially expressed gene products.

The subtraction-enhanced display technique that will be described here is a combination of the PCR-based subtractive hybridization and differential display on a polyacrylamide gel of the enriched products obtained after each subtraction step (Reference 5 and Figure 1). Because the subtraction simplifies the composition of the cDNA population, the differentially expressed products can be displayed on a single gel. The display gel monitors the progression of the elimination of the common products and the enrichment of the differentially expressed products. The latter fragments can readily be isolated from the display gel for sequence identification.

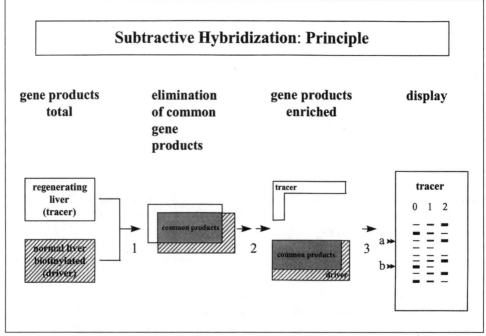

Figure 1. Principle of the subtractive hybridization procedure. [1] Formation of tracer–tracer, tracer–driver and driver–driver hybrids during the hybridization process; [2] subtractions in which common sequences are eliminated from the tracer; [3] display gel of starting material (lane 0) and of the tracer after 1 or 2 rounds of subtraction (lanes 1 and 2). [a] Examples of enriched fragment; [b] fragment eliminated during the subtractive hybridization procedure. See Procedures section for more detail.

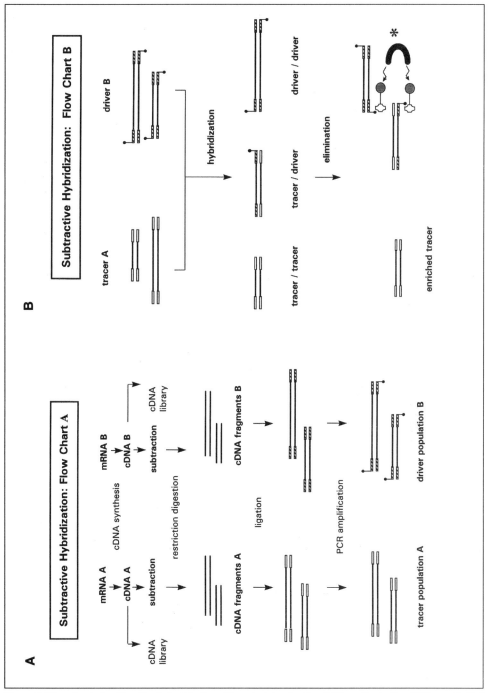

Figure 2. Flow chart of subtractive hybridization. Flow chart A: after the cDNA synthesis, the cDNAs are divided into two equal portions for the synthesis of a full-length cDNA library and for subtractive hybridization. The cDNAs for subtractive hybridization were fragmented and ligated to distinct primers/adaptors. Subsequently, the cDNA fragments were amplified using PCR for tracer and driver synthesis. The driver is synthesized with biotinylated primers (T: biotin residue). Flow chart B: subtractive hybridization. For the hybridization, tracer and driver are mixed in a ratio of 1:20, and the cDNA fragments are denatured and allowed to reanneal. After the hybridization, the driver hybrids are removed from the solution using SA-PMPs and a magnetic standard. The enriched tracer population is subsequently submitted to a round of subtraction. See Procedures section for more detail.

Because the subtraction step can be repeated, specific elimination of abundant cDNAs or other products previously identified by the display gel can be performed, greatly improving the sensitivity of the technique. Moreover, the population of enriched cDNA fragments can be used directly for the isolation of the corresponding full-length recombinant cDNAs (Hakvoort et al., in this volume).

BACKGROUND

The subtraction-enhanced differential display technique as described below was used for the isolation of factors involved in liver regeneration. The method is based on the subtractive hybridization procedure described by Wang and Brown (17) with modifications. The procedure consists of two reciprocal subtractions. In one subtraction, cDNA fragments derived from regenerating liver were used as tracer (i.e., the cDNA population of interest) and cDNA fragments derived from normal liver as driver (i.e., the cDNA population used for the elimination of the common sequences from the tracer population). In the other reciprocal subtraction, cDNA fragments derived from normal liver were used as a tracer and cDNA fragments from the regenerating liver as driver. Performing reciprocal subtractions allows the identification of both up- and down-regulated fragments because gene products that are down-regulated or absent in one tissue can be enriched in the reciprocal subtraction. This feature gives a more comprehensive overview of the gene expression patterns in the tissue of interest as compared to a single subtractive hybridization.

A chart of the subtraction procedure is given in Figure 2. mRNA from normal and regenerating (partially hepatectomized) rat liver was isolated and used to synthesize the initial cDNA populations. The cDNA populations were divided into two equal portions, one for the subtractive hybridization and one for the synthesis of a cDNA library from which full-length cDNA sequences could be isolated (Hakvoort et al., in this volume). The cDNAs to be used for subtractive hybridization were digested with frequently cutting enzymes into fragments of approximately 150–700 bp. The reduction of the size of the cDNA fragments presented several advantages: (*i*) a bias in favor of the smaller cDNAs during the PCR was largely avoided; (*ii*) with respect to the length of the products, the hybridization conditions for all cDNA fragments were comparable; (*iii*) multiple fragments of a single cDNA were generated. The last feature was important for the detection of differential expression of gene products sharing a high sequence homology (e.g., albumin [GenBank® Accession No. J00698], α-fetoprotein [m16111] and glutathione peroxidase I [x12367]). Fragments with less homology that reflect the difference between the two gene products can still be identified as up- or down-regulated products.

Distinct primers/adaptors were ligated to the cDNA fragments derived from the regenerating and normal liver, ensuring independent amplification of the tracer and driver populations. Then a first PCR amplification was performed to obtain the initial tracer and driver populations. The tracer was amplified with unmodified primers, whereas the driver was synthesized with primers containing a biotin residue at their 5′ end. The biotin formed a substrate during the extraction of the driver sequences during the subtractive procedure.

To remove abundant, common sequences, a short hybridization was performed first. The cDNA fragments present in the hybridization mixture were heat-denatured and reannealed subsequently at a specific hybridization temperature. During the reannealing, hybrids between the common fragments of tracer and driver can be formed. Because the rate of hybridization is dependent on the relative concentration (C_0t) of the fragments in the hybridization mixture, abundant sequences hybridize more rapidly compared to their more rare counterparts. After the completion of the short hybridization, tracer–driver and driver–driver hybrids were bound by their biotin residues to streptavidin-coated paramagnetic particles (SA-PMPs) and removed from the hybridization mixture using a magnetic standard. The short subtraction is immediately followed by a long hybridization step to remove the less abundant common sequences. Then the enriched products were amplified by PCR to prepare the driver population for the next round of the reciprocal subtraction. To avoid the introduction of PCR artifacts in the tracer population, the tracer was not further amplified for the subsequent subtraction rounds (see Results and Discussion). After a few rounds, the progression of the subtraction was assessed by amplification of the tracers obtained after each subtraction in the presence of radioactive nucleotides. The PCR products were subsequently analyzed on a polyacrylamide gel, on which the enriched products were visualized as distinct fragments with increasing intensity after each round of subtraction (5).

In the next section, we describe the method in detail and introduce several adaptations and modifications that resulted in a faster and more efficient method for obtaining differentially expressed gene products.

PROTOCOLS

Preparation of the cDNA Fragments as Substrate Material for the Subtraction

Materials and Reagents

- cDNA derived from normal and regenerating liver
- Restriction enzymes *Rsa*I (10 U/µL) and *Alu*I (10 U/µL)
- 10× restriction digestion buffer: 330 mM Tris-acetate, 100 mM magnesium acetate, 660 mM potassium acetate, 5 mM dithiothreitol (DTT)
- Primer sets: A21: 5′-CTCTTGCTTGAATTCGGACTA-3′ and
  A24: 5′-TAGGCCGAAGGCAAGCAAAGAGCAC-3′
  B21: 5′-AATTCAGGCCAAGTCGGCCGG-3′ and
  B24: 5′-CCGGCCGACTTGGCCTGAATTATA-3′
- 10 mM ATP
- T4 polynucleotide kinase (10 U/µL)
- 10× kinase buffer: 500 mM Tris-HCl, pH 7.6, 100 mM $MgCl_2$, 50 mM DTT, 0.5% bovine serum albumin (BSA)
- 20× standard saline citrate (SSC): 3 M NaCl, 0.3 M sodium acetate, pH 7.4
- 3 M sodium acetate (pH 5.2)
- 96% ethanol
- T4 DNA ligase (10 U/µL)
- 5× ligase buffer: 250 mM Tris-HCl, pH 7.8, 50 mM $MgCl_2$, 5 mM DTT, 5 mM ATP)
- Tris-buffered phenol
- Chloroform–isoamyl alcohol (24:1, vol/vol)
- TE buffer: 10 mM Tris-HCl, 1 mM EDTA, pH 7.5

- QIAquick® PCR Purification Kit (Qiagen, Chatsworth, CA, USA)

Procedures

RNA isolation and cDNA synthesis

Procedures for RNA isolation and cDNA synthesis can be found in previous chapters as well as in laboratory manuals (2,14). cDNAs from normal and regenerating rat liver (after partial hepatectomy) were synthesized using the Great Lengths™ cDNA Synthesis Kit (CLONTECH Laboratories GmbH, Heidelberg, Germany) according to the manufacturer's instructions. Both sets of cDNAs were divided into two equal portions. One portion was used for the subtractive hybridization procedure described below, and the other portion was used to generate cDNA expression libraries for the isolation of full-length clones of enriched products obtained using the subtraction procedure (Hakvoort et al., in this volume).

Restriction digestion

Both cDNA sets derived from the normal and the regenerating liver were digested with the frequently cutting restriction enzymes *Alu*I and *Rsa*I. The restriction digestion was performed in 2 reactions, one with *Alu*I alone and the other with *Alu*I and *Rsa*I as follows.

1. Mix 1 µg of cDNA and 10 µL of 10× restriction digestion buffer in 2 reaction tubes. Add 25 U of *Alu*I to one reaction tube and 10 U of *Alu*I plus 25 U of *Rsa*I to the other. Add H_2O to a final volume of 100 µL.
2. Incubate overnight at 37°C to ensure complete digestion of the cDNAs.
3. Inactivate the enzymes by incubation at 65°C for 15 min.
4. Remove the enzymes with a phenol–chloroform–isoamyl alcohol extraction.
5. Precipitate the purified cDNA fragments with 2.5 vol of 96% ethanol in the presence of 0.3 M sodium acetate at -20°C.
6. Keep the DNA in ethanol at -20°C until use.

These steps resulted in the creation of blunt-ended cDNA fragments of 150–700 bp, which were ligated to specific primers/adaptors to ensure PCR amplification.

Preparation of the primers/adaptors

To allow efficient ligation of the primers/adaptors to the cDNA fragments, primers A24 and B24 were phosphorylated using the following reaction procedure.

1. Mix together:
   2.5 nmol oligonucleotide primer (A24 and B24 separately)
   5 µL 10× kinase buffer
   5 µL 10 mM dATP
   20 U T4 polynucleotide kinase
   H_2O to 50 µL.
2. Incubate for 1 h at 37°C.

3. Stop the kinase reaction by heating the solution for 15 min at 65°C.
4. Precipitate the phosphorylated oligonucleotides with 2.5 vol of ethanol in the presence of 0.3 M sodium acetate overnight at -20°C.
5. Centrifuge for 10 min at 13 000 rpm in a Model 5415C Eppendorf® centrifuge (Brinkmann Instruments, Westbury, CT, USA) at 4°C, wash the pellet with 70% ethanol and dissolve the pellet in 25 µL TE buffer.

The primers/adaptors were made by annealing equimolar amounts of primer A21 with phosphorylated A24, and B21 with phosphorylated B24.

6. Mix together:
   2 nmol of primer A21 or B21
   2 nmol of primer A24 or B24
   2.5 µL of 20× SSC
   H$_2$O to a final volume of 50 µL.
7. Centrifuge briefly after mixing.
8. Heat the mixture for 10 min at 70°C and let the water bath cool down slowly to room temperature.
9. Precipitate the primers/adaptors with 250 µL of 96% ethanol in the presence of 0.3 M sodium acetate (pH 5.2) for at least 1 h at -20°C.
10. Centrifuge for 10 min at 4°C, rinse the pellet with 70% ethanol and dissolve the primers/adaptors in 20 mL TE buffer to a final concentration of approximately 100 nmol/mL.

Ligation of the primers/adaptors to the cDNA fragments

1. Recover the cDNA fragments by centrifugation for 10 min at 4°C, wash the pellets with 70% ethanol and dissolve the DNA in 20 µL TE buffer.
2. Establish the concentrations of the digested cDNA fragments (2 µL) by comparison with a known dilution series of DNA on an agarose gel containing ethidium bromide, either by estimation or with optical densitometry (Eagle Eye®; Stratagene GmbH, Heidelberg, Germany). The DNA samples will appear as a single band if the electrophoresis is very brief (maximum 5 min at 5 V/cm).
3. Ligate the primers/adaptors to the cDNA fragments. The ligation of the *Alu*I- and the *Alu*I/*Rsa*I-restricted cDNA fragments to the primers/adaptors is done separately. Make certain that the same primer/adaptor is ligated to the cDNA fragments derived from one tissue.
4. Set up the ligation reactions containing:
   0.5 µg of cDNA fragments
   6 nmol of primer/adaptor
   6 µL of 5× ligase buffer
   10 U of T4 DNA ligase
   H$_2$O to 30 µL.
5. Mix well and incubate overnight at 15°C.
6. Inactivate the ligase by heating the reaction for 15 min at 70°C.
7. Purify the DNA using a QIAquick PCR purification column following the manufacturer's instructions.

Note: Alternatively, the unligated primers/adaptors can be eliminated by isolation of the cDNA fragments from 150 bp and larger from a 1.5% agarose gel. The fragments are recovered from the agarose using electroelution (2,14).

8. Precipitate the DNA fragments overnight with 2.5 vol of 96% ethanol in the presence of 0.3 M sodium acetate (pH 5.2) at -20°C.
9. Recover the DNA fragments by centrifugation for 10 min at 4°C, rinse the pellet with 70% ethanol and dissolve in 20 µL TE buffer.
10. Establish the concentrations of the cDNA fragments as described above.
11. Mix 100 ng of *Alu*I-restricted cDNA fragments with 100 ng of *Alu*I/*Rsa*I-restricted cDNA fragments derived from the same tissue. The two cDNA mixtures derived from the regenerating and the normal liver are the template for the synthesis of tracer and driver.

PCR Amplification for Tracer and Driver Synthesis

Some general precautions should be taken when amplifying the tracer population prior to subtraction or analysis of the products on the display gel. If too much input DNA is added to the PCR mixture or if too many cycles of amplification are used, PCR artifacts (i.e., the formation of smaller fragments due to mispriming) can occur (M. Denijn, unpublished). The PCR conditions described below allow the synthesis of 100–150 ng of DNA per reaction. This protocol is written for SuperTaq® Polymerase (SphaeroQ, The Netherlands). If other primers or other brands of DNA polymerases are used, the PCR conditions might have to be adapted with respect to input DNA, number of cycles and temperature. The PCR conditions can be tested by the use of specific DNA fragments (e.g., pBluescript® [Stratagene GmbH] digested with *Hae*III) ligated to the primers/adaptors to be tested.

Materials and Reagents

- 10× SuperTaq buffer: 100 mM Tris-HCl, pH 9.0, 500 mM KCl, 15 mM $MgCl_2$, 1% Triton® X-100
- dNTP mixture: 10 mM each dNTP
- 2.5% BSA
- Primer (100 µM); for the synthesis of the driver, the primers should be biotinylated at their 5′ end (biotinylated primers are commercially available; e.g., Eurogentec, Liège, Belgium)
- SuperTaq DNA polymerase (5 U/µL)
- Initial cDNA populations
- Mineral oil (optional)
- PCR apparatus: PTC™-200 (MJ Research, Watertown, MA, USA)
- QIAquick PCR Purification Kit (spin columns)
- 70% and 96% ethanol
- 3 M sodium acetate (pH 5.2)

Procedures

Synthesis of the tracer

To synthesize the tracer cDNA, 10 PCRs, each containing approximately 1 pg of input DNA per reaction, should be performed to obtain approximately 1 µg of tracer DNA.

1. Make the following mixture:

10× SuperTaq PCR buffer	50 µL
10 mM dNTP	10 µL
2 µM primer	10 µL
2.5% BSA	1 µL
SuperTaq DNA polymerase (5 U/µL)	1 µL
10 pg DNA	x µL
H$_2$O	to 500 µL

2. Optional: 1 µCi of [^{32}P]dCTP can be added to monitor the tracer during the subtraction procedure.

3. Divide the mixture over 10 wells of a microplate or over 10 PCR tubes and perform the PCR amplification according to the following program.

   Note: Depending on the type of PCR apparatus, 50 µL of mineral oil should be added to the wells to prevent evaporation of the PCR mixture.

   Perform an initial denaturation of the DNA for 2 min at 92°C, then 23 cycles of denaturation for 1 min at 92°C, annealing for 30 s at 60°C and extension for 1 min at 72°C, with a final extension for 5 min at 72°C. Cool down to 4°C.

   Note: Depending on the primer, the PCR conditions with respect to input DNA and number of cycles and temperature should be adapted.

4. Purify the DNA using a QIAquick PCR purification column.

5. Pool the PCR mixtures, add 0.3 M sodium acetate (pH 5.2) (final concentration) and 2.5 vol of ethanol and allow precipitation for 1/2 h at -20°C.

6. Collect the DNA by centrifugation for 10 min at 4°C at 13 000 rpm in the Model 5415C Eppendorf centrifuge.

7. Wash the pellet with 70% ethanol.

8. Air dry the pellet.

9. Resuspend the pellet in 20 µL TE buffer.

10. Quantitate the yield of the PCRs by assessing the DNA concentration on an agarose gel containing ethidium bromide.

<u>Synthesis of the driver</u>

1. The PCR for the driver synthesis was performed as for the tracer synthesis, except that the driver is synthesized in the presence of primers with a biotin residue at their 5′ ends (same concentration as the nonbiotinylated primer). Also, 100 pg DNA template were added to the 500-µL mixture instead of 10 pg, and 25 cycles were performed instead of 23 to obtain sufficient driver material (see Results and Discussion). Per subtraction, 20 µg driver must be synthesized. For the short and the first long subtraction, the driver synthesized from the initial cDNA fragments is used. For the following subtractions, the tracer derived from the reciprocal subtraction is amplified in the presence of biotinylated primers to obtain the driver.

2. Purify the DNA using a QIAquick spin column.

3. Add sodium acetate (pH 5.2) to a final concentration of 0.3 M and 2.5 vol of ethanol. Precipitate the driver fragments for 30 min at -20°C.
4. Collect the DNA by centrifugation for 10 min at 4°C at 13 000 rpm.
5. Wash the pellet with 70% ethanol.
6. Air dry the pellet.
7. Resuspend the pellet in 20 µL TE buffer.
8. Quantitate the yield of the PCRs by assessing the DNA concentration on an agarose gel containing ethidium bromide.

Subtractive Hybridization

As was stated in the introduction, the subtractive hybridization consists of two reciprocal subtractions. We describe here the procedure for one tracer and one driver, but the reciprocal subtraction should be performed in parallel. In the procedure, multiple rounds of subtractive hybridization are performed until enrichment of fragments becomes visible on the display gel.

Materials and Reagents

- Tracer cDNA fragments
- Biotinylated driver cDNA fragments
- TE buffer: 10 mM Tris-HCl, 1 mM EDTA, pH 8.0
- 2× hybridization buffer: 1.5 M NaCl, 50 mM HEPES, pH 7.6, 10 mM EDTA, 0.2% sodium dodecyl sulfate (SDS)
- Mineral oil
- 20× SSC: 3 M NaCl, 0.3 M sodium citrate, pH 7.4
- 0.5× SSC: 1.2 mL per subtraction (mix 1.17 mL of sterile water with 30 µL of 20× SSC)
- SA-PMPs (Promega, Madison, WI, USA): 0.6 mL/20 µg driver. This is equivalent to one tube of the SA-PMPs in the Promega Kit System IV.
- Magnetic stand
- 3 M sodium acetate (pH 5.2)
- 96% and 70% ethanol
- PCR apparatus or water bath

Procedures

Preparation of the tracer–driver mixture

1. Mix 1 µg of tracer and 20 µg of driver cDNA fragments and precipitate with sodium acetate to a final volume of 0.3 M and 2.5 vol of 96% ethanol. Leave for ca. 30 min at -20°C.
2. Spin down for 10 min at room temperature.
3. Wash the pellet with 70% ethanol.
4. Air dry the pellet for ca. 5 min.
5. Dissolve the pellet in 5 µL TE buffer (pH 8.0) by pipetting or heating at 60°C. Make sure that the DNA is completely dissolved. No precipitation should be present after a centrifugation step of ca. 1 min.

The short subtractive hybridization

6. Bring the DNA solution into a PCR tube or into a well of a microplate.
7. Add 5 µL of 2× hybridization buffer, mix well and overlay with approximately 50 µL mineral oil.

   Note: At room temperature, SDS in the hybridization mixture precipitates in the presence of high salt, resulting in a cloudy solution. This can be prevented by heating the 2× hybridization mixture at 60°C immediately before use.
8. Denature the DNA for 10 min at 95°C.
9. Allow the solution to cool slowly to 68°C at a rate of 0.5°C/30 s and allow hybridization at 68°C for 4 h.

Preparation of the SA-PMPs

10. Resuspend the SA-PMPs by gently flicking the bottom of the tube until they are completely dispersed and then capture them by placing the tube in the magnetic stand until the SA-PMPs have collected at the side of the tube (±30 s). Carefully remove the supernatant.
11. Wash the SA-PMPs 3 times with 300 µL of 0.5× SSC, each time capturing them using the magnetic stand and removing the supernatant.
12. Resuspend the SA-PMPs in 100 µL 0.5× SSC.

Removal of the biotinylated hybrids using SA-PMPs

13. Quench the hybridization reaction by placing the tube on ice.
14. Mix 12.5 µL of 20× SSC and 477 µL of distilled water and add to the hybridization mixture to obtain a final concentration of 0.5× SSC.
15. At room temperature, add the diluted hybridization mixture to the SA-PMPs and mix gently.
16. Leave for 10 min at room temperature to allow binding of the streptavidin to the biotin residues.
17. Capture the SA-PMPs using the magnetic stand and carefully remove the supernatant without disturbing the SA-PMP pellet.
18. Transfer the supernatant to a microcentrifuge tube, centrifuge briefly to pellet any remaining SA-PMPs from the solution and transfer to a clean microcentrifuge tube.
19. Remove 25 µL of tracer solution and store at -20°C for subsequent amplification for the display gel.

The first long subtractive hybridization

20. Add 20 µg of driver DNA to the tracer solution and precipitate the DNA mixture by adding 0.3 M sodium acetate (final volume) and 2 vol of 96% ethanol and leave overnight at -20°C.
21. Repeat steps 2–18, but increase the hybridization time (step 9) to 48 h instead of 4 h.
22. Precipitate the tracer with 0.3 M sodium acetate (final volume) and 2 vol of

96% ethanol and leave overnight at -20°C.
23. Centrifuge the tracer populations at 13 000 rpm in the Model 5415C Eppendorf centrifuge for 10 min at 4°C. Dissolve the pellet in 100 µL of TE buffer (pH 8.0).
24. Remove 10 µL of the tracer solution, which is used for the driver synthesis for the next reciprocal subtraction and for analysis on the display gel. The remaining 90 µL tracer are used in the next long hybridization.

Note: For the synthesis of the driver, make a 1000-fold dilution of the DNA fragments and amplify 1 µL per reaction.

The following subtractive hybridization rounds

The next subtraction rounds consist only of long hybridization steps.

25. For the subsequent hybridizations, the driver is amplified using the tracer obtained from the previous reciprocal subtraction round, as described in the "Synthesis of the driver" section. The tracer is not further amplified to avoid potential PCR artifacts.
26. The 90 µL of the tracer obtained after subtraction (step 25) are mixed with 20 µg of newly synthesized driver and precipitated with sodium acetate to a final volume of 0.3 M and 2.5 vol of 96% ethanol. Leave for ca. 30 min at -20°C.
27. Repeat steps 21–24.

The Display Gel and Subsequent Recovery of Enriched cDNA Fragments

Enriched cDNA fragments were amplified in the presence of [^{32}P]dCTP and visualized on a denaturing 6% polyacrylamide gel. For polyacrylamide gel electrophoresis, we refer to Sambrook et al. (14) or Ausubel et al. (2).

Materials and Reagents

- [^{32}P]dCTP (>3000 Ci/mmol; 10 µCi per PCR)
- The starting tracer cDNA population and enriched tracer populations obtained from successive rounds of subtraction
- Loading buffer: 95% formamide, 20 mM EDTA, 0.025% xylene cyanol, 0.025% bromophenol blue
- 6% denaturing polyacrylamide gel
- TE buffer: 10 mM Tris-HCl, 1 mM EDTA, pH 7.5
- Glycogen (20 µg/µL)
- QIAquick PCR purification spin columns

Procedures

Display gel

1. The PCR is performed as described for the tracer synthesis, except that to increase the incorporation of radioactivity, the concentration of dNTPs was lowered to 40 µM instead of 200 µM, and the volume of the reaction is

reduced to 25 µL instead of 50 µL.

2. The PCR products obtained are precipitated with 0.3 M sodium acetate (final concentration) and 2 vol of 96% ethanol, left at -20°C for 1 h and spun down for 10 min at room temperature.
3. The pellets are washed with 70% ethanol, dried in a Speed-Vac® centrifuge (Savant Instruments, Holbrook, NY, USA) for 5 min and resuspended in 20 µL of loading buffer.
4. For 1 µL, the amount of radioactivity incorporated is established by scintillation counting and mixing equal amounts of radioactivity to 2 µL of loading buffer, with a final volume of 6 µL.
5. Allow the polyacrylamide gel to prerun for 1 h.
6. Prior to loading on the polyacrylamide gel, the cDNA fragments are denatured for 5 min at 80°C.
7. The samples are loaded on the polyacrylamide gel according to the following scheme: tracer A0, tracer A1,...tracer A4,...tracer B4,...tracer B1,...tracer B0, where A0 and B0 are the initial tracer materials, and Ax and Bx are the tracer materials obtained after the different subtraction rounds.
8. The gel is run for approximately 3 h at 60–70 W, fixed and dried as for a normal sequencing gel. Subsequently, the dried gel is tagged with landmarks (stickers labeled with radioactive ink) to allow the localization of the display after autoradiography.
9. Allow overnight autoradiography with X-ray film.

<u>Recovery of the amplified fragments</u>

10. After the autoradiography, the developed X-ray film is positioned on the display gel using the landmarks. The enriched gene products are cut out of the gel (through the X-ray film) with new scalpel knives for each fragment.
11. The gel slices are put into microcentrifuge tubes, and the DNA is recovered according to the following procedure.
    a. Add 100 µL of TE buffer (pH 7.5) to each polyacrylamide gel slice.
    b. Elute the DNA by boiling the gel slices for 10 min.
    c. Centrifuge briefly to precipitate the polyacrylamide and transfer the supernatant into a fresh microcentrifuge tube.
    d. Add 20 µg of glycogen and precipitate in the presence of 0.3 M sodium acetate and 2.5 vol of 96% ethanol.
    Note: Alternatively, 25 µL of the DNA solution can be amplified without precipitation.
12. Recover the DNA by centrifugation for 10 min at 4°C and rinse the pellet with 70% ethanol.
13. Dissolve the DNA in 10 µL TE buffer (pH 7.5).
14. Amplify the fragments in a single PCR using the conditions for the tracer amplification. Use 4 µL of TE buffer (step 13) as template for PCR. The number of cycles was increased to 30 to ensure the synthesis of sufficient material for the cloning strategy, and the extension step is omitted.
    Note: During the extension step, *Taq* DNA polymerase often adds an addition-

al dAMP at the 5' end of the PCR fragments, which prevents blunt-end ligation in the steps described below. Alternatively, when the extension step is performed, cloning of the fragments should be done into vectors especially designed for the cloning of PCR fragments, such as the TA Cloning® vector (Invitrogen, Leek, The Netherlands).
15. Purify the cDNA fragments using a QIAquick PCR purification spin column.
   Note: Authors' observations are that when less than 1 µL of the PCR mixture is used in the subsequent ligation step, purification of the DNA fragments is not necessary.
16. The amount of DNA synthesized is determined on a 1.5% agarose gel containing ethidium bromide.

Cloning of the Enriched cDNA Fragments

All the techniques described below are standard molecular biology techniques, and protocols can be found in laboratory manuals (2,14)

The cDNA fragments (ca. 500 ng) were ligated into 100 ng of the pBluescript vector, which was linearized with the restriction enzyme *Eco*RV. The authors' observations were that this cloning strategy works fine, although higher cloning efficiencies can be obtained using other vectors (e.g., the TA Cloning vector or pZErO™ [Invitrogen]). The constructs were transformed into DH5α™ cells (Life Technologies, Gaithersburg, MD, USA) by chemical transformation. The transformed cells were plated on LB plates containing ampicillin. The cultures were incubated overnight, and positive colonies were selected by blue/white color identification in the presence of 5-bromo-4-chloro-3-indolyl-β-D-galactopyranoside (X-gal) and isopropyl β-D-thiogalactopyranoside (IPTG). From each construct, about 6 white clones were analyzed by restriction digestion, and the fragment length was controlled on a 1.5% agarose gel. Each clone with a different length was sequenced. The sequences were compared to the data available in public databases.

RESULTS AND DISCUSSION

The subtractive hybridization presented here was used to isolate gene products involved in liver regeneration. cDNA derived from the regenerating and normal liver was used as starting material for the subtractive hybridization. First, tracer and driver material was prepared from cDNA fragments, which were modified to allow a specific PCR amplification. Tracer and driver fragments were amplified using distinct primers to avoid the removal of primer/adaptor sequences as in the original procedure (17). For the PCR amplification of the tracers (used for the isolation of the up-regulated gene products if insufficient starting material [<1 µg] was available), we started with as little as 1 pg of cDNA fragments per amplification reaction, which was amplified in 23 cycles. We noticed that if too much starting DNA (i.e., more than 15 pg DNA) was added per reaction or if too many cycles were performed, the PCR was prone to produce artifacts because of mispriming (M. Denijn, unpublished). This small amount of input cDNA, however, contains sufficient material to obtain a complete representation of mRNAs with an abundance of 20 copies per cell, as was calculated according to Klickstein (7). The following formula: $[P = 1 - (1 - n/T)^B]$, in which P is probability to find a specific mRNA, n is number of copies of the specific mRNA (n = 20), T is total number of all mRNAs in a cell [estimated at 500 000 mRNA molecules per cell (7)], and B is the number of mRNAs needed to identify the specific mRNA with a P of 0.99, was used (note that 1 - n/T is the probability of one mRNA not being the specific one). We calculated that B corresponded to 1.2×10^5 mRNAs. Assuming that the average length of the mRNA molecules is 1500 bases, at least $1500 \times 330 \times 2 \times 1.2 \times 10^5/6.02 \times 10^{23}$ = 197 fg of cDNA are needed. In this calculation, 330 is the average molecular weight of a nucleotide, 2 is the factor to calculate the molecular weight of the cDNA made from the mRNA, and 6.02×10^{23} is Avogadro's constant.

In contrast, for the synthesis of the driver material, 10 pg of starting DNA were added to each PCR, and two additional cycles were performed as compared to the synthesis of the tracer to obtain sufficient driver material. We reasoned that if PCR artifacts occurred in the driver population, they were not as critical as for the tracer population because the driver fragments are eliminated after subtraction. Performing the PCR with little input DNA and fewer cycles than described originally (17) also prevents normalization of the relative amounts of abundant fragments during amplification (12). If a fragment is present at a high concentration in the PCR,

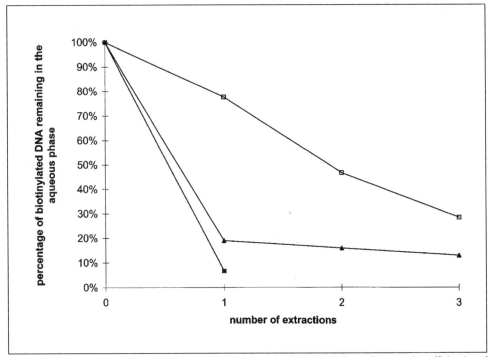

Figure 3. Extraction efficiency of biotinylated DNA fragments. This graph shows the extraction efficiencies of cDNA fragments containing one biotin residue per strand (□) (■) or multiple biotin residues per strand (▲) and extracted either with SA-PMPs or with an organic extraction procedure. cDNA fragments were tagged with [^{32}P]dCTP during the amplification. The extraction efficiency of the radioactive cDNAs was estimated by measuring the amount of radioactivity remaining after the extraction procedure. The extraction efficiency of cDNAs labeled either with biotinylated primers [one biotin per single strand (□)] or with photobiotin [multiple biotins per single strand (▲)] using a streptavidin-organic extraction was compared to that of cDNAs labeled with biotinylated primers and extracted with SA-PMPs [one biotin per strand; (■)]. For the organic extraction, 3 extraction steps were performed, each comprising a 15-min incubation with 2 mg/mL streptavidin at room temperature and a phenol–chloroform extraction. For the SA-PMPs, a single extraction was performed.

re-hybridization of the complementary strands will interfere with the PCR amplification, diminishing the relative concentration of the highly abundant cDNA fragments after PCR as compared to the initial cDNA population (12). If such a normalization occurred in one population only, it could alter the initial relative concentrations between tracer and driver. This in turn could affect the outcome of the subtractive hybridization.

The subtraction procedure started with a single short subtraction to remove the highly abundant common cDNA fragments, whereas the subsequent subtraction steps consisted of long hybridizations that aimed at the elimination of the less abundant common sequences (17). In the original method (15,17), the driver cDNA fragments were biotinylated using photobiotin. After hybridization, the biotinylated hybrids were removed by binding to streptavidin followed by an organic extraction. In the procedure described above, the extraction was simplified by the incorporation of a single biotin residue per single-stranded DNA during the PCR amplification, using biotinylated primers and by carrying out the extraction of driver hybrids in a single step using SA-PMPs. Using this technique, a better extraction efficiency was obtained because 93% of the driver hybrids were removed in a single extraction instead of 70% or 85% obtained after three organic extractions for driver labeled with a single biotin residue or for photobiotinylated driver, respectively (Figure 3). After each long subtraction, the driver for the next subtraction round was synthesized by PCR amplification of tracer material obtained from the reciprocal subtraction. Indeed, the up-regulated genes in one cDNA population are eliminated in the reciprocal cDNA population after subtraction, making cDNA fragments obtained after subtraction better drivers compared to the initial material (17). In contrast, the tracer was not amplified to avoid artificial cDNA fragments as a result of PCR artifacts.

For the isolation of factors involved in liver regeneration, four rounds of subtraction were performed, after which the progression of the subtraction was examined using a display gel (5). Because the up-regulated products have increased in their relative concentration after each subtraction cycle, they can be identified as discrete bands that increase in intensity (Figure 4). Also, a direct comparison could be made between the gene products in the regenerating and the normal liver, which allows the detection of up- and down-regulated fragments, provided that the primers used for tracer and driver amplification have the same length (5).

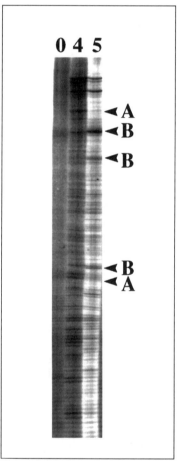

Figure 4. Display gel of the enriched cDNA fragments derived from regenerating liver mRNA. Lanes 0, 4 and 5 display the initial cDNA fragments and the enriched products obtained after 4 and 5 rounds of subtraction. In lane 5, the enriched population of fragments obtained after selective subtractive elimination with albumin cDNA, fibrinogen cDNA and mitochondrial DNA are visualized. The arrowheads indicate examples of gene products that were eliminated after the selective subtraction (A) and up-regulated gene products that are visualized after the selective elimination (B).

These up-regulated fragments were easily recovered from the display gel, amplified and sequenced. In addition to several fragments with no known homology with sequences published in databases, we found markers such as α_2-macroglobulin, albumin, mitochondrial DNA, hemopexin and vitamin D-binding protein. These gene products were abundant, as could be interpreted from the intensity of the bands on the display gel. A fifth subtraction round was performed using selected cDNA sequences of the identified abundant gene products as a driver. The abundant cDNA fragments were efficiently removed as shown in Figure 4, lane 5. By performing this elimination step, other less abundant up-regulated fragments, previously masked by the intensity of the dominant fragments, were revealed. The manipulation of the subtraction process by removing known (abundant) tracer-specific sequences from the tracer population increases the sensitivity of the technique. This feature proved to be a major advantage of the subtractive procedure presented here.

The subtractive hybridization technique presented here offers several advantages as compared to other techniques described to isolate differentially expressed gene products. First, the gene pool to be analyzed is enriched for differentially expressed genes by the extraction of common cDNA fragments present in the initial cDNA population. During the procedure, distinct primers are used to amplify tracer and driver population, so that additional manipulation of tracer and driver is avoided. Also, the use of biotinylated primers containing a single biotin residue in combination with SA-PMPs allows a simple and efficient removal of driver hybrids. The extraction procedure is easily monitored by the display of the distinct enrichment steps on a polyacrylamide gel. Multiple elimination rounds, which increase the sensitivity of the technique, are possible. Moreover, the subtractive hybridization can be performed in combination with the triple-helix method for the isolation of full-length cDNA sequences (Hakvoort et al., in this volume) and is a powerful tool for efficient isolation and characterization of differentially expressed gene products as a full-length cDNA.

REFERENCES

1. **Ariazi, E.A. and M.N. Gould.** 1997. Identifying differential gene expression in monoterpene-treated mammary carcinomas using subtractive display. J. Biol. Chem. *271*:29286-29294.
2. **Ausubel, F.M., R. Brent, R.E. Kingston, D.D. Moore, J.G. Seidman, J.A. Smith and K. Struhl.** 1994. Current Protocols in Molecular Biology. John Wiley & Sons, New York.
3. **Diatchenko, L., Y.F. Lau, A.P. Campbell, A. Chenchik, F. Moqadam, B. Huang, S. Lukyanov, K. Lukyanov et al.** 1996. Suppression subtractive hybridization: a method for generating differentially regulated or tissue-specific cDNA probes and libraries. Proc. Natl. Acad. Sci. USA *93*:6025-6030.
4. **Gurskaya, N.G., L. Diatchenko, A. Chenchik, P.D. Siebert, G.L. Khaspekov, K.A. Lukyanov, L.L. Vagner, O.D. Ermolaeva, S.A. Lukyanov and E.D. Sverdlov.** 1996. Equalizing cDNA subtraction based on selective suppression of polymerase chain reaction: cloning of Jurkat cell transcripts induced by phytohemagglutinin and phorbol 12-myristate 13-acetate. Anal. Biochem. *240*:90-97.
5. **Hakvoort, T.B.M., A.C.J. Leegwater, F.A. Michiels, R.A.F.M. Chamuleau and W.H. Lamers.** 1994. Identification of enriched sequences from a cDNA subtraction-hybridization procedure. Nucleic Acids Res. 22:878-879.
6. **Hubank, M. and D.G. Schatz.** 1994. Identifying differences in mRNA expression by representational difference analysis of cDNA. Nucleic Acids Res. 22:5640-5648.
7. **Klickstein, L.B.** 1994. Construction of a complete cDNA library, p. 5.8.1. *In* F.M. Ausubel, R. Brent, R.E. Kingston, D.D. Moore, J.G. Seidman, J.A. Smith and K. Struhl (Eds.), Current Protocols in Molecular Biology. John Wiley & Sons, New York.
8. **Lavery, D.J., L. Lope-Molina, F. Fleury-Olela and U. Schibler.** 1997. Selective amplification via biotin and restriction mediated enrichment (SABRE), a novel selective amplification procedure for detection of differentially expressed mRNAs. Proc. Natl. Acad. Sci. USA *94*:6831-6836.
9. **Liang, P., L. Averboukh and A.B. Pardee.** 1993. Distribution and cloning of eukaryotic mRNAs by means

of differential display: refinements and optimization. Nucleic Acids Res. *21*:3269-3275.
10. **Liang, P. and A.B. Pardee.** 1992. Differential display of eukaryotic messenger RNA by means of the polymerase chain reaction. Science *257*:967-971.
11. **Liang, P. and A.B. Pardee.** 1995. Recent advances in differential display. Curr. Opin. Immunol. *7*:274-280.
12. **Mathieu-Daudé, F., J. Welsh, T. Vogt and M. McClelland.** 1996. DNA rehybridization during PCR: the "C_0t effect" and its consequences. Nucleic Acids Res. *24*:2080-2086.
13. **McClelland, M., F. Mathieu-Daudé and J. Welsh.** 1995. RNA fingerprinting and differential display using arbitrarily primed PCR. Trends Genet. *11*:242-246.
14. **Sambrook, J., E.F. Fritsch and T. Maniatis.** 1989. Molecular Cloning: A Laboratory Manual, 2nd ed. CSHL Press, Cold Spring Harbor, NY.
15. **Sive, H.L. and T. St. John.** 1988. A simple subtractive hybridization technique employing photoactivatable biotin and phenol extraction. Nucleic Acids Res. *16*:10937.
16. **Velculescu, V.E., L. Zhang, B. Vogelstein and K.W. Kinzler.** 1995. Serial analysis of gene expression. Science *270*:484-487.
17. **Wang, Z. and D.D. Brown.** 1991. A gene expression screen. Proc. Natl. Acad. Sci. USA *88*:11505-11509.
18. **Welsh, J., K. Chada, S.S. Dalal, R. Cheng, D. Ralph and M. McClelland.** 1992. Arbitrarily primed PCR fingerprinting of RNA. Nucleic Acids Res. *20*:4965-4970.
19. **Zhang, L., W Zhou, V.E. Velculescu, S.E. Kern, R.H. Hruban, S.R. Hamilton, B. Vogelstein and K.W. Kinzler.** 1997. Gene expression profiles in normal and cancer cells. Science *276*:1268-1272.

Section IV

Obtaining Full-Length cDNAs and Homologous cDNAs

19 Enriched Full-Length cDNA Expression Library by RecA-Mediated Affinity Capture

Theodorus B.M. Hakvoort, Jacqueline L.M. Vermeulen and Wouter H. Lamers
Department of Anatomy and Embryology, Academic Medical Center, University of Amsterdam, Amsterdam, The Netherlands

OVERVIEW

Multiple specific full-length cDNAs can be isolated simultaneously in one experiment using the triple helix-mediated affinity capture method. The approach is based on the formation of triple-helical structures between single-stranded (ss)DNA and double-stranded (ds)DNA molecules mediated by the recombinase A (RecA) protein. We report that the combination of the subtractive hybridization method (to enrich a population of cDNA fragments for up- or down-regulated gene products) and the RecA-mediated affinity capture method yields a rapid and powerful procedure to prepare specific, enriched full-length cDNA expression libraries. These enriched libraries can be used directly for secondary screening assays based on function.

BACKGROUND

The cellular fingerprinting of gene products can result in the identification of a set of marker molecules that reflect the condition of the cell. Knowledge of these specific molecules can provide us with information that is essential for unraveling the biochemical complexity of metabolic events. During recent years, various techniques have been introduced to study the composition of the transcriptome. In these comparative studies, the differential expression of the genes in the cells is highlighted, and the up- and down-regulated gene products are taken as leads in the identification of fundamental cellular aspects.

Several methods have been designed to analyze the gene products within a cell or tissue. Examples are the simultaneous analysis of the entire population of gene products as in Velculescu et al. (14) and the differential display procedure (9) or by eliminating common products and identifying the remaining, enriched products as in subtractive hybridization (6,15) or suppression subtractive hybridization (2). (See also chapters by Denijn et al. and Diatchenko et al. in this volume.) In all these procedures, the identification of differential expression of genes is based on characterization of fragments of the gene products (cDNA).

Such an inventory does not, however, allow one to discriminate between the gene products that are responsible for the change in the cellular phenotype and those that result from it. To distinguish between these two groups, a functional test with the full-length cDNA is necessary. The feasibility of successfully executing such a laborious test, such as a cell proliferation assay, is largely dependent on the complexity of the population of gene products being studied. Ideally, this secondary screen is performed not as a one-by-one analysis of the clonal cDNAs but on the total enriched cDNA population corresponding with the cDNA fragments or subclasses (e.g., based on specified size range) of these cDNAs.

The isolation of specific recombinant plasmids from a cDNA library can be done in various ways, such as colony lifting and phage screening. Both procedures are the options of choice when only a few plasmid cDNAs have to be obtained but have their limitations when larger numbers of recombinant plasmids need to be isolated. It should be noted that the differentially expressed gene products detected by serial analysis of gene expression (SAGE) and differential display analysis have to be isolated as single fragments and therefore are less amenable to batchwise processing. This is in contrast to the subtractive hybridization and the suppression subtractive hybridization methods that can be streamlined into a batchwise secondary screening procedure based on functionality.

In principle, all specific full-length cDNAs corresponding to the population of the enriched cDNA fragments can be isolated simultaneously in a single experiment using the triple helix-mediated affinity capture method. The approach is based on the formation of triple-helical structures that can then be selectively isolated. These triple-helical structures are stable complexes between ssDNA and dsDNA molecules that, because of their sequence similarity, are held together by Hoogsteen hydrogen bonding. The *Escherichia coli* RecA protein facilitates the parallel positioning of the ssDNA fragment in the major groove of the dsDNA molecules and stabilizes the complex (7,8,11,12). Initially, typical triple helix motifs were used to form triple-helical structures. We extended this approach using cDNA fragments without any sequence constraint. We report that the combination of: *(i)* the subtractive hybridization method (to enrich a population of cDNA fragments for up- or down-regulated gene products) and *(ii)* the RecA-mediated formation of triple-stranded complexes between enriched cDNA fragments and double-stranded recombinant plasmids containing the full-length cDNAs yields a rapid and powerful procedure to prepare specific, enriched cDNA libraries. These enriched libraries can be analyzed for differential expression and can be used directly for the obligatory secondary screening based on function. Part of this method has been published (4), but a detailed description of the procedure, including cautionary notes, is given in this chapter.

PROTOCOLS

Materials and Reagents

- 10× RecA medium: 20 mM $CoCl_2$, 80 mM $MgCl_2$, 300 mM Tris-HCl, pH 8.0
- Buffer 1: 10 mM Tris-HCl, pH 7.5, 1 mM EDTA, 300 mM NaCl
- Buffer 2: 10 mM Tris-HCl, pH 7.5, 1 mM EDTA, 50 mM NaCl
- Buffer 3: 30 mM Tris-HCl, pH 8.8, 0.1 mM EDTA
- ATP-γ-S (20 µmol/200 µL; Boehringer Mannheim, Indianapolis, IN, USA)
- Bovine serum albumin (BSA) at 5 µg/µL in TE buffer
- 2 M NaCl
- 10% sodium dodecyl sulfate (SDS)
- Proteinase K at 20 mg/mL, freshly prepared
- Salmon sperm DNA at 10 mg/mL, sonicated
- RecA protein at 2 mg/mL (New England Biolabs, Beverly, MA, USA)
- Butanol-2
- Glycogen at 20 mg/mL (Boehringer Mannheim)
- 100 mM phenylmethylsulfonyl fluoride (PMSF)
- 100 mM dNTPs (Amersham Pharmacia Biotech, Piscataway, NJ, USA)
- 1 mM digoxigenin (DIG)-11-dUTP, alkali-stable (Boehringer Mannheim)
- Anti-DIG magnetic particles (Boehringer Mannheim)
- Magnetic stand
- Water baths, set at 37° and 65°C
- TE buffer: 10 mM Tris-HCl, 1 mM EDTA, pH 8.0
- XL1-Blue MRF' electrocompetent cells (Stratagene, La Jolla, CA, USA)
- QIAquick® Spin PCR Column (Qiagen, Chatsworth, CA, USA)
- 3 M NaOAc, pH 5.2
- Ethanol

Isolation of Differentially Expressed cDNA Fragments

A population of up- and down-regulated products (from here on referred to as enriched cDNA fragments) was obtained by the subtractive hybridization method (3,15) using hepatic cDNA from regenerating and normal tissue. For a detailed description of the procedure, see Denijn et al. in this volume.

Preparation of Digoxigenin-Labeled, Enriched cDNA Fragments

The enriched cDNA fragments (ranging in size from 200 to 700 bp) were labeled with DIG-11-dUTP. In the polymerase chain reaction (PCR) amplification, a 20–25-fold excess of dTTP over DIG-11-dUTP is used. The nucleotide concentration in the PCR amplification step is 200 µM NTP, in which the dUTP is set at 192 µM and supplemented with 8 µM DIG-11-dUTP.

Note: Instead of using DIG-11-dUTP during the PCR, one can also use 5'-end DIG-labeled oligonucleotide primers. Theoretically, the high binding affinity of the antibody for the DIG moiety should be sufficient to capture the cDNA fragment with only one DIG residue. In the subtractive hybridization procedure, the end-labeled biotin cDNA fragments were effectively removed from the mixtures (see Denijn et al.

in this volume).

The enriched DIG-labeled cDNA fragments are purified using a QIAquick Spin PCR Column (or equivalent) and recovered in TE buffer. The concentration was determined both spectrophotometrically and by ethidium bromide staining.

Note: It is essential to purify the DIG-labeled cDNA products because of the interference of the free DIG-11-dUTP with the anti-DIG binding step later on in the procedure.

Preparation of Full-Length cDNA Library

High-quality poly(A+) RNA was isolated from normal rat liver and used to prepare cDNA.

Note: It is essential to start with high-quality RNA preparations, because the preparation of full-length cDNA is dependent on the quality of the starting mRNA population.

The Great Lengths™ cDNA Synthesis Kit (CLONTECH Laboratories, Palo Alto, CA, USA) was used for this purpose with modification. The first-strand synthesis was initiated with an anchored oligo(dT) primer with a *Not*I site (5′-AAGCGGC-CGCT$_{25}$V) in which V = G, C or A in equal ratios.

Note: The anchor is needed for uniform first-strand synthesis, and the *Not*I site is present for cloning purposes.

The cDNA was unidirectionally inserted into the multiple cloning site of a eukaryotic expression vector like pcDNA3 (Invitrogen, Carlsbad, CA, USA), using a *Bst*XI linker and the *Not*I restriction site. Recombinant plasmids were propagated in XL1-Blue MRF′ cells by electrotransformation and cultivation on agar plates. Plasmid cDNA was isolated from 400 000 transformants using the alkaline lysis method (1) and taken up into TE buffer. RNA was removed by RNase treatment followed by phenol extraction and ethanol precipitation.

Note: The plasmid DNA preparation should be of good quality and clean because part of it will be used to transfect cells at the end of this procedure.

DNA concentration was calculated from the absorbance measured at 260 nm. Nucleotide sequence analysis of randomly selected clones showed that most of the known cDNA inserts contained the ATG start codon (albumin: 1.9 kb; fibrinogen: 1.6 kb; glutamine synthetase: 1.4- and 2.8-kb form; contrapsin-like protease inhibitor: 2.0 kb).

Triple Helix Affinity Capture

The method can be divided into four parts (Figure 1). Details on the formation of filaments and helical structures, the capturing of the structures and the final release of the recombinant plasmids are listed below.

RecA Protein Filament Formation and Single-Strand Binding

DIG-labeled double-stranded cDNA fragments (25–250 ng) were denatured and added to 10 µg RecA protein according to the procedure given below.

1. Add 2 µL BSA in TE buffer to a microcentrifuge tube.
2. Add 25–250 ng DIG-labeled double-stranded cDNA fragments.
3. Adjust volume to 80 µL with water.
4. Insert tube for 5 min in boiling water.
5. Immediately after step 4, place the tube in ice water for 5 min.
6. Add 10 µL from the 10× RecA medium.
7. Add 5 µL RecA protein.

    Note: We did not optimize the RecA protein to the DIG-labeled double-stranded cDNA fragment (see Results and Discussion).
8. Mix very carefully.

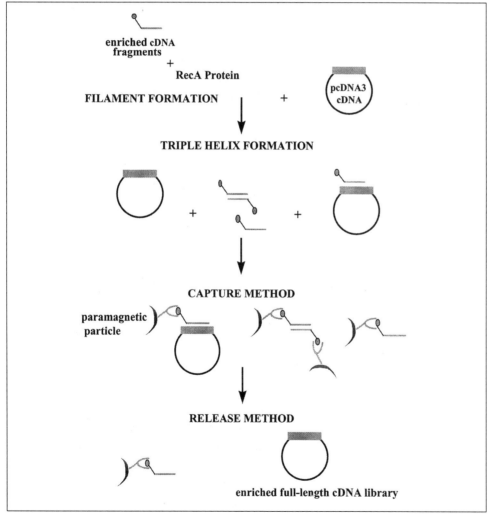

Figure 1. Schematic diagram of the isolation of differentially expressed full-length recombinant cDNAs. Enriched ssDNA fragments were fitted with a DIG hapten and used to generate triple-helical structures with double-stranded recombinant full-length cDNA in the presence of the RecA protein. The formed complexes were captured using immobilized anti-DIG antibodies, and the enriched full-length cDNAs were released from the single-stranded cDNA fragment. See Protocols section for details; the events labeled refer to corresponding steps in the procedure.

9. Incubate mixture for 1 min at 37°C.
10. Add 1.5 µL from the nonhydrolyzable ATP-γ-S.
11. Incubate tube for an additional 10 min at 37°C.

Formation of Triple Helices

12. Add 250–2500 ng of a full-length cDNA expression library to the mixture from step 11. Final volume is now 100 mL.
    Note: The average size of the cDNA expression library used in this method was 8 kbp.
13. Place tube at 37°C for another 20 min.

Capture of the Triple-Helical Structures

14. Add an excess of small nonrelated DNA fragments (5–10 µg), e.g., pBluescript® (Stratagene) restricted with *Msp*I and phenol-extracted.
15. Incubate tube for 3 min at 37°C.
16. Raise salt concentration to 50 mM NaCl by adding 2.5 µL 2 M NaCl.
17. Incubate tube for 3 min at 37°C.
18. Destroy the RecA protein filament structures by adding 1 µL proteinase K.
19. Add 2 µL 10% SDS.
20. Incubate tube for 7 min at 37°C.
21. Meanwhile, prepare the magnetic anti-DIG particles.
    a. Remove 10 µL magnetic anti-DIG beads from the stock solution and place them into a microcentrifuge tube.
    b. Collect particles to one side of the tube using a magnetic stand.
    c. Remove supernatant carefully.
    d. Add 100 µL buffer 1 to the beads and mix suspension.
    e. Repeat steps b, c and d four more times.
    f. Replace the supernatant with 10 µL buffer 1.
       Note: The binding specifications of the anti-DIG beads were >10 and >150 pmol for DIG-labeled double-stranded cDNA (1.5 kb) and DIG-labeled oligonucleotides, respectively. Therefore, 10 µL beads can bind up to 1000 and 15 000 fmol DIG-labeled probe. Assuming an average size of 400 nucleotides (nt) for the enriched cDNA fragment, this would translate into 90–900 fmol (step 2) of DIG-labeled single-stranded cDNA probe. These probes will have at most 4 DIG-labeled dUTPs per molecule when a 25-fold excess of dTTP over DIG-11-dUTP is used.
22. Add 1.5 µL PMSF to the tube from step 20.
    Note: The function of the protease inhibitor is to prevent degradation of the anti-DIG antibodies.
23. Add 10 µL magnetic anti-DIG particles from step 21f.
24. Mix suspension and place on lab table for 10 min at room temperature.
25. Place in magnetic stand and collect particles towards one side.

26. Remove medium from the beads.
27. Add 500 µL medium supplemented with 5 µg salmon sperm DNA and 5 µL PMSF.
28. Leave mixture on lab table for 2 min at room temperature.
29. Repeat steps 25–28 five times.

    Note: *(i)* Special care should be taken not to lose the beads during the many washing procedures. In our experience, turning the tube back and forth slowly in the magnetic stand causes the particles to concentrate in one spot. When pipetting off the supernatant, one can see them conveniently; and *(ii)* during the sequential washing steps, we occasionally observed the formation of clumps of particles. Mixing the suspension minimized this; however, these clumped particles did not have a negative effect on the full-length cDNA capture method.

30. Remove buffer 1 from beads.
31. Add 500 µL buffer 2 to beads.
32. Leave mixture on lab table for 2 min.
33. Remove buffer 2 using magnetic stand.
34. Repeat the sequence of steps 31–33 four times.

Release of the Plasmid cDNAs

35. Add 500 µL buffer 3 to the beads, prewarmed at 65°C.
36. Place tube from step 35 in water bath at 65°C for 10 min.
37. Collect the supernatant in a fresh microcentrifuge tube using the magnetic stand.
38. Add 500 µL buffer 3 to the beads, prewarmed to 65°C, and mix.
39. Place back at 65°C for an additional 10 min.
40. Collect the supernatant and combine it with the supernatant from step 37.
41. Reduce the volume of the eluate by butanol-2 extraction.

    a. Add equal volume of butanol-2.

    b. Mix thoroughly.

    c. Centrifuge for 5 min in a microcentrifuge tube.

    d. Remove the upper (organic) phase.

    e. Repeat steps a–d three to four times until volume is reduced to 100–200 µL.

42. Any remaining magnetic beads are removed from the solution by a short centrifugation step (15 000× *g*, 1 min); the supernatant is transferred to a fresh tube.
43. Add 5 µL glycogen.
44. Add 1/10 vol 3 M NaOAc.
45. Add 2 1/2 vol ethanol.
46. Place tube in ethanol/dry ice for one hour or at -20°C overnight.
47. Collect precipitate by centrifugation in a microcentrifuge tube for 30 min at maximum speed in the cold (4°C).
48. Carefully remove supernatant.

49. Add 500 µL 70% ethanol.
50. Centrifuge for 10 min.
51. Carefully remove supernatant.
52. Add 100 µL pure ethanol.

    Note: This step can be omitted. Its function is only to reduce the drying time on the lab table dramatically.

53. Centrifuge for 5 min.
54. Carefully remove supernatant.
55. Air dry pellet.

    Note: Only a short time is needed. Alternatively, one could use a vacuum centrifuge.

56. Resuspend precipitate in 10 µL TE buffer.

Propagation of the Enriched Full-Length Plasmid cDNA

We usually introduced the population of enriched cDNA plasmids into super electrocompetent cells in aliquots of 2–3 µL and plated them out directly onto Hybond®-N filters (FMC BioProducts, Rockland, ME, USA) (1). This serves two purposes. First, one can make copies for colony screening directly and simply and isolate the more rare gene products from even a normal library in a very efficient manner. This is only true if the particular gene product is not absent in one population. These latter gene products would require a specific cDNA library. The isolated gene products can be used to generate cRNA probes for localization studies of mRNA in tissues. Second and more important, the bacterial cells can be scraped from the plate or filter in portions and used to isolate plasmid cDNA (and once amplified, compared to the RecA-captured material). The isolated plasmids can be used in transfection assays for a secondary screening assay based on function to discriminate between the gene products that are responsible for the change in cellular phenotype and those that result from it.

RESULTS AND DISCUSSION

The DNA strand-exchange protein RecA is characterized by its ability to bind and pair DNA with no sequence constraints. For the high-affinity DNA binding state of RecA protein, a nucleoside triphosphate cofactor is required, and for this the nonhydrolyzable ATP-γ-S analog is often used (10). In our experimental setup, the conditions are such that only the binding and pairing activity of the RecA protein can occur and not strand exchange. In other words, the RecA protein functions to facilitate the formation of triple-helical structures, and this intermediate is the basis of our method. After that, the triple helices are selectively isolated and finally broken, yielding the captured recombinant plasmids.

We have applied the RecA-mediated affinity capture procedure to obtain an enriched full-length cDNA expression library from differentially expressed genes in regenerating and tumorous liver tissue. Differentially expressed cDNA fragments were isolated as described by Denijn et al. (in this volume) and used, as described in this chapter, to isolate the corresponding full-length cDNAs from a normal expression cDNA library. Figure 2 shows a colony blotting experiment in which the frequency of a novel gene product (R5-25-5) in a normal expression library is compared with that in the obtained enriched library. In the normal library, 8000 transformants were plated on a Hybond-N filter and screened with the R5-25-5 cDNA fragment isolated from the display gel as probe that was labeled with the random-prime method. The same probe was used for a filter on which 8000 transformants from the enriched library were plated. The number of positive colonies in the enriched library was 50-fold higher than that in the normal population. This ratio matched very well with that calculated for the enrichment during the subtractive hybridization (the ratio of prevalence of the cDNA fragment of this gene product in the enriched versus the initial cDNA fragment population [45-fold]) and therefore suggests a very efficient conversion step.

The experiment shown in Figure 2 is an example in which the frequency of a specific gene product in the population of the isolated enriched cDNA library is displayed. We also performed the same experiment with only one cDNA fragment to isolate the corresponding full-length cDNA from a library. In this particular experiment, we used 25 ng of the 3' end (400 nt) of the glutamine synthetase gene as the DIG-labeled probe and again the normal expression library (see Protocols section). Restriction analysis with isolated plasmid DNA preparations from randomly chosen clones showed that all the isolated clones contained a glutamine synthetase cDNA insert. Colony screening experiments in which a random prime-labeled 700-bp fragment of the 5' end of the gene product (nonoverlapping with the DIG-labeled glutamine synthetase fragment in the capture method) was used that showed that all colonies were glutamine synthetase-positive and therefore probably nearly full-length (results not shown).

Knowing the principle of the capture approach, we can evaluate those steps that are important for the outcome of the experiment and therefore are discussed below.

The RecA filament can bind ssDNA molecules. In addition, it has an affinity for

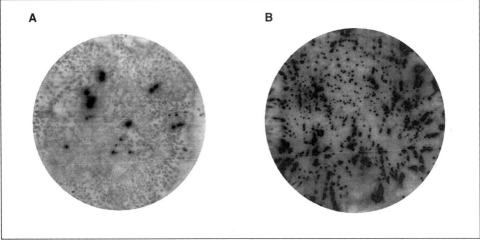

Figure 2. Colony hybridization experiment showing the frequency of a selected gene product in an expression library. (A) Normal expression library. (B) RecA-mediated affinity capture-enriched expression library. The enriched library is a population of cDNAs corresponding to the isolated differentially expressed cDNA fragments (see Protocols section and chapter by Denijn et al. in this volume for details). The probe used was a single enriched 300-bp cDNA fragment isolated from the display gel and labeled with the random-prime method. (An equal number of colonies were plated.)

dsDNA but to a much lesser extent (13). An excessive binding of dsDNA molecules to the RecA filaments should not interfere with the availability of the dsDNA in solution for the homologous base pairing with the ssDNA on the filaments. However, it should also be noted that the binding of the dsDNA to the RecA filaments enhances the formation of triple-helical structures by bringing them into close contact. Hence, an optimum ratio of ssDNA to dsDNA to the RecA protein has to be established (5). To minimize the dsDNA binding, either the RecA protein concentration has to be lowered or the binding sites on the filaments have to be saturated with nonrelated short DNA molecules. This should be added immediately after step 10 in our protocol. We did not optimize this step in the protocol. The natural recombinase activity of the protein is blocked by ATP-γ-S, so no strand exchange will take place. After the formation of the triple-helical structures, the RecA filaments are no longer needed. They are ruptured and destroyed by protease treatment. Special care should be taken not to break the triple helix at this stage of the procedure (factors that can destabilize the complex are high temperature, high pH and low salt concentration). The triple helices are isolated using the DIG moieties attached on the ssDNA fragments and an immobilized antibody. The DIG haptens in the cDNA fragments can also be substituted with biotin residues (preferentially by 5′ end-labeled oligonucleotide primers [see Denijn et al. in this volume]). When biotin moieties are used, an immobilized streptavidin should be used for capturing the triple-helical structures. The specificity of the conversion of the cDNA fragment into full-length cDNAs is better when DIG-labeled haptens are used (unpublished).

The magnetic beads have been suggested to preferentially bind small DNA molecules in a nonspecific manner. The final step in the procedure is the release of the captured recombinant plasmids from the beads. Therefore, special attention should be paid to minimizing the binding of other nonhomologous plasmids. These plasmids will also be released from the beads. To reduce the binding of nonspecific plasmid DNA to the beads, an excess of small DNA fragments unrelated to our cDNAs were added in step 14 of the procedure before the magnetic beads were added. These small DNA fragments will saturate the nonspecific binding sites on the beads. Because they are linear DNA fragments, they are not propagated in transfection assays.

The procedure as described in this chapter allows one to screen a complete library of cDNAs for a subset of full-length cDNAs in a simple step. The selection is done in liquid media, allowing the screening of a complete library in a single test tube. Conventional methods do rely on screening of libraries on solid media or multiple agar plates (colony and plaque screening), and they are rather time-consuming. Furthermore, the entire procedure does not require the preparation of single-stranded plasmid DNA and subsequent conversion to dsDNA, as is often required in other methods. The presented RecA-mediated affinity capture procedure provides a one-step conversion of isolated enriched cDNA fragments into the corresponding enriched full-length cDNA library. The availability of such a subset of differentially expressed full-length cDNAs makes secondary screening assays for function manageable.

REFERENCES

1. Ausubel, F.M., R. Brent, R.E. Kingston, D.D. Moore, J.G. Seidman, J.A. Smith and K. Struhl. 1994. Current Protocols in Molecular Biology. Wiley Interscience, New York.
2. Diatchenko, L., Y.C. Lau, A.P. Campbell, A. Chenchik, F. Moqadam, B. Huang, S. Lukyanov et al. 1996. Suppression subtractive hybridization: a method for generating differentially regulated or tissue-specific cDNA probes and libraries. Proc. Natl. Acad. Sci. USA 93:6025-6030.

3. **Hakvoort, T.B.M., A.C.J. Leegwater, F.A.M. Michiels, R.A.F.M. Chamuleau and W.H. Lamers.** 1994. Identification of enriched sequences from a cDNA subtraction-hybridization procedure. Nucleic Acids Res. *22*:878-879.
4. **Hakvoort, T.B.M., J.A.A. Spijkers, J.L.M. Vermeulen and W.H. Lamers.** 1996. Preparation of a differentially expressed, full-length cDNA expression library by RecA-mediated triple-strand formation with subtractively enriched cDNA fragments. Nucleic Acids Res. *24*:3478-3480.
5. **Honigberg, S.M., B.J. Rao and C.M. Radding.** 1986. Ability of RecA protein to promote a search for rare sequences in duplex DNA. Proc. Natl. Acad. Sci. USA *83*:9586-9590.
6. **Hubank, M. and D.G. Schatz.** 1994. Identifying differences in mRNA expression by representational difference analysis of cDNA. Nucleic Acids Res. *22*:5640-5648.
7. **Ito, T., C.L. Smith and C.R. Cantor.** 1992. Sequence-specific DNA purification by triplex capture. Proc. Natl. Acad. Sci. USA *89*:495-498.
8. **Ji, H. and L.M. Smith.** 1993. Rapid purification of double-stranded DNA by triple-helix-mediated affinity capture. Anal. Chem. *65*:1323-1328.
9. **Liang, P. and A.B. Pardee.** 1995. Recent advances in differential display. Curr. Opin. Immunol. *7*:274-280.
10. **Menetski, J.M., D.G. Bear and S.C. Kowalczykowsky.** 1990. Stable DNA heteroduplex formation catalyzed by the *Escherichia coli* RecA protein in the absence of ATP hydrolysis. Proc. Natl. Acad. Sci. USA *87*:21-25.
11. **Rigas, B., A.A. Welcher, D.C. Ward and S.M. Weissman.** 1986. Rapid plasmid library screening using RecA-coated biotinylated probes. Proc. Natl. Acad. Sci. USA *83*:9591-9595.
12. **Roman, L.J. and S.C. Kowalczykowsky.** 1989. Formation of heteroduplex DNA promoted by the combined activities of *Escherichia coli* recA and recBCD proteins. J. Biol. Chem. *264*:18340-18348.
13. **Story, R.M., I.T. Weber and T.A. Steitz.** 1992. The structure of the *E. coli* recA protein monomer and polymer. Nature *355*:318-325.
14. **Velculescu, V.E., L. Zhang, B. Vogelstein and K.W. Kinzler.** 1995. Serial analysis of gene expression. Science *270*:484-487.
15. **Wang, Z. and D.D. Brown.** 1991. A gene expression screen. Proc. Natl. Acad. Sci. USA *88*:11505-11509.

20 Gene-Capture PCR

Renato Mastrangeli and Silvia Donini
Istituto di Ricerca Cesare Serono SpA, Ardea, Rome, Italy

OVERVIEW

Gene-capture polymerase chain reaction (GC-PCR) is a novel homologous gene cloning strategy that was successfully used to isolate the full-length mouse counterpart of the human lymphocyte activation gene 3 (LAG-3) (12,17). This technique does not require sequence information on the 5' or 3' region of the mRNA of interest and permits gene isolation to be sensitive, fast, simple, specific and at relatively low cost. In principle, this technique should have wide use and could be applied to the isolation of homologous genes, gene families, gene variants and protein-encoding DNA sequences when limited knowledge of the amino acid sequence exists.

GC-PCR as used by the authors will be described (12). Potential applications and possible improvements will be discussed.

BACKGROUND

Isolation of homologous genes from related species is necessary to obtain information about human genes when they are known in other species and vice versa, and to define evolutionary paths and study conserved inter-/intra-species related genes (7). Conventional procedures to clone a homologous gene include the construction of a tissue- or cell-specific cDNA library and library screening with probes derived from various regions of the known gene of interest. Given that no information on homology is a priori available, this approach is time-consuming and could lead to the isolation of a large number of false-positive clones. Alternatively, homologous genes can be cloned by PCR but only when the homology is high enough to allow the design of specific primers. Screening procedures among homologous sequences derived from different animals and/or eukaryotic models are required to quickly obtain useful data and information on functionally important genes and their encoded products. The yeast genome sequence has been recently completed (1,13) and found to have nearly 31% potential genes sharing homology with mammals (1). Yeast manipulation is easy and cheap, and at least 71 human genes have been found to complement yeast mutations (1). Mouse models are commonly used for the functional characterization of new proteins by performing tissue distribution analysis, in vitro and in vivo biological tests, analysis of expression regulation and generation of gene knock-out and transgenic animals. A paradigm on human-mouse synteny has also been established (5). Comparison of protein sequences from different species is also relevant to define possible functionally conserved motifs and residues, which could help understanding of species specificity and rational design of agonist–antagonist peptidomimetics and muteins. The genomic era demands fast tools to study genes involved in human diseases and their counterparts in animal models, which are essential for the acceleration of discovery of novel therapeutic targets and development of new therapies.

Principle of GC-PCR

GC-PCR combines two powerful techniques: *(i)* the magnetic separation of DNA molecules, which allows easy and fast isolation of complexes bound to magnetic beads from a reaction mixture when a magnet is applied and *(ii)* PCR amplification which, because of its high sensitivity, allows the amplification and cloning of minute amounts of cDNA.

The strategy of GC-PCR as depicted in Figure 1 relies on the selection of amplifiable single-stranded cDNA molecules by multiple capture steps starting from total RNA or from DNA libraries.

Each capture step consists of single-stranded cDNA hybridization with a homologous capture probe followed by magnetic separation of hybridized (captured) single-stranded cDNA molecules. Selection is performed by sequential repeated capture steps with different homologous capture probes on single-stranded cDNA molecules released in a previous capture step. The capture probe is a single-stranded DNA sequence linked to magnetic beads and designed on a suitable region of the gene of interest. The capture probe can be a synthetic oligonucleotide or a cDNA molecule prepared by PCR. Amplifiable single-stranded cDNA to be subjected to GC-PCR must contain suitable sequences at the 5′ and 3′ ends to allow the PCR amplification of captured molecules.

Compared to traditional library screening, sequential hybridization with different probes is performed in liquid phase on single-stranded cDNA molecules in a microcentrifuge tube, thus allowing fast hybridization reactions, the processing of many samples such as libraries from different species and a 50% decrease in the library complexity, which results in a reduction of background. This cloning approach should speed up homologous gene screening between different species or within a gene family.

PROTOCOLS

Materials and Reagents

- TRIZOL® Reagent (Life Technologies, Gaithersburg, MD, USA)
- Oligonucleotide Purification Cartridge (OPC®) (PE Applied Biosystems, Foster City, CA, USA)
- LC Biotin-ON™ Phosphoramidite (CLONTECH Laboratories, Palo Alto, CA, USA)

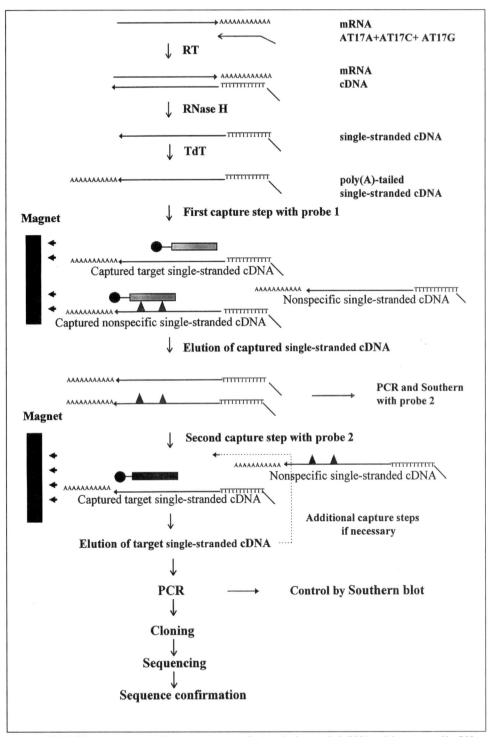

Figure 1. GC-PCR strategy. Total RNA is reverse-transcribed to single-stranded cDNA and then removed by RNase H treatment. Single-stranded cDNA is poly(A)-tailed with TdT and subjected to two rounds of hybridization with biotinylated homologous probes bound to streptavidin-coated beads, followed by magnetic separation of the hybridizing molecules. The selected cDNA is then amplified by PCR with oligo(dT) mixture and anchor primer A, controlled by Southern blotting, cloned and sequenced.

- Eluent 1: 0.1 M triethylammonium acetate (TEAA), pH 7.0, 5% acetonitrile
- Eluent 2: 100% acetonitrile
- Aquapore® Butyl Brownlee™ Column (220 × 4.6 mm; PE Applied Biosystems)
- Amicon® Centricon®-SR3, Microcon®-100 (Millipore, Beverly, MA, USA)
- Thermostable DNA polymerases: *Taq* (Advanced Biotechnologies Ltd., Leatherhead, Surry, UK) and *Pfu* (Stratagene, La Jolla, CA, USA)
- 2.5 mM dNTP, 6.5 mM dNTP, 2.5 mM dATP
- OLIGO™ 5.0 Primer Analysis Software (National Bioscience, Plymouth, MN, USA)
- 0.2× TE buffer: 2 mM Tris-HCl, 2 mM EDTA, pH 8.0
- TEN buffer: 10 mM Tris-HCl, 0.1 mM EDTA, 1 M NaCl, pH 8.0
- Streptavidin-coated M280 Dynabeads® (10 mg/mL), magnetic particle concentrator for Eppendorf microcentrifuge tubes (MPC®-E; Dynal AS, Oslo, Norway)
- 0.15 M NaOH
- RNase H RNase inhibitor (Boehringer Mannheim GmbH, Mannheim, Germany)
- SUPERSCRIPT™ II RNase H⁻ Reverse Transcriptase and 5′ RACE System for Rapid Amplification of cDNA Ends (both from Life Technologies): terminal deoxyribonucleotidyl transferase (TdT), SUPERSCRIPT, *E. coli* RNase H, 10× synthesis buffer, 0.1 M dithiothreitol (DTT) and 10 mM dNTP
- 20× standard saline citrate (SSC): 3 M NaCl, 0.3 M sodium citrate
- 10% (wt/vol) sodium dodecyl sulfate (SDS)
- Hybridization buffer 1: 6× SSC, 0.1% (wt/vol) SDS
- Hybridization buffer 2: 5× SSC, 0.1% (wt/vol) SDS, 1% *N*-lauroylsarcosine, 0.5% (wt/vol) blocking reagent
- Elution buffer: 10 mM Tris-HCl, 1 mM EDTA, pH 8.0
- 25 mM $MgCl_2$
- Dimethyl sulfoxide (DMSO)
- Positively charged nylon membrane, blocking reagent, PCR digoxigenin (DIG) labeling mixture, horseradish peroxidase (HRP)-conjugated anti-DIG antibodies and DNA DIG-labeled molecular weight markers (all from Boehringer Mannheim GmbH)
- Washing buffer: 1× SSC, 0.1% (wt/vol) SDS
- ECL™ reagents and Hyperfilm® ECL (both from Amersham Pharmacia Biotech, Piscataway, NJ, USA)
- Prism™ Cycle Sequencing Ready Reaction Kit (PE Applied Biosystems)
- Model 392 Automatic DNA Synthesizer and Model 373A Automated DNA Sequencer (both from PE Applied Biosystems)
- HPLC BioLC 4500 (Dionex, Sunnyvale, CA, USA)
- Thermal Cycler The Protocol™ (AMS Biotechnology Europe Ltd., Lugano, Switzerland)

Oligonucleotides

Oligonucleotides used as primers are listed in Table 1.

The primer mixture (10 pmol/μL each of AT17G, AT17A and AT17C) is referred to as oligo(dT) mixture and allows single-stranded cDNA fragments to be obtained with homogeneous 5′ ends (6,10). Oligo(dT) mixture introduces *Bam*HI and *Acc*III restriction sites with the anchor (underlined sequences). *Sfi*I and *Not*I restriction sites could be introduced in the anchor to facilitate the cloning. This primer mixture must be prepared in diethyl pyrocarbonate (DEPC)-treated water to avoid possible RNA degradation during the reverse transcription (RT) reaction.

F1, F2, F3, R1, R2 and R3 oligonucleotide sequences are designed from the known gene of interest. Bio-F1, Bio-F2 and Bio-F3 are biotinylated oligonucleotides to be used as forward primers. R1, R2 and R3 are nonbiotinylated oligonucleotides to be used as reverse primers. Primer sets Bio-F1/R1, Bio-F2/R2 and Bio-F3/R3 should amplify different DNA regions of the gene of interest and will be used to synthesize capture probes 1, 2 and 3, respectively. Bio-F2/R2 and Bio-F3/R3 will also be used to obtain the DIG-labeled probes 2 and 3 to be used in control Southern blotting experiments. Oligonucleotides can be synthesized by any automatic DNA synthesizer with a dimethoxytrityl-On (DMT-On) procedure and purified using OPCs or analogous devices following the manufacturer's instructions. Alternatively, they can be purchased from any commercial oligonucleotide synthesis service.

Biotinylated oligonucleotides are synthesized by the DMT-Off procedure using LC Biotin-ON phosphoramidite, which is incorporated at the 5′ end of the oligonucleotides. LC Biotin-ON has a 14-atom linking arm that reduces steric hindrance and facilitates streptavidin binding. Biotinylated oligonucleotides are purified according to the manufacturer's instructions or by reverse-phase high-performance liquid chromatography (RP-HPLC).

RP-HPLC purification

A reverse-phase column such as the Aquapore butyl Brownlee column (220 × 4.6 mm) is used with eluents 1 and 2 at a flow rate of 1 mL/min and the UV detector set at 260 nm. A linear gradient between 5% and 45% eluent 2 for 20 min is applied. Biotinylated oligonucleotides are collected after the elution of nonbiotinylated material (Figure 2). The recovered biotinylated oligonucleotides are then ultrafiltered in a Centricon-SR3 (with 3 washings in 1 mL H_2O) to remove salts and acetonitrile.

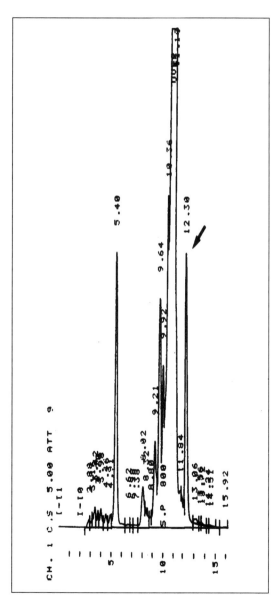

Figure 2. RP-HPLC chromatogram of crude biotinylated 22-mer oligonucleotide.

Table 1. Primers Used in GC-PCR

Poly(A)-Specific Sequences	
5'-GGCCCTGGATCCGGACCTAA(T)$_{17}$G-3'	AT17G
5'-GGCCCTGGATCCGGACCTAA(T)$_{17}$A-3'	AT17A
5'-GGCCCTGGATCCGGACCTAA(T)$_{17}$C-3'	AT17C
5'-GGCCCTGGATCCGGACCTAA-3'	Primer A
Gene-Specific Sequences	
5' Bio- [F1 sequence] 3'	Bio-F1
5' Bio- [F2 sequence] 3'	Bio-F2
5' Bio- [F3 sequence] 3'	Bio-F3
5' [R1 sequence] 3'	R1
5' [R2 sequence] 3'	R2
5' [R3 sequence] 3'	R3

Note: Figure 2 shows the RP-HPLC profile of a crude synthetic mixture in which the yield of biotinylated product was low. The arrow shows the biotinylated material that can be easily purified by RP-HPLC even in such a critical situation.

Total RNA Extraction

Extract total RNA from cells or tissues using TRIZOL reagent according to the manufacturer's instructions or by other conventional procedures.

Synthesis and Immobilization of Biotinylated Single-Stranded Capture Probes

Biotinylated probes are prepared by PCR on distinct DNA regions of the gene of interest using biotinylated forward (Bio-F) and nonbiotinylated reverse (R) primer sets (Figure 3, Step 1).

PCR Procedure

1. PCR mixture:
   x µL DNA template (150 fmol)
   9 µL 10× *Taq* Buffer IV (Advanced Biotechnologies Ltd.)

3 μL Bio-F primer (10 μM)
3 μL R primer (10 μM)
8 μL dNTP (2.5 mM)
y μL H$_2$O
90 μL total volume

2. Add mineral oil.
3. Denature PCR mixture at 95°C for 5 min, then maintain the temperature at 80°C.
4. Perform hot start by adding 10 μL of *Taq* DNA polymerase (2.5 U in 1× *Taq* buffer) at 80°C.
5. Perform 25 PCR cycles of denaturation at 96°C for 30 s, annealing at X°C for 30 s and extension at 72°C for 3 min.
6. Perform final extension at 72°C for 5 min.
7. Store at 4°C.

Annealing is performed at a suitable temperature (X°C) for each primer pair (optimal annealing temperature for each primer pair can be obtained using software such as OLIGO 5.0 Primer Analysis Software). The hot-start procedure (2) should be used to prevent mispriming and primer dimerization. The biotinylated PCR products are controlled and purified by agarose gel electrophoresis or anion-exchange (AE)-HPLC. However, if a unique product is obtained, the PCR mixture can be directly ultrafiltered using a Microcon-100 microconcentrator with 500 μL TEN buffer four times to remove the residual biotinylated primer (which could affect the next binding step to magnetic beads) and to equilibrate the biotinylated probe in 100 μL TEN buffer.

Biotinylated single-stranded capture probes are prepared according to Dynal instructions, as briefly described below.

Immobilization of biotinylated double-stranded probe to magnetic beads

One hundred microliters of magnetic beads (1 mg) (streptavidin-coated M-280 Dynabeads [10 mg/mL]) are washed four times with 100 μL TEN using the MPC-E. The biotinylated probe (90 μL) is added to the beads and incubated at room temperature for 30 min (Figure 3, step 2). After probe immobilization, the beads are washed four times with 100 μL TEN.

To control the immobilization efficiency, 10 μL of biotinylated double-stranded probe before binding and 10 μL of supernatant of the binding reaction are loaded on an agarose gel. A decrease of ethidium bromide signal in the post-binding supernatant as compared to the one before binding indicates that most of the probe is immobilized (Figure 4).

Denaturation of immobilized double-stranded probe

The immobilized double-stranded probe is then denatured with 0.15 M NaOH for 10 min at room temperature (Figure 3, step 3). After the magnetic separation and three 100-μL washes with TEN, the single-stranded capture probe is ready for the downstream applications. The material can be stored for days at 4°C. For long-term storage, 0.02% sodium azide should be added.

Synthesis of Amplifiable Single-Stranded cDNA Starting from Total RNA

Amplifiable single-stranded cDNA for GC-PCR must contain suitable sequences at the 5′ and 3′ ends for further PCR amplification. Different procedures can be used, e.g., TdT tailing (6), single-strand ligation of cDNA (SLIC) (4) and the switch mech-

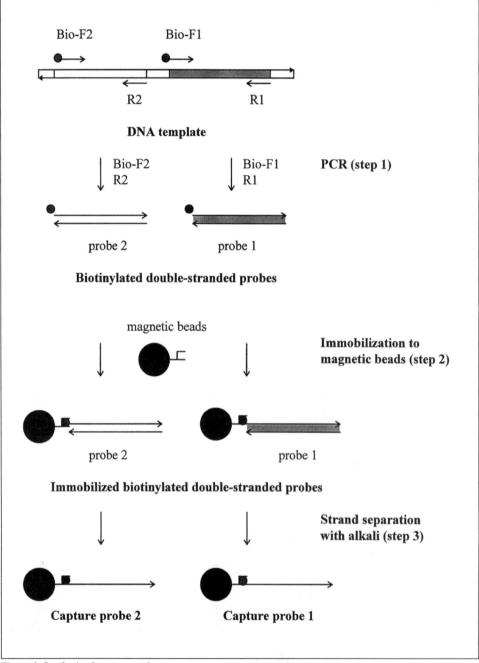

Figure 3. Synthesis of capture probes.

anism at the 5' end of RNA template (SMART™) technology using the SMART oligonucleotide during the RT reaction (SMART Kit; CLONTECH). Each procedure has its own advantages in terms of simplicity, cost and rapidity.

We used the simplest and least expensive TdT homopolymeric tailing with dATP (6). By this procedure, the same primer mixture for RT and PCR was used. The disadvantage was that the PCR product must be cloned to obtain the DNA sequence. Possible problems with the inverted repeat at the 5' and 3' ends, such as the suppression PCR effect (3), were not observed, probably because of the high primer concentration and the similar stability between intramolecular and intermolecular annealing.

The RT reaction is a critical step, especially for long retrotranscripts (6). The total RNA preparation should be of good quality, and the reaction conditions should favor the full-length single-stranded cDNAs. Several discrete products might derive from the same transcript if the RT reaction is not completed. Many commercial kits are now available to perform long-distance RT-PCR; the SMART kit is one of them and should be chosen accordingly.

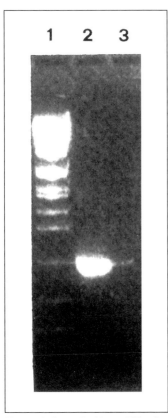

Figure 4. Control of immobilization efficiency of biotinylated double-stranded probe on magnetic beads. Biotinylated double-stranded probe (713 bp) before (lane 2) and after (lane 3) binding reaction. Molecular weight marker VII (Boehringer Mannheim) is shown in lane 1.

Reverse Transcription Procedure

We recommend to follow all indications to avoid RNA degradations described in the 5' RACE instruction manual or in any current RNA isolation procedure.

1. Place the following components in a 0.5-mL PCR tube:

    x µL   total RNA (5 µg)
    2 µL   oligo(dT) mixture (10 µM each primer)
    y µL   DEPC-treated H$_2$O
    10 µL   total volume

2. Incubate at 65°C for 10 min, then chill on ice.
3. Add the following components:

    10 µL   5× SUPERSCRIPT II buffer
    5 µL   0.1 M DTT
    5 µL   10 mM dNTP
    1 µL   RNase inhibitor (10–50 U)
    2 µL   SUPERSCRIPT II (400 U)
    17 µL   DEPC-treated H$_2$O
    50 µL   total volume

4. Incubate at 37°C for 90–120 min (42°C might be preferred).
5. Stop the reaction at 90°C for 5 min, then chill on ice for 10 min.
6. Add 1 µL RNase H (1 U).
7. Incubate at 37°C for 30 min to degrade RNA.

8. Single-stranded cDNA is ultrafiltered using Microcon-100 microconcentrators to remove residual primers, dNTP, degraded RNA and possibly cDNA molecules shorter than 300 bases. This step avoids unwanted TdT reactions such as different nucleotide incorporation or primer tailing. Ultrafiltration is performed with 0.2 mL of 0.2× TE buffer at 500× g for 5 min followed by four ultrafiltration steps with 0.5 mL of 0.2× TE buffer at 500× g for 12 min. Final volume is adjusted to 50 µL with 0.2× TE buffer.

   Note: Compared to Centricon, Microcon might experience leakage because of membrane damage or mechanical defects. Be careful not to damage the membrane, especially when the sample reservoir is disconnected from the Eppendorf tube during multiple washing steps. Store all eluates that should be concentrated with a new device in case of leakage.

9. Store ultrafiltered single-stranded cDNA at -20°C.

Single-Stranded cDNA Poly(A) Tailing with TdT

1. Place 16 µL ultrafiltered single-stranded cDNA (corresponding to 1.6 µg total RNA) in a 0.5-mL PCR tube.
2. Incubate at 70°C for 5 min and chill on ice. By this treatment, single-stranded cDNA with 3′ ends free from secondary structures is obtained.
3. Add 1 µL 10× synthesis buffer (from 5′ RACE system kit), 2 µL dATP (2.5 mM) and 1 µL TdT (10 U) for a total volume of 20 µL.
4. Incubate for 10 min at 37°C.
5. Inactivate TdT for 10 min at 70°C.
6. Store poly(A)-tailed single-stranded cDNA at -20°C.

   Note: Bovine serum albumin (BSA) at a final concentration of 50 µg/mL deriving from the synthesis buffer is present in the final mixture.

   Poly(A)-tailed single-stranded cDNA can be amplified without further purification. Four microliters of the mixture could be used to control the reaction. A DNA smear larger than 300 bp is expected. In our RT-PCR conditions, a smear between 500 and 2700 bp with higher intensity in the 700–1150-bp region was observed on the ethidium bromide-stained gel.

Capture Step

Procedure

1. Equilibrate magnetic beads linked to probe 1 at 0.25 mg beads/100 µL by washing 3 times in 100 µL hybridization buffer.
2. Add 10 µL poly(A)-tailed single-stranded cDNA (equivalent to 0.8 µg total RNA) and mix gently.
3. Incubate the mixture for 10 min at 70°–90°C to denature possible secondary structures on both single-stranded probe and poly(A)-tailed single-stranded cDNA.
4. Incubate the mixture for 4 h at 45°C under gentle agitation to maintain the beads in suspension. Incubation temperature should be adjusted depending on

the required stringency.

5. Hybrids between capture probe 1 and poly(A)-tailed single-stranded cDNA are retrieved from the solution by the MPC-E magnet. The beads are washed 3 times at room temperature with 100 µL of hybridization buffer and then resuspended in 300 µL of the same buffer.
6. Beads are divided in three 100-µL aliquots (each equivalent to 0.26 µg total RNA).
7. Wash each aliquot 5 times with 100 µL of hybridization buffer at different temperatures to determine the optimal selection conditions, i.e., 45°, 55° and 65°C. Alternatively, the salt concentration can be decreased to increase the stringency.
8. Release the captured poly(A)-tailed single-stranded cDNA from the beads by elution with 100 µL of a low-salt buffer (i.e., 10 mM Tris-HCl, 1 mM EDTA, pH 8.0) for 5 min at 90°C.
9. Add BSA (50 µg/mL final concentration) to the recovered poly(A)-tailed single-stranded cDNA to avoid DNA loss.
10. Thirty microliters of each recovered poly(A)-tailed single-stranded cDNA are amplified by PCR and controlled by Southern blotting with DIG-labeled probe 2 as described below. This control confirms the presence of the putative homologous gene and provides information on the complexity of captured single-stranded cDNA. In addition, by performing Southern blotting under various stringencies, one can get useful information on the conditions required for the next capture step.
11. The next capture step is performed as described above (steps 1–10) except that magnetic beads linked to biotinylated single-stranded probe 2 (0.25 mg beads/200 µL) are used. Capture probe 2 is hybridized to poly(A)-tailed single-stranded cDNA captured under optimal hybridization conditions, as derived from the results of step 10. Mix the following components:

138 µL	captured single-stranded cDNA (in low-salt buffer of step 8)
60 µL	20× SSC
2 µL	10% SDS
200 µL	total hybridization volume

12. Incubate for 4 h at a suitable temperature according to the results of the previous Southern blotting. In our study, the hybridization was performed at 65°C. After hybridization, the reaction mixture was divided into two 100-µL aliquots, which were washed at 65°C with either 6× or 1× SSC.
13. Thirty microliters of each recovered poly(A)-tailed single-stranded cDNA are amplified by PCR and controlled by Southern blotting with DIG-labeled probe 3 as described below. This control should confirm the isolation of the putative homologous gene and indicate whether other capture steps are required. In our study, two capture steps were sufficient to isolate murine LAG-3.
14. If necessary, perform an additional capture step with biotinylated single-stranded probe 3.

As a control of the capture step efficiency, PCR and Southern blotting of captured single-stranded cDNA and desalted noncaptured single-stranded cDNA should be performed as depicted in Figure 5.

PCR of Captured Amplified Single-Stranded cDNA

PCR must be performed in conditions suitable to obtain full-length molecules, especially in the critical first cycle when amplification fails if the full-length double-stranded cDNA is not obtained. Captured poly(A)-tailed single-stranded cDNA is amplified by PCR with *Taq* DNA polymerase according to the method of Frohman (6) with minor modifications.

PCR Procedure

1. PCR mixture:
    - 30.0 µL poly(A)-tailed single-stranded cDNA
    - 0.5 µL oligo(dT) mixture (10 µM each primer)
    - 8.0 µL 10× *Taq* buffer IV (without MgCl$_2$)
    - 8.0 µL DMSO
    - 21.0 µL MgCl$_2$ (25 mM)
    - 8.0 µL dNTP (6.25 mM)
    - 75.5 µL total volume
2. Add mineral oil.
3. Denature PCR mixture at 95°C for 5 min, then adjust to 80°C.
4. Perform hot start by adding 4 µL of *Taq* DNA polymerase (2.5 U in 1× *Taq* buffer) at 80°C.
5. Start the first PCR cycle of denaturation at 96°C for 30 s, annealing at 48°C for 2 min and extension at 72°C for 30 min.
6. Perform additional cycles of denaturation at 96°C for 4 min and 80°C for 5 min.

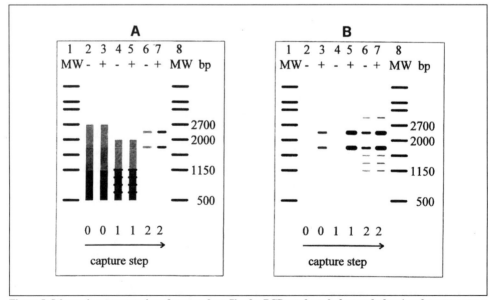

Figure 5. Schematic representation of expected profiles for PCR products before and after 1 or 2 capture steps. (A) Ethidium bromide-stained gel and (B) Southern blotting with DIG-labeled probe 3. (+) Induced cells; (-) noninduced cells; (MW) DIG-labeled molecular weight markers.

Add 6 µL of primer A (10 µM) at 80°C.
7. Perform 35 PCR cycles of denaturation at 96°C for 30 s, annealing at 60°C for 30 s and extension at 72°C for 4 min.
8. Perform a final extension at 72°C for 10 min.
9. Store at 4°C.

Note: A buffer composition of 6.5 mM $MgCl_2$, 10% DMSO and 1.5 mM dNTP was found by Frohman to give high amplification yield of cDNA.

In step 7, the extension time at 72°C (ca. 1 min/kb) will depend on the expected PCR fragment length (length might be deduced from previous Northern blot experiment).

As an alternative, to avoid anchor primer A mispriming, all PCR cycles can be performed with oligo(dT) mixture, which should be used at a final concentration of 600 nM. Under these conditions, PCR specificity should be improved. In this case, do not perform step 6. Two or three additional cycles might be necessary if PCR products are not visible on an ethidium bromide-stained gel.

In addition, AmpliTaq Gold™ (Perkin-Elmer, Norwalk, CT, USA), which is activated by high temperature, TaqStart™ Antibody (CLONTECH) or HotWax™ Mg^{2+} Beads (Invitrogen, Carlsbad, CA, USA) allow to avoid the manual hot-start procedure (step 4).

The PCR products are controlled by Southern blotting with a DIG-labeled probe whose sequence is not present in the capture probes. PCR products are blunt-ended with *Pfu* DNA polymerase, purified and cloned into a plasmid vector.

Southern Blotting of Captured cDNA

Ten microliters of amplified cDNA are subjected to suitable agarose gel electrophoresis and blotted onto a positively charged nylon membrane. The blot is hybridized with a DIG-labeled probe (50 µL) overnight at 45°C in hybridization buffer 2.

For probe synthesis, a suitable 5′ region of the gene of interest, which is external to the capture probes, is first amplified. The PCR product is ultrafiltered using Microcon-100 microconcentrators to remove primers, dNTP and enzyme, and 15 ng are subjected to 25 cycles of asymmetric PCR with 20 pmol of reverse primer R, 10 µL of PCR DIG-labeling dNTP mixture and 2.5 U *Taq* DNA polymerase in 100 µL final volume. Store the reaction mixture at -20°C until required.

Blots are washed 4 times with 1× SSC, 0.1% SDS at a suitable temperature depending on the requested stringency. A priori, because the degree of homology is unknown, various stringency conditions should be tested (the OLIGO program provides melting temperatures at several concentrations for a given probe). Hybrid detection is performed with HRP-conjugated anti-DIG antibodies according to the manufacturer's instructions and ECL reagents.

Cloning of Captured cDNA

Amplified cDNA is ligated to a linearized plasmid (i.e., *Sma*I-digested pBluescript® [Stratagene]). The ligation mixture is transformed into *E. coli* strain XL1-Blue. However, SURE® strains are available from Stratagene that allow the cloning of unstable DNA structures such as inverted repeats as well as ultracompetent cells suitable for transformation of plasmid library (i.e., XL10-Gold cells). Recombinant plasmids are controlled by restriction analysis, and one positive clone is sequenced.

DNA Sequence and Analysis

Plasmid DNA is sequenced to confirm the cloned cDNA to be the correct one. Universal primers present on the plasmid vector upstream and downstream of the cloned insert can be used to start DNA sequencing. A gene-walking strategy can then be used to complete the sequencing. Some available primers designed on the known gene can be used for sequencing if the 3′ end perfectly matches the homologous gene. We usually perform cycle sequencing using the DyeDeoxy Terminator chemistry and the automated DNA sequencer (both from Perkin-Elmer).

As the isolated sequence is amplified with *Taq* DNA polymerase and a modified PCR buffer that increases the yield but also increases the polymerase error rate (6), misincorporation as well as single-base deletion might be possible (personal observation). Therefore, two additional independent amplifications with primers derived from the cloned sequence must be performed under optimal conditions to minimize misincorporations. Sequencing of these PCR products will confirm the correctness of the sequence. Different enzymes such as *Pfu* or *Pwo* (Boehringer Mannheim) might be used to perform the additional amplification under increased fidelity conditions. Generally speaking, sequencing on 3 different independent RT-PCR products is sufficient to unequivocally determine the natural DNA sequence.

Expected Results

As described above, results of PCR and Southern blotting of single-stranded cDNA derived from RNA and the various capture steps will provide information on the GC-PCR performance. In case of induced genes, comparison of induced (+) vs. noninduced (-) captured cDNA might help to further evaluate GC-PCR performance (Figure 5).

Southern blotting must be preferentially performed using a DIG-labeled probe comprising a region not included in the capture probes. If the gene of interest comprises 3 regions (e.g., 1, 2 and 3) and assuming that the capture probes correspond to regions 1 and 2, the DIG-labeled probe should match region 3 to assure that all regions (1, 2 and 3) are present in the captured single-stranded cDNA. In addition, using this approach, the background signal due to nonspecific captured cDNA will be avoided.

Figure 5 shows the expected band profile of an ethidium bromide-stained gel and the corresponding Southern blot.

RESULTS AND DISCUSSION

Under our RT-PCR conditions, the noncaptured material showed a smear between 500 and 2700 bp with higher intensity in the region of 700–1150 bp, as determined by ethidium bromide staining. Two capture steps were sufficient to isolate murine LAG-3 from less than 0.8 µg of total RNA (Figure 6). Two major 2.2- and 1.8-kb bands were observed, corresponding to the full-length and incomplete cDNAs, respectively.

After the first capture step, a smear is still observed by ethidium bromide staining, but the homologous gene is clearly visible by Southern blotting in activated thymocytes only (Figure 6, A and B, compare lanes 6 and 7). After the second capture step, the ethidium bromide-stained gel clearly showed only the bands of interest (Figure 6A, lane 2). In addition, by Southern blotting, these bands were even found in non-activated thymocytes (Figure 6B, lane 1). Because nonactivated thymocytes and T cells do not express LAG-3 at detectable levels, as determined by Northern blotting (12,17), this result demonstrates the potential of the strategy to isolate even genes expressed at very low levels.

In our study, the homology of capture probes 1 and 2 (713 and 290 nucleotides long, respectively) with the gene of interest was found to be 79% and 82%, respectively. Total homology between human and mouse encoding regions was found to be 79%.

Additional bands were observed by Southern blotting (Figure 6B, lanes 1 and 2). They could derive either from unprocessed mRNA molecules, variants, incomplete reverse transcripts, single-stranded PCR products or even additional highly homologous genes.

Under the PCR conditions used to maximize the yield (6), several misincorporations were observed. Two additional PCRs were performed with *Taq* and *Pfu* DNA polymerases, individually, using anchored specific primer under optimal reaction conditions to avoid misincorporations. Both sequences were found identical, suggesting

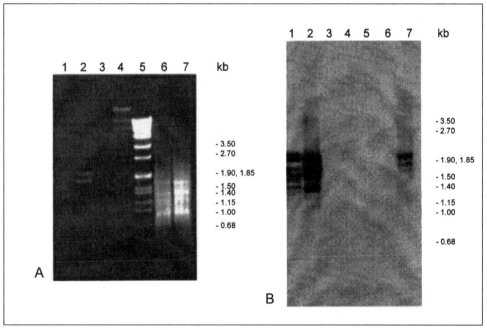

Figure 6. **Southern blot analysis of captured cDNA.** Poly(A)-tailed single-stranded cDNA from nonactivated (lanes 1 and 6) and ConA-activated (lanes 2 and 7) murine thymocytes was captured with biotinylated single-stranded hLAG-3 probe 1 (lanes 6 and 7) and then with single-stranded hLAG-3 probe 2 (lanes 1 and 2) bound to magnetic beads. Captured single-stranded cDNA was amplified by PCR with oligo(dT) mixture and anchor primer A and analyzed by agarose gel electrophoresis (A) and Southern blotting with DIG-labeled single-stranded hLAG-3 probe 2 (B). PCR without DNA template (negative control) (lane 3) and molecular weight markers (MW II DIG-labeled and MW VII from Boehringer Mannheim, lanes 4 and 5) are also shown. (From Reference 12, copyright 1996 by Academic Press).

identity with the natural sequence. Anchored specific primers were used to facilitate the gene cloning into an expression vector for recombinant protein production.

Potential Application of GC-PCR

Isolation of Gene Family Members

When two or more conserved domains are present, such as in G-coupled receptors (8,14,16), tyrosine kinase receptors (18) and metalloproteinases (15), they could be used to design consensus capture probes. The same capture probes could be used to select different genes of the family specifically expressed by a given cell or tissue. Many cell- or tissue-specific gene families can be contemporaneously analyzed and directly cloned to obtain the corresponding expression library. In this case, gene selection does not require perfect 3′-end matching of primers as in the PCR approach. Partial DNA sequencing of the library clones will provide expressed sequence tags (ESTs) that will be informative on expression levels for the gene family.

Gene Isolation from Protein Sequence

When part of the protein sequence is known, its encoding gene could be isolated using two or more oligonucleotide capture probes. They might be designed following the protocol of Lathe (9), thus obtaining a homologous probe with about 80%–85% homology with the target and/or by introducing modified bases such as 5-nitroindole as a universal base analogue able to display equal base pairing affinity for all natural bases (11) in a position of high degeneracy.

Biotin and nitroindole can be introduced directly in the capture probes during automated synthesis using LC Biotin-ON and Universal Base Phosphoramidites (CLONTECH).

Gene Isolation from a DNA Library

Unidirectional cDNA libraries can also be screened by GC-PCR. In this case, one should perform a PCR amplification of the library with a biotinylated universal forward and nonbiotinylated universal reverse primer set. The amplified cDNA library is bound to magnetic beads and denatured by alkali. The supernatant is neutralized to obtain amplifiable single-stranded cDNA to be used in GC-PCR.

Alternatively, a library cloned in a plasmid containing the F1 origin might be used to produce single-stranded cDNA. After two or more capture steps, single-stranded cDNA might be converted to double-stranded cDNA, which is transformed into *E. coli*. Colonies should contain the captured genes. This approach is used by the GeneTrapper™ System (Life Technologies), in which a 3′ biotinylated oligonucleotide is used to capture the complementary target. By the gene-capture approach, this system can also be adapted to the isolation of homologous genes, thus increasing the potential of GeneTrapper system.

Improvement of GC-PCR

Many improvements to this technique can be envisaged, deriving from continuous improvement of commercial reagents and general procedures for molecular biology, such as long-distance RT-PCR, RT reaction improvement to obtain full-length retrotranscript and high-fidelity and processivity enzymes.

Many commercial kits are now available containing enzyme mixtures that increase PCR fidelity and yield.

SLIC technology and a different TdT tailing should avoid the cloning procedure, and direct sequencing of the isolated fragment can be performed if an acceptable homogeneity of the PCR product is achieved.

The SMART oligonucleotide included in the SMART PCR cDNA Library Construction Kit might introduce a strong improvement in GC-PCR strategy. This oligonucleotide allows the selective amplification of full-length cDNAs. Furthermore, the SMART kit is optimized for long-distance RT-PCR.

For the isolation of rare genes, an amplification/denaturation step on the released material between each capture step could also be considered.

Miniaturization and automation provide a tool to accelerate GC-PCR. By automated multiple capture steps and using SMART technology, one should be able to directly and quickly isolate the full-length gene of interest without any Southern blot control because of the higher selection obtained with a higher number of capture steps. Oligonucleotides used as capture probes will shorten hybridization time (by minutes), and the multiple capture step process will be greatly speeded up.

ACKNOWLEDGMENTS

We thank Ruben Papoian for supporting this work and for reviewing the manuscript and Emilia Micangeli for expert technical assistance.

REFERENCES

1. **Botstein, D., S.A. Chervitz and J.M. Cherry.** 1997. Yeast as a model organism. Science *277*:1259-1260.
2. **Chou, Q., M. Russell, D.E. Birch, J. Raymond and W. Bloch.** 1992. Prevention of pre-PCR mis-priming and primer dimerization improves low-copy-number amplifications. Nucleic Acids Res. *20*:1717-1723.
3. **Diatchenko, L., Y.F. Lau, A.P. Campbell, A. Chenchik, F. Moqadam, B. Huang, S. Lukyanov, K. Lukyanov et al.** 1996. Suppression subtractive hybridization: a method for generating differentially regulated or tissue-specific cDNA probes and libraries. Proc. Natl. Acad. Sci. USA *93*:6025-6030.
4. **Edwards, J.B.D.M., J. Delort and J. Mallet.** 1991. Oligodeoxyribonucleotide ligation to single-stranded cDNAs: a new tool for cloning 5' ends of mRNAs and for constructing cDNA libraries by in vitro amplification. Nucleic Acids Res. *19*:5227-5232.
5. **Friedrich, G.A.** 1996. Moving beyond the genome projects—does the future of genomics-based drug discovery lie with the mouse? Nature Biotechnol. *14*:1234-1237.
6. **Frohman, M.A.** 1993. Rapid amplification of complementary DNA ends for generation of full-length complementary DNAs: thermal RACE. Methods Enzymol. *218*:340-356.
7. **Huard, B., R. Mastrangeli, P. Prigent, D. Bruniquel, S. Donini, N. El-Tayar, B. Maigret, M. Dreano and F. Triebel.** 1997. Characterization of the major histocompatibility complex class II binding site on LAG-3 protein. Proc. Natl. Acad. Sci. USA *94*:5744-5749.
8. **Kunapuli, P. and J.L. Benovic.** 1993. Cloning and expression of GRK5: a member of the G protein-coupled receptor kinase family. Proc. Natl. Acad. Sci. USA *90*:5588-5592.
9. **Lathe, R.** 1985. Synthetic oligonucleotide probes deduced from amino acid sequence data. Theoretical and practical considerations. J. Mol. Biol. *183*:1-12.
10. **Liang, P., W. Zhu, X. Zhang, Z. Guo, R.P. O'Connell, L. Averboukh, F. Wang and A.B. Pardee.** 1994. Differential display using one-base anchored oligo-dT primers. Nucleic Acids Res. *22*:5763-5764.
11. **Loakes, D. and D.M. Brown.** 1994. 5-Nitroindole as an universal base analogue. Nucleic Acids Res. *22*:4039-4043.
12. **Mastrangeli, R., E. Micangeli and S. Donini.** 1996. Cloning of murine LAG-3 by magnetic bead bound

homologous probes and PCR (gene-capture PCR). Anal. Biochem. *241*:93-102.
13. **Mewes, H.W., K. Albermann, M. Bahr, D. Frishman, A. Gleissner, J. Hani, K. Heumann, K. Kleine et al.** 1997. Overview of the yeast genome. Nature *387*:7-65.
14. **Rohrer, L., F. Raulf, C. Bruns, R. Buettner, F. Hofstaedter and R. Schule.** 1993. Cloning and characterization of a fourth human somatostatin receptor. Proc. Natl. Acad. Sci. USA *90*:4196-4200.
15. **Sato, H., T. Takino, Y. Okada, J. Cao, A. Shinagawa, E. Yamamoto and M. Seiki.** 1994. A matrix metalloproteinase expressed on the surface of invasive tumour cells. Nature *370*:61-65.
16. **Savarese, T.M. and C.M. Fraser.** 1992. In vitro mutagenesis and the search for structure-function relationships among G protein-coupled receptors [Review]. Biochem. J. *283*:1-19.
17. **Triebel, F., S. Jitsukawa, E. Baixeras, S. Roman-Roman, C. Genevee, E. Viegas-Pequignot and T. Hercend.** 1990. LAG-3, a novel lymphocyte activation gene closely related to CD4. J. Exp. Med. *171*:1393-1405.
18. **Wilks, A.F.** 1991. Cloning members of protein-tyrosine kinase family using polymerase chain reaction. Methods Enzymol. *200*:533-546.

21 Improved Technique for Walking in Uncloned Genomic DNA

Stephen S. Chen, Alex Chenchik, Konstantin A. Lukyanov and Paul D. Siebert
Gene Cloning and Analysis Group, CLONTECH Laboratories, Palo Alto, CA, USA

OVERVIEW

We describe an improved DNA walking technique that facilitates the cloning of unknown genomic DNA sequences adjacent to a known sequence, such as a cDNA. A key innovation known as suppression polymerase chain reaction (PCR) allows efficient amplification without the background and reproducibility problems that have plagued other methods for PCR-based walking in complex genomes. In combination with long-distance PCR, this allows individual steps of up to 6 kb in genomic DNA. The resulting PCR products can be easily cloned and are usually pure enough to allow restriction mapping without cloning. This method is valuable for rapidly finding promoters and regulatory elements from sequences obtained from cloned cDNAs, for determining the exon–intron boundaries of genes and for walking upstream or downstream from sequence-tagged sites generated from genome studies.

BACKGROUND

Several PCR-based methods are available for walking from a known region to an unknown region in cloned or uncloned genomic DNA. The methods are of three types: *(i)* inverse PCR (14), *(ii)* randomly primed PCR (16) and *(iii)* adaptor ligation PCR (9,10,17,18). However, these methods have not been widely used for walking in uncloned genomic DNA because they are either complicated or inefficient. They have recently been improved (5,8); however, walks in uncloned DNA have been limited to distances of <1 kb.

We have investigated the application of long and accurate PCR (1,3) to walking in uncloned genomic DNA. We initially examined the use of unpredictably primed PCR (18), a new method based on randomly primed PCR, and vectorette PCR (10), which is based on adaptor ligation. As a model system, we attempted to walk upstream from

Gene Cloning and Analysis by RT-PCR
Edited by Paul Siebert and James Larrick
© 1998 BioTechniques Books, Natick, MA

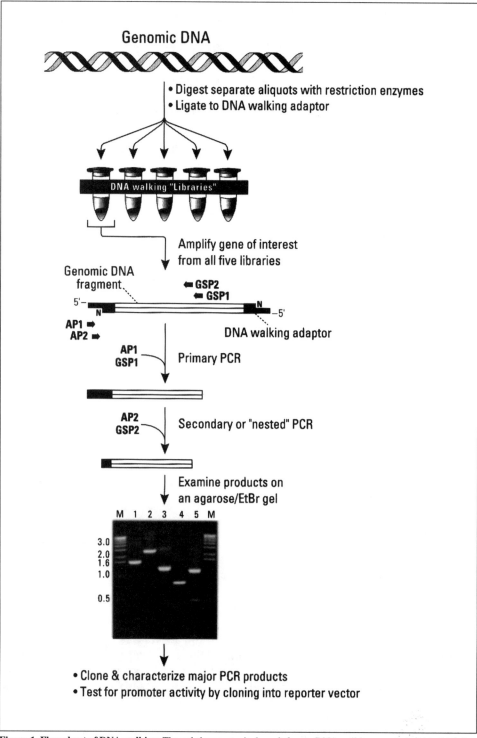

Figure 1. Flow chart of DNA walking. The gel shows a typical result from a DNA walking experiment in which five DNA libraries were used. Lane 1: *Eco*RV library. Lane 2: *Sca*I library. Lane 3: *Dra*I library. Lane 4: *Pvu*II library. Lane 5: *Ssp*I library. Lane M: 1-kb DNA size markers.

exon 1 of the human tissue-type plasminogen activator (tPA) gene that has been previously characterized (4).

In our hands, both unpredictably primed PCR and vectorette PCR generated multiple products even after nested PCR was performed. Although upon analysis, some of the PCR products were found to be derived from successful walks, the presence of multiple PCR products would complicate further characterization without Southern blot hybridization or extensive cloning.

We have improved on the adaptor ligation method by combining vectorette PCR with a recently developed method termed suppression PCR (11). A walking adaptor is ligated to the ends of DNA fragments generated by digestion of human genomic DNA with *Eco*RV, *Sca*I, *Dra*I, *Pvu*II and *Ssp*I separately. Following adaptor ligation, a small amount of the DNA is used as a template for PCR using adaptor primers and gene-specific primers as illustrated in Figure 1. The enzymes used were selected because they have six-base recognition sites and generate blunt ends.

The sequences of the walking adaptor and primers are shown in Figure 2. One end of the adaptor is blunt so that it will ligate to both ends of any DNA fragment generated by restriction enzymes that yield blunt ends. The adaptor also contains several rare restriction enzyme sites for *Mlu*I, *Sal*I (staggered ends) and *Srf*I/*Sma*I (blunt ends) to allow cloning into commonly used vectors such as pBluescript® (Stratagene, La Jolla, CA, USA).

The vectorette feature of the adaptor is the presence of an amine group on 3′ end of the lower strand. This blocks polymerase-catalyzed extension of the lower adaptor strand, preventing the generation of the primer binding site unless a defined, distal gene-specific primer extends the DNA strand opposite the upper strand of the adaptor. The suppression PCR technology uses an adaptor primer that is shorter than the adaptor and is capable of hybridizing to the outer primer binding site. If any PCR

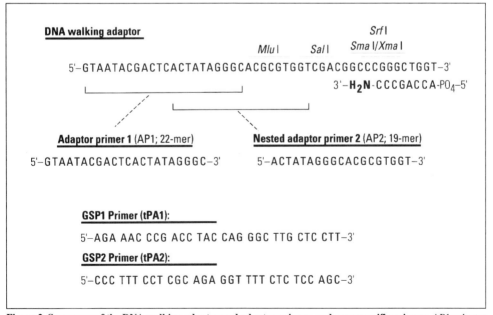

Figure 2. Sequences of the DNA walking adaptor and adaptor primers and gene-specific primers. AP1: adaptor primer 1. AP2: adaptor primer 2. GSP1: gene-specific primer 1. GSP2: gene-specific primer 2. tPA: human tissue-type plasminogen activator gene.

products are generated that contain double-stranded adaptor sequences at both ends (because of nonspecific DNA synthesis), the ends of the individual DNA strands will form panhandle structures following every denaturation step because of the presence of inverted terminal repeats (Figure 3). These structures are more stable than the primer–template hybrid and therefore will suppress exponential amplification. However, when a distal gene-specific primer extends a DNA strand through the adaptor, the extension product will contain the adaptor sequence only on one end and thus cannot form the panhandle structure. PCR amplification can then proceed normally.

The use of touchdown PCR (6) in our model system also improved the specific amplification. Touchdown PCR involves using an annealing/extension temperature that is several degrees higher than the melting temperature (T_m) of the primers during the initial PCR cycles. Although primer annealing is less efficient at this higher temperature, it is much more specific. The higher temperature also enhances the sup-

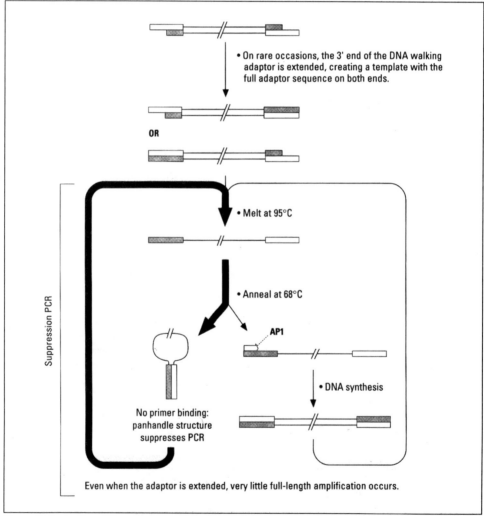

Figure 3. Illustration of the suppression PCR effect. Structure of PCR products formed by nonspecific amplification. The sequences on the ends of the products contain inverted terminal repeats and will form panhandle structures that suppress PCR.

pression PCR effect, allowing a critical amount of gene-specific product to accumulate. The annealing/extension temperature is then reduced to slightly below the primer T_m for the remaining PCR cycles, permitting efficient, exponential amplification of the gene-specific product.

PROTOCOLS

Protocol for Oligonucleotide Synthesis

Materials and Reagents

- 3′-DMT-C6-Amine-ON CPG (CLONTECH Laboratories, Palo Alto, CA, USA)
- 5′-Phosphate-ON™ (CLONTECH)
- Oligonucleotide Cartridge Purification (OCP™) Columns (CLONTECH)

Procedure

All oligonucleotides were synthesized on an ABI 394 DNA Synthesizer (PE Applied Biosystems, Foster City, CA, USA). The lower adaptor strand was synthesized with a 3′-end amine group and a 5′-end phosphate. Both modifications were incorporated directly into the lower adaptor strand during its synthesis using 3′-DMT-C6-Amine-ON CPG and 5′-Phosphate-ON. The adaptor oligonucleotides, both upper and lower strands, were purified by denaturing polyacrylamide gel electrophoresis. PCR primers were purified using OCP Columns.

Protocol for Preparation of the Adaptor-Ligated DNA

Materials and Reagents

- Human genomic DNA (0.1 µg/µL; CLONTECH)
- Restriction enzymes and buffers (Life Technologies, Gaithersburg, MD, USA)
  - *Eco*RV (10 U/µL)
  - 10× *Eco*RV restriction buffer
  - *Sca*I (10 U/µL)
  - 10× *Sca*I restriction buffer
  - *Dra*I (10 U/µL)
  - 10× *Dra*I restriction buffer
  - *Pvu*II (10 U/µL)
  - 10× *Pvu*II restriction buffer
  - *Ssp*I (10 U/µL)
  - 10× *Ssp*I restriction buffer
- Phenol–chloroform–isoamyl alcohol (25:24:1) (Sigma Chemical, St. Louis, MO, USA)
- Chloroform–isoamyl alcohol (24:1) (Sigma)
- 3 M sodium acetate, pH 4.5
- Glycogen (10 µg/µL) (molecular biology grade; Boehringer Mannheim GmbH, Mannheim, Germany)
- 95% ethanol
- 80% ethanol

- TE buffer 1: 10 mM Tris-HCl, pH 7.5, 0.1 mM EDTA
- TE buffer 2: 10 mM Tris-HCl, pH 7.4, 1 mM EDTA
- T4 DNA ligase (1 U/µL; Life Technologies)
- 5× DNA ligase reaction buffer (Life Technologies): 250 mM Tris-HCl, pH 7.6, 50 mM $MgCl_2$, 5.0 mM ATP, 5.0 mM dithiothreitol, 25% (wt/vol) polyethylene glycol 8000
- 25 µM DNA walking adaptor
- Deionized H_2O
- 1-kb ladder DNA size markers (Life Technologies)

Procedure

Digestion of genomic DNA

1. Label five 1.5-mL tubes: DL1, DL2, DL3, DL4, DL5.
2. For each reaction, combine the following in a separate 1.5-mL tube:

    25 µL genomic DNA (0.1 µg/µL)

    8 µL restriction enzyme

    10 µL restriction enzyme buffer

    57 µL deionized H_2O

    Mix gently by inverting tube. Do not vortex mix. Vigorous mixing will shear genomic DNA.
3. Incubate at 37°C for 2 h in an air incubator.
4. Vortex mix the reaction at slow speed for 5–10 s. Return to 37°C overnight (16–18 h).
5. From each reaction tube, remove 5 µL and run on an ethidium bromide (EtdBr)-stained 0.5% agarose gel to determine whether digestion is complete. Also save an additional 5 µL of each sample to run on the gel used in step 17 of the "Purification of DNA" section.

Purification of DNA

1. To each reaction tube add an equal volume (90 µL) of phenol–chloroform–isoamyl alcohol (25:24:1).
2. Vortex mix at slow speed for 5–10 s.
3. Spin briefly to separate the aqueous and organic phases.
4. Using a pipet, transfer the upper (aqueous) layer into a fresh 1.5-mL tube. Discard the lower (organic) layer.
5. To each tube add an equal volume (90 µL) of chloroform–isoamyl alcohol (24:1).
6. Vortex mix at slow speed for 5–10 s.
7. Spin briefly to separate the aqueous and organic phases.
8. Using a pipet, transfer the upper (aqueous) layer into a fresh 1.5-mL tube. Discard the lower (organic) layer.
9. To each tube add 2 vol (180 µL) of ice-cold 95% ethanol, 1/10 vol (9 µL) of 3 M NaOAc (pH 4.5) and 20 µg of glycogen.

10. Vortex mix at slow speed for 5–10 s.
11. Centrifuge at 20 200× g for 10 min.
12. Decant supernatant and wash pellet in 100 µL of ice-cold 80% ethanol.
13. Centrifuge at 20 200× g for 5 min.
14. Decant supernatant and air dry the pellet.
15. Dissolve pellet in 20 µL of TE buffer 1.
16. Vortex mix at slow speed for 5–10 s.
17. From each reaction tube, remove 1 µL and run on an EtdBr-stained 0.5% agarose gel to determine the approximate quantity of DNA after purification.

Ligation of genomic DNA to DNA walking adaptors

For each library construction, set up a total of six ligation reactions.
1. From each tube, transfer 4 µL of digested, purified DNA to a fresh 0.5-mL tube. To each, add the following:
    1.9 µL DNA walking adaptor (25 µM)
    1.6 µL 5× DNA ligase reaction buffer
    0.5 µL T4 DNA ligase (1 U/µL)
2. Incubate at 16°C overnight.

    Note: A PCR thermal cycler holds a very constant temperature and is recommended in place of a water bath for this reaction.

3. To stop the reactions, incubate at 70°C for 5 min.
4. To each tube add 72 µL of TE buffer 2.
5. Vortex mix at slow speed for 10–15 s.
6. Store at -20°C.

Protocol for PCR Amplification

Materials and Reagents

- GeneAmp® PCR tubes (Perkin-Elmer, Norwalk, CT, USA)
- Advantage® Genomic Polymerase Mix (CLONTECH): 10× *Tth* PCR buffer (400 mM Tris-HCl [pH 9.3 at 25°C], 150 mM KOAc), 25 mM Mg(OAc)$_2$, 50× Advantage *Tth* Polymerase Mix
- 50× dNTP Mix (10 mM each) (Amersham Pharmacia Biotech, Piscataway, NJ, USA)
- 10 µM tPA gene-specific primer 1 (GSP1)
- 10 µM tPA gene-specific primer 2 (GSP2)
- 10 µM adaptor primer 1 (AP1)
- 10 µM adaptor primer 2 (AP2)

Procedure

PCR amplifications were performed using a long-distance thermostable DNA polymerase mixture of *Tth* (2.5 U/µL; Perkin-Elmer) and Vent® (2 U/µL; New England Biolabs, Beverly, MA, USA) at a ratio of 20:1 (vol/vol) (3). Before use, 2 μL (11 μg) of TthStart™ Antibody (CLONTECH) were mixed with 10 μL of the *Tth*/Vent enzyme mixture to facilitate a hot-start PCR. We also used a commercially available enzyme mixture, Advantage *Tth* Polymerase Mix or r*Tth* XL (2 U/μL; Perkin-Elmer)/TthStart, with similar results.

1. Prepare enough primary PCR master mixture for all five reactions plus one additional tube.

    Combine the following reagents in a 0.5-mL tube:

per 6 reactions	per 1 reaction	
226.2 μL	37.8 μL	deionized H_2O
30.0 μL	5.0 μL	10× *Tth* PCR buffer
6.0 μL	1.0 μL	dNTPs (10 mM each)
13.2 μL	2.2 μL	$Mg(OAc)_2$ (25 mM)
6.0 μL	1.0 μL	AP1 (10 μM)
6.0 μL	1.0 μL	GSP1 (10 μM)
6.0 μL	1.0 μL	50× Advantage *Tth* Polymerase Mix
294.0 μL	49.0 μL	Total volume

    Vortex mix well (without introducing bubbles) and briefly spin the tube in a microcentrifuge.

2. Add 49 μL of the primary PCR master mixture to the appropriately labeled tubes.
3. Add 1 μL of each DNA library to the appropriately labeled tubes.
4. Overlay the contents of each tube with 1 drop of mineral oil and place caps firmly on tubes.
5. Briefly spin tubes in a microcentrifuge.
6. Commence thermal cycling in a Perkin-Elmer DNA Thermal Cycler 480 using the following two-step cycle parameters: 7 cycles of denaturation at 94°C for 25 s and extension at 72°C for 3 min, then 32 cycles at 94°C for 25 s and extension at 67°C for 3 min, and then a final extension at 67°C for an additional 7 min after the final cycle.

    Note: Do not use a three-step cycling program (e.g., 95°C denaturation, 60°C annealing, 68°C extension).

7. Analyze 8 μL of the primary PCR products on an EtdBr-stained 1.5% agarose gel, along with DNA size markers such as a 1-kb ladder and electrophorese at 150 V in 0.5× TBE buffer (0.045 M Tris-borate, 0.001 M EDTA, pH 8.0). If no product is detected, perform 5 additional cycles. Expected results of primary PCR: in all lanes, predicted banding patterns should be observed. Be aware, however, that there might be some smearing in some lanes or multiple banding patterns in some lanes, ranging in size from about 500 bp to 5 kb.
8. Using a clean 0.5-mL tube for each sample, dilute 1 μL of each primary product into 49 μL of deionized H_2O.
9. Prepare a secondary PCR master mixture for all five reactions plus one additional tube.

    Combine the following reagents in a 0.5-mL tube:

per 6 reactions	per 1 reaction	
226.2 μL	37.8 μL	deionized H_2O
30.0 μL	5.0 μL	10× *Tth* PCR buffer

6.0 µL	1.0 µL	dNTPs (10 mM each)
13.2 µL	2.2 µL	Mg(OAc)$_2$ (25 mM)
6.0 µL	1.0 µL	AP2 (10 µM)
6.0 µL	1.0 µL	GSP2 (10 µM)
<u>6.0 µL</u>	<u>1.0 µL</u>	50× Advantage *Tth* Polymerase Mix
294.0 µL	49.0 µL	Total volume

Vortex mix well (without introducing bubbles) and briefly spin the tube in a microcentrifuge.

10. Add 49 µL of the secondary PCR master mixture to the appropriately labeled tubes.
11. Add 1 µL of each diluted primary PCR product (from step 8) to the appropriately labeled tubes.
12. Overlay the contents of each tube with 1 drop of mineral oil and place caps firmly on tubes.
13. Briefly spin tubes in a microcentrifuge.
14. Commence thermal cycling in a Perkin-Elmer DNA Thermal Cycler 480 using the following two-step cycle parameters: 5 cycles of denaturation at 94°C for 25 s and extension at 72°C for 3 min, then 20 cycles of denaturation at 94°C for 25 s and extension at 67°C for 3 min, and a final extension at 67°C for an additional 7 min after the final cycle.

Note: Do not use a three-step cycling program (e.g., 95°C denaturation, 60°C annealing, 68°C extension).

15. Analyze 8 µL of the primary PCR products on an EtdBr-stained 1.5% agarose gel, along with DNA size markers such as a 1-kb ladder and electrophorese at 150 V in 0.5× TBE. If the amplification product is not detected, perform 5 additional cycles.

Protocol for Restriction Mapping of DNA Walking PCR Products

Materials and Reagents

- 0.5× TBE buffer
- EtdBr-stained 1.5% agarose gel
- Restriction enzymes and buffers (Life Technologies)
  - *Bam*HI (10 U/µL)
  - 10× *Bam*HI restriction buffer
  - *Pvu*II (10 U/µL)
  - 10× *Pvu*II restriction buffer

Procedure

DNA walking PCR products are generally clean enough to allow simple restriction mapping without cloning.

<u>Digestion of PCR products with *Bam*HI</u>

1. Label five 0.5-mL tubes for five PCR products.
2. Prepare enough restriction enzyme master mixture for all five reactions plus one

additional tube. Combine the following reagents in a 0.5-mL tube:

per 6 reactions	per 1 reaction	
6.0 µL	1.0 µL	10 U/µL *Bam*HI
7.2 µL	1.2 µL	10× *Bam*HI restriction buffer
13.2 µL	2.2 µL	Total volume

Vortex mix well and briefly spin the tube in a microcentrifuge.

3. Add 2.2 µL of master mixture to the appropriately labeled tubes.
4. Add 10 µL of each PCR product to the appropriately labeled tubes.
5. Vortex mix well and briefly spin the tube in a microcentrifuge.
6. Incubate all reactions at 37°C for 4 h in an air incubator.

Digestion of PCR products with *Pvu*II

Perform the same experiment described above with *Pvu*II instead of *Bam*HI.

Visualization of digested PCR products

Analyze 5 µL of each digested PCR product on an EtdBr-stained 1.5% agarose gel along with DNA size markers such as a 1-kb ladder, and electrophorese at 150 V in 0.5× TBE for about 30 min.

RESULTS AND DISCUSSION

The result of an experiment in which we walked upstream from exon 1 of the tPA gene is shown in Figure 4. In all cases, single major PCR products were generated that had the following sizes: 1.8, 4.5, 0.9, 1.5 and 3.9 kb. The sizes of the three smaller PCR products were predicted from the known sequence of the tPA gene (4). Not enough upstream sequence was available in GenBank® to predict the sizes of the two larger PCR products.

We then mapped the PCR products by digestion with *Bam*HI and *Pvu*II. The restriction enzyme pattern obtained is shown in Figure 5. The release of common DNA fragments at 1.6 kb with *Bam*HI and 1.4 kb with *Pvu*II from the larger of the two PCR products was predicted from the known sequence. These data thus verified the gene specificity of the walk for all five adaptor-ligated DNA libraries. The results also show that provisional restriction maps of the

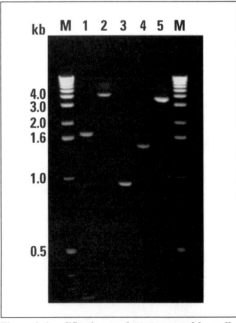

Figure 4. Amplification products generated by walking upstream from exon 1 of the human tPA gene. PCR products in lanes 1–5 were obtained from the *Eco*RV, *Sca*I, *Dra*I, *Pvu*II and *Ssp*I adaptor-ligated DNA libraries, consecutively. Lanes M: 1-kb DNA size markers.

unknown sequence can be obtained without cloning.

We have found that the high specificity of the method described is primarily due to the suppression effect of the adaptor–adaptor primer combination, although in some cases, also having an amine group on the lower strand of the adaptor (vectorette blocking) yielded better results.

We have used the same adaptor-ligated DNA to walk upstream from 5′-end coding regions of the human transferrin receptor gene (15) with similar clean results. In this case, the walks obtained ranged in length from 0.2 to 6 kb. We have also applied this method to mouse interleukin 6 (20), mouse tumor necrosis factor (TNF) (19), rat interleukin 6 (13) and wheat acetyl coenzyme A (CoA) carboxylase (7), walking either upstream or downstream on those genes. In all cases, we achieved the expected results.

TROUBLESHOOTING

The following information can be useful in optimizing or troubleshooting the method.

Genomic DNA Preparation

The starting DNA must be very pure and have a high average molecular weight, so a higher-quality preparation than the minimum suitable for Southern blotting or conventional PCR is required. Before constructing libraries, check the size and purity of

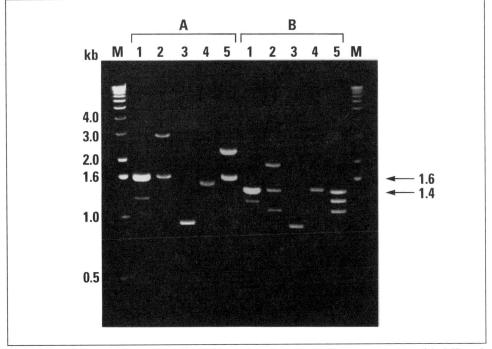

Figure 5. Restriction mapping of the DNA walking PCR products. Restriction enzyme digestion of the PCR products shown in Figure 4. Lanes 1–5 correspond to the products obtained from the *Eco*RV, *Sca*I, *Dra*I, *Pvu*II and *Ssp*I adaptor-ligated DNA libraries, consecutively. (A) Digestion with *Bam*HI. (B) Digestion with *Pvu*II. Lane M: 1-kb DNA size markers.

the genomic DNA on an EtdBr-stained 0.5% agarose gel in 0.5× TBE, along with DNA size markers such as a λ/*Hin*dIII. Genomic DNA should be visibly larger than the largest molecular weight (50 kb) marker.

Primer Design

There are several suggestions for designing gene-specific primers. In general, the gene-specific primers should not cross intron–exon boundaries and should be derived from sequences as close to the end of the known sequence as possible. For walking upstream from a cDNA sequence, the primer should be as close to the 5′ end as possible. Ideally, the primers should be derived from the first exon of the gene. Gene-specific primers should be 26–30 nucleotides in length and have a GC content of 40%–60%. Even if the T_m seems high, do not design primers shorter than 26 bp. Primers should not be able to fold back and form intramolecular hydrogen bonds, and sequences at the 3′ ends of the primers should not be able to anneal to the 3′ ends of the adaptor primers. There should be no more than three Gs and Cs in the last six positions at the 3′ end of the primer. The sequence of the AP1 and nested primer AP2 should not overlap. If overlapping primers must be used, the 3′ end of the nested primer should have as much unique sequence as possible.

DNA Digestion and DNA Purification

The DNA concentration should be the same before and after phenol–chloroform extraction. Run samples of the DNA on an agarose gel before and after purification. If the intensity of EtdBr staining is 50% less after purification, the DNA should be concentrated by ethanol precipitation or by placing tubes in a rotating evaporator. Then the DNA should be resuspended in a smaller volume of water.

One of the possible causes of nonspecific PCR products is incomplete restriction digestion of the DNA. If this is the case, repeat the DNA digestion and check again. To determine whether digestion is complete, remove 5 µL of the digested DNA mixture and run on an EtdBr-stained 0.5% agarose gel. If the DNA is completely digested, a long smear of DNA beween 0.5 and 20 kb will be seen on the gel. If this is not the case, the DNA digestion either failed or is incomplete.

Normally, if the DNA is completely digested, a single major band should be observed after secondary PCR. However, multiple bands can result, depending on the species used (e.g., some plants are multiploid) or from genes that belong to multigene families.

Adaptor–DNA Ligation

If no PCR product is detected after secondary PCR from all five libraries, the adaptor–DNA ligation probably failed. Two major reasons for this problem are poor quality of the adaptor and low efficiency of the T4 DNA ligase. Before repeating the adaptor–DNA ligation step, check the quality of adaptor oligonucleotides and obtain fresh T4 DNA ligase.

PCR Amplification

We recommend starting with the cycle parameters suggested in the protocol. If no amplification product is detected even after increasing the number of primary cycles from 32 to 37, reduce all annealing/extension temperatures by 2°C (i.e., 72° to 70°C and 67° to 65°C) and reduce the length of the incubation at 94°C.

In our experience, 20% of cases contain GC-rich sequences that prevent their amplification by standard PCR techniques. Because these GC-rich sequences possess strong secondary structure that resists denaturation and prevents primer annealing, PCR often fails to yield any product. If this is the case, repeat the PCR amplification with the addition of a final concentration of 5% DMSO and 1 M N,N,N-trimethylglycine (betaine) (2,12) in the primary and secondary PCRs. Betaine destabilizes DNA secondary structure and effectively makes AT and GC base pairs equally stable. This allows efficient amplification of virtually all of the GC-rich sequences that are resistant to PCR by standard techniques. We have also used a commercially available reagent, the Advantage GC Genomic Polymerase Mix (CLONTECH), with similar results.

Generally, the simplified touchdown PCR program suggested in this chapter can significantly improve the amplification by increasing specificity. However, secondary products can occasionally be generated in some libraries for certain genes.

ACKNOWLEDGMENTS

The authors wish to thank Stephaine Trelogan for excellent critical review of the manuscript, Rose Deng and Josephine Mo for their timely synthesis of oligonucleotides and Christopher Scola for the artwork.

REFERENCES

1. **Barnes, W.M.** 1994. PCR amplification of up to 35-kb DNA with high fidelity and high yield from λ bacteriophage templates. Proc. Natl. Acad. Sci. USA *91*:2216-2220.
2. **Baskaran, N., R.P. Kandpal, A.K. Bhargava, M.W. Glynn, A. Bale and S.M. Weissman.** 1996. Uniform amplification of a mixture of deoxyribonucleic acids with varying GC content. Genome Res. *6*:633-638.
3. **Cheng, S., C. Fockler, W.M. Barnes and H. Russell.** 1994. Effective amplification of long targets from cloned inserts and human genomic DNA. Proc. Natl. Acad. Sci. USA *91*:5695-5699.
4. **Degen, S.J., B. Rajput and E. Reich.** 1986. The human tissue plasminogen activator gene. J. Biol. Chem. *261*:6972-6985.
5. **Dominguez, O. and C. Lopez-Larrea.** 1994. Gene walking by unpredictably primed PCR. Nucleic Acids Res. *22*:3247-3248.
6. **Don, R.H., P.T. Cox, B.J. Wainwright, K. Baker and J.S. Mattick.** 1991. 'Touchdown' PCR to circumvent spurious priming during gene amplification. Nucleic Acids Res. *19*:4008.
7. **Gornicki, P., J. Podkowinski, L.A. Scappino, J. DiMaio, E. Ward and R. Haselkorn.** 1994. Wheat acetyl-coenzyme A carboxylase: cDNA and protein structure. Proc. Natl. Acad. Sci. USA *91*:6860-6864.
8. **Iwahana, H., T. Tsujisawa, R. Katashima, K. Yoshimoto and M. Itakura.** 1994. PCR with end trimming and cassette ligation: a rapid method to clone exon-intron boundaries and an upstream sequence of genomic DNA based on a cDNA sequence. PCR Methods Appl. *4*:19-25.
9. **Jones, D.H. and S.C. Winistorfer.** 1993. Genome walking with 2- to 4-kb steps using panhandle PCR. PCR Methods Appl. *2*:197-203.
10. **Lagerstrom, M., J. Parik, H. Malmgren, J. Stewart, U. Pettersson and U. Landegren.** 1991. Capture PCR: efficient amplification of DNA fragments adjacent to a known sequence in human and YAC DNA. PCR Methods Appl. *1*:111-119.
11. **Luk'ianov, S.A., N.G. Gurskaia, K.A. Luk'ianov, V.S. Yarabykin and E.D. Sverdlov.** 1994. Highly-effective subtractive hybridization of cDNA. Bioorg. Khim. *20*:701-704.
12. **Melchior, W.B., Jr. and P.H. Vun Vippel.** 1973. Alteration of the relative stability of dA.dT and dC.dG base pairs in DNA. Proc. Natl. Acad. Sci. USA *70*:298-302.
13. **Northemann, W., T.A. Braciak, M. Hattori, F. Lee and G.H. Fey.** 1989. Structure of the rat interleukin 6 gene and its expression in macrophage-derived cells. J. Biol. Chem. *264*:16072-16082.
14. **Ochman, H., A.S. Gerber and D.L. Hartl.** 1988. Genetic applications of an inverse polymerase chain reaction. Genetics *120*:621-623.

15. **Owen, D. and L.C. Kuhn.** 1987. Noncoding 3' sequences of the transferrin receptor gene are required for mRNA regulation by iron. EMBO J. *6*:1287-1293.
16. **Parker, J.D., P.S. Rabinovitch and G.C. Burmer.** 1991. Targeted gene walking polymerase chain reaction. Nucleic Acids Res. *19*:3055-3060.
17. **Riley, J., R. Butler, D. Ogilvie, R. Finniear, D. Jenner, S. Powell, R. Anand, J.C. Smith and A.F. Markham.** 1990. A novel, rapid method for the isolation of terminal sequences from yeast artificial chromosome (YAC) clones. Nucleic Acids Res. *18*:2887-2890.
18. **Rosenthal, A., D. Stephen and C. Jones.** 1990. Genomic walking and sequencing by oligo-cassette mediated polymerase chain reaction. Nucleic Acids Res. *18*:3095-3096.
19. **Shakhov, A.N. and S.A. Nedospasov.** 1987. Molecular cloning of genes coding for tumor necrosis factors: complete nucleotide sequence of the genomic copy of TNF-alpha in mice. Bioorg. Khim. *13*:701-705.
20. **Tanabe, O., S. Akira, T. Kamiya, G.G. Wong, T. Hirano and T. Kishimoto.** 1988. Genomic structure of the murine IL-6 gene. High degree conservation of potential regulatory sequences between mouse and human. J. Immunol. *141*:3875-3881.

Section V

Ancillary PCR Methods

22. Generation and Use of High-Quality cDNA from Small Amounts of Total RNA by SMART™ PCR

Alex Chenchik, York Y. Zhu, Luda Diatchenko, Roger Li, Jason Hill and Paul D. Siebert
Gene Cloning and Analysis Group, CLONTECH Laboratories, Palo Alto, CA, USA

OVERVIEW

A simplified procedure is described for the selective amplification of the poly(A+) RNA fraction of total RNA. The first cDNA strand is primed from total RNA using an oligo(dT) primer in the presence of a second template-switching (TS) oligonucleotide. When reverse transcriptase reaches the 5' end of the mRNA, the enzyme continues replication using the TS oligonucleotide as a second template. The resulting single-stranded cDNA contains the complete 5' end of the mRNA as well as sequences complementary to the TS oligonucleotide. The cDNA is then amplified by polymerase chain reaction (PCR) using an oligo(dT) and TS primer. The representation of the starting mRNA population in the amplified cDNA was confirmed through analysis of 45 randomly chosen genes. This technique can be used to generate high-quality cDNA from nanogram quantities of total RNA for cDNA library construction, cDNA subtraction and "virtual" Northern blots.

BACKGROUND

Many gene cloning and analysis methods such as cDNA library construction and cDNA subtraction require large quantities of mRNA (5–50 µg). However, there are many situations in which it is not possible to obtain such quantities of mRNA. Examples of this include biopsy samples, unstable cell lines, tissues at different developmental stages and studies of small collections of cells. In such cases, amplification of cDNA in a sequence-independent manner can be performed (12). Generation of cDNAs with a high representation of mRNA sequences is critical for cloning and analyzing both novel genes and genes that are transcribed at low levels. However, with current techniques, a large portion of the amplified cDNA is not derived from the mRNA.

Amplification of the cDNA requires universal primer binding sites at both cDNA ends. An arbitrary sequence can easily be attached to the 5′ end of the cDNA by priming the reverse transcription (RT) reaction with an anchored oligo(dT) primer. Currently, there are three ways to generate universal PCR priming sites (anchors) on the 3′ end of the cDNA: *(i)* homopolymer tailing on the 3′ end of the first-strand cDNA (1,4,10), *(ii)* single-stranded anchor ligation to the 5′ end of mRNA (15,18) or to the 3′ end of the first-strand cDNA (2) and *(iii)* double-stranded adaptor ligation to the 5′ end of double-stranded cDNA (11). Each of these approaches requires several inefficient and complex manipulations.

We describe a simple sequence-independent cDNA amplification method. The method combines, in one step, first-strand cDNA synthesis and attachment of anchor sequences at the 5′ and 3′ ends of single-stranded cDNA. This is achieved by utilizing two intrinsic properties of Moloney murine leukemia virus (MMLV) reverse transcriptase: *(i)* the ability to add nontemplated nucleotides at the end of an RNA strand and *(ii)* the ability to switch to a second template. Both properties are utilized during the retroviral life cycle (7,14,17,19). For example, it has been shown that recombinogenic template switches at the 5′ ends of RNA template are highly mutagenic (17,19).

In the course of our studies, we found that these additional nontemplated nucleotides can be used to promote the reverse transcriptase to switch to a second short acceptor template (TS oligonucleotide) and copy it. The anchored single-stranded DNA can then be amplified by PCR. We find that the method generates a high yield of full-length double-stranded cDNA products with a high representation of the starting mRNA population. We present a detailed protocol for conducting this method, called switch mechanism at the 5′ end of RNA templates (SMART™) PCR (CLONTECH Laboratories, Palo Alto, CA, USA). Several applications of this method are described, including cDNA library construction, cDNA subtraction and Northern blot analysis.

PROTOCOLS

Materials and Reagents

- MMLV Reverse Transcriptase (200 U/μL) or SUPERSCRIPT™ II Reverse Transcriptase (200 U/μL) (both from Life Technologies, Gaithersburg, MD, USA)
- TS oligonucleotide: 5′-d(AAGCAGTGGTAACAACGCAGAGTACGC)r(GGG)-3′
- cDNA synthesis primer: 5′-AAGCAGTGGTAACAACGCAGAGTACT$_{(30)}$VN-3′ (N = A, G, C or T; V = A, G or C)
- PCR primer: 5′-AAGCAGTGGTAACAACGCAGAGT-3′
- 5× first-strand cDNA synthesis buffer: 250 mM Tris-HCl, pH 8.0, 375 mM KCl, 30 mM MgCl$_2$
- dNTP mixture (10 mM each; Amersham Pharmacia Biotech, Piscataway, NJ, USA)

- Dithiothreitol (DTT; 20 mM)
- Advantage™ KlenTaq Polymerase Mix and Buffer (CLONTECH): a mixture of KlenTaq-1 DNA polymerase (3), Deep Vent™ DNA polymerase (New England Biolabs, Beverly, MA, USA) and TaqStart™ antibody
- Deionized H_2O
- Mineral oil (Sigma Chemical, St. Louis, MO, USA)
- 1× TE buffer (pH 7.6): 10 mM Tris-Cl (pH 7.6), 1 mM EDTA (pH 8.0)
- DNA size markers (1-kb DNA ladder; Life Technologies)
- 1× TAE electrophoresis buffer: 40 mM Tris-acetate, pH 8.3, 1 mM EDTA
- Model 480 Thermal Cycler (Perkin-Elmer, Norwalk, CT, USA) (all cycling parameters were optimized). Perkin-Elmer GeneAmp® PCR Systems 2400 and 9600 have also been tested. For the latter, the denaturation time needs to be reduced from 30 to 10 s. For different types of thermal cyclers, the cycling parameters must be optimized.

Preparation of RNA

For purification of total RNA, we recommend the acid guanidinium thiocyanate–phenol–chloroform method (6).

cDNA Synthesis

First-Strand cDNA Synthesis Procedure

1. Combine the following reagents in a sterile 0.5-mL reaction tube.

1–3 µL	RNA sample (0.05–1 µg total RNA)
1 µL	cDNA synthesis primer (10 µM)
1 µL	TS oligonucleotide (10 µM)
____	deionized H_2O*
5 µL	Total volume

   *Adjust volume to 5 µL with deionized H_2O.

2. Mix contents and spin the tube briefly in a microcentrifuge.
3. Incubate the tube at 72°C for 2 min.
4. Spin the tube briefly in a microcentrifuge and keep the tube at room temperature.
5. Add the following to the reaction tube.

2 µL	5× first-strand buffer
1 µL	DTT (20 mM)
1 µL	dNTP (10 mM)
1 µL	MMLV or SUPERSCRIPT II reverse transcriptase (200 U/µL)
10 µL	Total volume

6. Mix by gently pipetting and spin in the tubes briefly in a microcentrifuge.
7. Incubate the tube at 42°C for 1 h in an air incubator.

8. Dilute the first-strand cDNA synthesis mixture by addition of 40 µL of 1× TE buffer.
9. Heat the tube at 72°C for 5 min in a thermal cycler.
10. First-strand reaction mixture can be stored at -20°C for up to three months.

Procedure for Amplification of cDNA

It is important to optimize the number of PCR cycles. We have found that the optimal number of PCR cycles varies with different templates, amounts of starting total RNA and type of thermal cycler. Once optimized, repeat the amplification in a scale large enough for your application. In our experience, each 100-µL PCR typically yields 2–5 µg of double-stranded cDNA.

1. Depending on the amount of total RNA used for first-strand synthesis, separate into aliquots the appropriate volume of first-strand reaction mixture into a 0.5-mL reaction tube. Enough H_2O is used to adjust the reaction volume to 100 µL.

Total RNA (µg)	Volume of diluted cDNA for PCR (µL)
1.0	1
0.5	2
0.25	4
0.1	10
0.05	10

2. In the test tube from step 1 combine the following components.

74 µL	deionized H_2O
10 µL	10× KlenTaq PCR Buffer
2 µL	dNTP mixture (10 mM each)
2 µL	5′ PCR primer (10 µM)
2 µL	50× Advantage KlenTaq Polymerase Mix
100 µL	Total volume

3. Overlay the reaction mixture with 2 drops of mineral oil, cap the tube and place it in a preheated (95°C) thermal cycler.
4. Commence thermal cycling using the following program: initial denaturation at 95°C for 1 min, followed by 15 cycles of denaturation at 95°C for 30 s and annealing at 65°C for 7 min.
5. Determine the optimal number of PCR cycles (Figure 1).
   a. Transfer 15 µL from the 15-cycle PCR to a clean microcentrifuge tube (for agarose–ethidium bromide [EtdBr] gel analysis).
   b. Run three additional cycles (for a total of 18) with the remaining 85 µL of the PCR mixture.
   c. Transfer 15 µL from the 18-cycle PCR to a clean microcentrifuge tube (for agarose–EtdBr gel analysis).
   d. Run three additional cycles (for a total of 21) with the remaining 70 µL of the PCR mixture.
   e. Transfer 15 µL from the 21-cycle PCR to a clean microcentrifuge tube (for agarose–EtdBr gel analysis).
   f. Run three additional cycles (for a total of 24) with the remaining 55 µL of the

PCR mixture.

6. Electrophorese 5 µL of each aliquot removed alongside 0.1 µg of 1-kb DNA size markers on a 1.2% agarose–EtdBr gel in 1× TAE buffer.

Figure 1 shows a typical gel profile of double-stranded cDNA synthesized from the human placental RNA. In general, cDNA synthesized from mammalian total RNA should appear on a 1.2% agarose–EtdBr gel as a moderately strong smear of 0.5–6 kb with some distinct bands. The number and position of the bands obtained will be different for each particular total RNA used. Furthermore, cDNA prepared from some mammalian tissue sources (e.g., human brain, spleen and thymus) might not display bright bands because of the very high complexity of the poly(A+) RNA. For nonmammalian species and for some tissues like pancreas and thyroid gland, the size distribution might be smaller.

Choosing the optimal number of PCR cycles ensures that the double-stranded cDNA will remain in the exponential phase of amplification. When the yield of PCR products stops increasing with more cycles, the reaction has reached its plateau. We find that overcycled cDNA is very poor template for application of this method. Undercycling, on the other hand, results in a lower yield of the PCR product. The optimal number of cycles for the experiment is one cycle fewer than is needed to reach the plateau.

Figure 1 provides an example of how the analysis should proceed. In this experiment, the PCR reached its plateau after 18 cycles (i.e., the yield of PCR products stopped increasing). After 21 and 24 cycles, a smear appeared in the high-molecular-weight region of the gel, indicating that the reaction was overcycled. Because the plateau was reached at 18 cycles, the optimal number of cycles for this experiment would be 17.

7. Repeat amplification of cDNA (steps 1–4) following the optimal number of PCR cycles (step 5). Each 100-µL PCR typically yields 2–5 µg of double-stranded cDNA.

8. When the cycling is completed, analyze a 5-µL sample of each test tube alongside 0.1 µg of 1-kb DNA size markers on a 1.2% agarose–EtdBr gel in 1× TAE buffer.

9. Add 2 µL of 0.5 M EDTA to each tube to terminate the reaction.

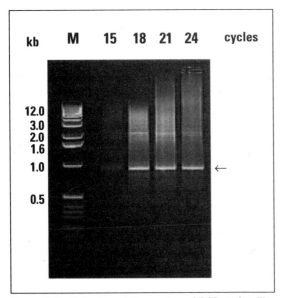

Figure 1. Optimization of the number of PCR cycles. Five microliters of each PCR product were electrophoresed on a 1.2% agarose–EtdBr gel in 1× TAE buffer following the indicated number of PCR cycles. The optimum number of cycles determined in this experiment was 17. Lane M: 1-kb DNA ladder. The arrow indicates a strong band at 900 bp, which is typically seen when amplifying human placental RNA.

cDNA Library Construction

Following amplification, the double-stranded cDNA is digested with proteinase K and then treated with T4 DNA polymerase to generate blunt ends. A double-stranded adaptor with an *Eco*RI sticky end is ligated at both ends of the double-stranded cDNA using T4 DNA ligase. Following adaptor ligation, the double-stranded cDNA is phosphorylated at the *Eco*RI sites, size-fractionated and ligated into suitable *Eco*RI-digested, dephosphorylated vector. In this study, the double-stranded cDNA was cloned nondirectionally into λgt11 vector. Alternatively, if the appropriate restriction sites are included in the oligo(dT) primer and TS oligonucleotide, the double-stranded cDNA can also be cloned directionally into any suitable vector (20,22).

RESULTS AND DISCUSSION

Model System to Examine Polynucleotide Addition and Template Switching

The substrates that were used to monitor the blunt-end addition (7,17) consisted of complementary oligodeoxyribonucleotides and oligoribonucleotides annealed to form the following blunt-ended duplexes.

S1: 5′-d(ACAACGCAGAATGGTCTCTCCC)-3′
3′-r(CACCATTGTTGCGTCTTACCAGAGAGGG)-5′

S1-CAP: 5′-d(ACAACGCAGAATGGTCTCTCCC)-3′
3′-r(CACCATTGTTGCGTCTTACCAGAGAGGG$_{ppp}$G^{me})-5′

S2: 5′-d(ACAACGCAGAATGGTCTCTCCT)-3′
3′-r(CACCATTGTTGCGTCTTACCAGAGAGGA)-5′

S2-CAP: 5′-d(ACAACGCAGAATGGTCTCTCCT)-3′
3′-r(CACCATTGTTGCGTCTTACCAGAGAGGA$_{ppp}$G^{me})-5′

S3: 5′-d(ACAACGCAGAATGGTCTCTCCG)-3′
3′-r(CACCATTGTTGCGTCTTACCAGAGAGGC)-5′

S3-CAP 5′-d(ACAACGCAGAATGGTCTCTCCG)-3′
3′-r(CACCATTGTTGCGTCTTACCAGAGAGGC$_{ppp}$G^{me})-5′

S4: 5′-d(ACAACGCAGAATGGTCTCTCCA)-3′
3′-r(CACCATTGTTGCGTCTTACCAGAGAGGT)-5′

S5: 5′-d(ACAACGCAGAATGGTCTCTGTC)-3′
3′-r(CACCATTGTTGCGTCTTACCAGAGACAG)-5′

S6 5′-d(ACAACGCAGAATGGTCTCTCTG)-3′
3′-r(CACCATTGTTGCGTCTTACCAGAGAGAC)-5′

$_{ppp}$G^{me} is 7-methylguanosine triphosphate group, the mRNA 5′-terminal cap structure (18).

To analyze the products of the template-switching reaction, we use the same substrates as for monitoring the blunt-end addition reaction and one of the following TS oligonucleotides.

TS-oligo-rG: 5′-d(CTAATACGACTCACTATAGGGC)r(G)-3′

TS-oligo-rG2: 5′-d(CTAATACGACTCACTATAGGGC)r(GG)-3′

TS-oligo-rG3:   5′-d(CTAATACGACTCACTATAGGGC)r(GGG)-3′
TS-oligo-rG5:   5′-d(CTAATACGACTCACTATAGGGC)r(GGGGG)-3′
TS-oligo-rU3:   5′-d(CTAATACGACTCACTATAGGGC)r(UUU)-3′
TS-oligo-dG3:   5′-d(CTAATACGACTCACTATAGGGCGGG)-3′
R-TS-oligo: 5′-r(CTAATACGACTCACTATAGGGCGGG)-3′

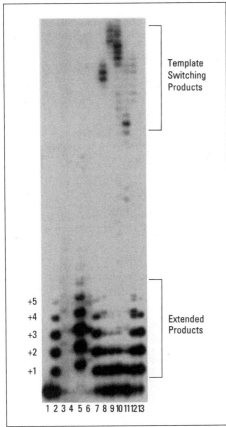

Figure 2. Addition of nontemplated nucleotides and template-switching reaction. A blunt-ended RNA–DNA duplex: S1 (Table 1) was used as a model substrate for this study. ^{32}P-labeled, blunt-ended RNA–DNA duplex (lane 1) was extended by MMLV reverse transcriptase in the presence of four unlabeled dNTPs (lane 2). Unlabeled RNA–DNA duplex was extended in the presence of three unlabeled dNTPs and one ^{32}P-labeled dNTP: dATP (lane 3), dGTP (lane 4), dCTP (lane 5) and dTTP (lane 6). The ^{32}P-labeled duplex was extended in the presence of all four unlabeled dNTPs and one of the following TS oligonucleotides (see text for sequence): TS-oligo-rG (lane 7), TS-oligo-rG2 (lane 8), TS-oligo-rG3 (lane 9), TS-oligo-rG5 (lane 10), R-TS-oligo (lane 11), TS-oligo-dG3 (lane 12) or TS-oligo-rU3 (lane 13). +1, +2, +3, +4, +5: position of extended products with 1, 2, 3, 4 and 5 additional nontemplated nucleotides; the upper brackets show the position of anchored oligodeoxyribonucleotides generated in the course of template switching.

All oligonucleotides were synthesized using standard phosphoramidite chemistry or by T7 RNA polymerase from double-stranded DNA templates (oligoribonucleotides with cap structure at the 5′ ends) and purified by gel electrophoresis. To analyze extended products and products of the template-switching reaction, oligodeoxyribonucleotides were 5′ end-labeled with ^{32}P by T4 polynucleotide kinase.

For the blunt-end nucleotide addition assay, 1 pmol of every two complementary oligonucleotides was annealed to the other in a volume of 5 µL of deionized water by heating the mixture at 70°C, followed by cooling on ice for 2 min. For the analysis of template switching, the annealing mixture additionally contained 1 pmol of the TS oligonucleotides. Blunt-end extension/template-switching reactions were carried out under same the conditions as described for first-strand cDNA synthesis. To determine the preference of nucleotides incorporated in the blunt-end extension assay, each reaction contained an unlabeled deoxyribonucleotide and 5 µCi of one of the four ^{32}P-labeled nucleotides. All reactions were terminated by the addition of dye–formamide solution and analyzed by electrophoresis on a 15% denaturing 7 M urea polyacrylamide gel run in 1× TBE buffer (45 mM Tris-borate, 1 mM EDTA, pH 8.0).

Principle of SMART PCR

DNA Template Switching

In the course of retroviral replication, a template-switching reaction occurs when MMLV reverse transcriptase reaches the 5′ end of genomic RNA. The literature suggests that reverse transcriptase can search for potential acceptor template sites when cDNA synthesis is interrupted (14,17,19). To efficiently catalyze template switching, the acceptor template should contain a homologous region with the 3′ end of the cDNA. As a result, this switching reaction takes place at the 3′ end of the first-strand cDNA. To develop a simple method of adding universal anchor sequences to the 3′ ends of the cDNA for PCR, we utilized another intrinsic property of MMLV reverse transcriptase. Like many DNA polymerases, when MMLV reverse transcriptase reaches the 5′ end of the DNA–RNA duplex, the enzyme will add more nontemplated nucleotides because of its intrinsic terminal transferase activity (7,17).

As shown in Figure 2 and Table 1, for several model blunt-ended RNA–DNA hybrid templates, wild-type MMLV reverse transcriptase usually adds 2–5 more nontemplated nucleotides to the 3′ end of the cDNA. We estimate that there is more than a 50% yield of the extended products. We find that, in accordance with previously published data (17), the presence of a cap structure at the 5′ end of mRNA does not significantly influence the preference of addition of these nontemplated nucleotides (data not shown). As shown in Table 1, the preference for addition of dNTPs is mainly determined by substrate specificity of the reverse transcriptase rather than complementary interaction between 5′-end mRNA sequence and nontemplated nucleotides. In general, MMLV reverse transcriptase shows the following preference: dCTP, dATP>>dGTP, dTTP. The consensus sequence of the extended product is 5′-d(Ac-Ac-Ca-Ca-Ca)-3′. The RNase H⁻ mutant of MMLV reverse transcriptase (SUPERSCRIPT II reverse transcriptase) shows a similar preference, but the efficiency of adding extended sequences is slightly lower than wild-type reverse transcriptase (data not shown).

We further explored the possibility of using this rather conservatively extended sequence at the 3′ end of cDNA to promote template switching. If an additional oligonucleotide (TS oligonucleotide) is present in the first-strand synthesis mixture, it could anneal with the extended sequence, and then reverse transcriptase could switch templates and replicate this oligonucleotide. We tested several TS oligonucleotides that could interact with the extended sequence and found that to achieve the highest efficiency of the template switching, the TS oligonucleotide should contain at least three consecutive rG nucleotides at the 3′ end. Replacement of any or all of the rGs for other ribonucleotides (rA, U or rC) or with dGs significantly reduces the efficiency of the template switching (Figure 2). We believe that the hydrogen bonding interaction of the $r(G)_3$ stretch of the TS oligonucleotide with dC-rich extended sequence of cDNA promotes template switching. Other types of complementary interactions such as oligo(U)–oligo(dA) are less efficient at template switching, probably because of their lower annealing stability. Favorable interactions of the oligo(rG) sequence with the active site of MMLV reverse transcriptase might also contribute to the efficiency of the template switching. As shown in Table 1, template switching does not require a perfect match because even the dA-rich extended sequences can be effectively anchored by a TS oligonucleotide containing $r(G)_3$.

Amplification of the cDNA

To be able to amplify the cDNA by PCR, it is first necessary to attach primer binding sites to each end. How this is achieved is illustrated in Figure 3. A primer

Table 1. Preference of Nontemplated dNTP Addition and Efficiency of Template-Switching Reaction for Model Blunt-Ended Substrates

Blunt-Ended Substrate	Nontemplated Extension Product	Efficiency of Template Switching
S1: 5'-d---------CCC-3' 3'-r---------GGG-5'	t a a 5'-d---------CCCCCCCC 3'-r---------GGG	+++
S2: 5'-d---------CCT-3' 3'-r---------GGA-5'	t 5'-d---------CCTAACCCC 3'-r---------GGA	+
S3: 5'-d---------CCG-3' 3'-r---------GGC-5'	g g t c c 5'-d---------CCGAAAAA 3'-r---------GGC	+
S4: 5'-d---------CCA-3' 3'-r---------GGT-5'	5'-d---------CCAAA 3'-r---------GGT	+
S5: 5'-d---------GTC-3' 3'-r----------CAG-5'	CaAAA 5'-d---------GTCCACCC 3'-r----------GGC	++

S6:
5'-d----------CTG-3'
3'-r----------GAC-5'
	t t t
	         g g g t
	        g A a a c
	5'-d---------CTGAACCCG
	3'-r----------GAC
		+

Consensus Extension Sequence:	c c a a a 5'-d---------NNNAACCC-3' 3'-r----------NNN-5'	

r: oligoribonucleotide
d: complementary oligodeoxyribonucleotide
Column 1: for all blunt-ended substrates, only differences in the structure of blunt ends are shown.
Column 2: structure of the extended products. The most effectively incorporated dNTP is shown by upper-case letters; alternatively, less efficiently incorporated nucleotides are shown above by lower-case letters.
Column 3: efficiency of template switching catalyzed by TS-oligo-rG3; "+++" is 30%–50%, "++" is 10%–30%, and "+" is 5%–10% yield of anchored template-switching oligonucleotide products.

binding site is attached to the 5' end of the cDNA by priming the RT reaction with an oligo(dT)-linked arbitrary sequence. This primer anneals to the poly(A) tail of the mRNA. A primer binding site is attached to the 3' end of the cDNA by template switching with an arbitrary sequence incorporated into the TS oligonucleotide.

Unlike other PCR-based cDNA amplification methods, no ligation, tailing or any intervening purification steps are required before the amplification step. For the amplification step, we performed long PCR, which uses a combination of thermostable DNA polymerases (3,16). The use of long PCR allows amplification of nearly full-sized cDNAs (in our experiments up to 6 kb) with an error rate significantly lower than that of conventional PCR (3).

Quality of cDNA Amplified by SMART Technology

Size Distribution and rRNA Impurities

To examine the efficiency of the SMART PCR technology, we compared the size distribution and banding pattern of double-stranded cDNA generated from human skeletal muscle total RNA by the SMART method to double-stranded cDNA generated from human skeletal muscle poly(A+) RNA by the conventional method. The results are shown in Figure 4. In both cases, the size distribution of cDNA ranged from about 0.5 to 6 kb, with little bias in the amplification of short cDNAs. Additionally, the cDNA banding patterns were the same whether the cDNA was synthesized from poly(A+) RNA or only 50 ng of total RNA. This demonstrates that the SMART PCR method allows selective amplification of cDNA from the poly(A+) fraction of total RNA. To further characterize the quali-

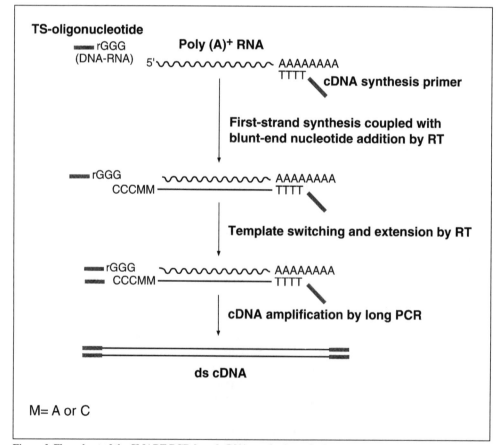

Figure 3. Flow chart of the SMART PCR-based cDNA synthesis method.

Generation and Use of High-Quality cDNA

ty of the double-stranded cDNA, we constructed a cDNA library from the amplified cDNA. A GenBank® and European Molecular Biology Laboratory (EMBL) database search of 50 randomly selected cDNA clones showed that more than 90% of them were of mRNA origin. Furthermore, none of the clones were found to have homology to ribosomal RNA (data not shown).

Gene Representation

To further examine the quality of SMART PCR cDNA, we used PCR to test for the presence of 45 different genes (including cytokines, receptors and housekeeping genes). Two cDNA populations were tested, one generated by SMART PCR starting from 50 ng human skeletal muscle total RNA and another synthesized by the conventional method of Gubler and Hoffman (13) from 5 µg of human skeletal muscle poly(A+) RNA (Table 2). These genes were chosen for their different levels of expression and their classification as different functional groups. In both populations of cDNA, we found the presence of 42 out of the 45 (93%) specific cDNAs. We conclude from this experiment that there is little or no difference in the representation of genes in the cDNAs generated by the SMART method and by the conventional method.

In comparison with other PCR-based cDNA amplification methods, the higher gene representation obtained is probably due to several factors: *(i)* there are fewer manipulations, which improves yields and thus requires fewer cycles of amplification; *(ii)* the template-switching reaction catalyzed by MMLV might be more efficient than single-stranded oligonucleotide tailing or oligonucleotide ligation; and *(iii)* the use of long PCR (3,17) extends the size of the cDNA that can be amplified.

Of course, we cannot exclude the possibility that the amplification step of the SMART procedure changes the relative concentration of particular cDNAs (e.g., cDNAs with GC-rich sequences or very long cDNAs).

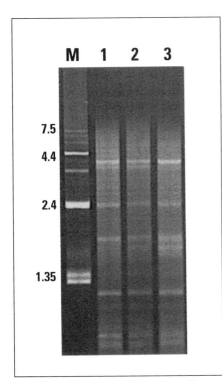

Figure 4. Gel electrophoresis patterns of human skeletal muscle double-stranded cDNAs. Lane M: RNA ladder; lane 1: double-stranded cDNA generated by the conventional method of Gubler and Hoffman (13) starting with 500 ng of poly(A+) RNA; lane 2: double-stranded cDNA generated by the SMART PCR method starting from 500 ng of total RNA; and lane 3: double-stranded cDNA generated starting from 50 ng of total RNA. Five microliters of the PCR products were analyzed on a 1.1% agarose EtdBr gel, along with the indicated RNA ladder.

cDNA Quality vs. the Quantity of Starting Total RNA

To examine the sensitivity of the SMART PCR method, we generated cDNA from a series of RNA dilutions. As shown in Figure 5 (lanes 1–4), the size distribution and pattern of the double-stranded cDNA amplified from 1 µg to 1 ng of total RNA are practically indistinguishable from each other. Howev-

315

Table 2. Comparison of Gene Representation in cDNAs Generated by SMART PCR vs. Conventional Method

Human Gene	Conventional cDNA [from 5 µg poly(A+) RNA]	SMART PCR cDNA (from 50 ng total RNA)
β-actin	+	+
β-myosin	+	+
$β_2M$	+	+
c-fms	+	+
c-fos	+	+
c-jun	+	+
c-kit	+	+
c-myc	+	+
EGF	+	+
EGFR-3	+	+
GAPD	+	+
G-CSF	+	+
GM-CSF	−	+
IGF2	+	+
IGF1R	+	+
IFNA1	+	+
IFNB1	+	+
IFNG	+	+
IL1A	+	+
IL1B	+	+
IL2	+	+
IL3	−	+
IL4	+	+
IL5	+	+
IL6	+	+
IL7	+	+
IL8	+	+
IL10	+	+
IL11	+	−
IL2RA	+	+
IL2RB	+	−
IL4R	+	+
IL6R	+	+
NFKB1	+	+
p53	+	+
PDGFA	+	+
TFR	+	+
TGFA	+	+
TGFB1	+	+
TGFB2	+	+
TNF	+	+
LTA	+	+
TNFR1	+	+
TNFR2	−	−
Positive	93.3% (42/45)	93.3% (42/45)

Gene-specific RT-PCR primers (45 pairs) were used for this test. One microliter of each double-stranded cDNA sample was amplified as described in the "Protocols" section, except that 35–40 cycles of PCR were performed for low-copy genes (cytokine and cytokine receptor genes), and 25–30 cycles were performed for the more highly abundant housekeeping genes. PCR products were analyzed on a 1.1% agarose gel. "+" and "-" indicate the presence and absence of positive PCR signals, respectively.

er, when the quantity of total RNA was less than 1 ng, the amplification products became shorter with more distinct bands. We also tested the cDNAs for the presence of interferon-β1 and interferon-γ, which are low-abundance mRNAs, and glyceraldehyde-3-phosphate dehydrogenase (*GAPD*) and β-myosin, which are high-abundance mRNAs. As shown in Figure 5, all of these gene sequences were detected in the cDNA amplified from 1 ng or more of total RNA. Lower quantities of starting RNA give unreproducible amplification of the low-abundance mRNAs. Based on the data (Figure 5), we suggest that a minimum of 50 ng of total RNA be

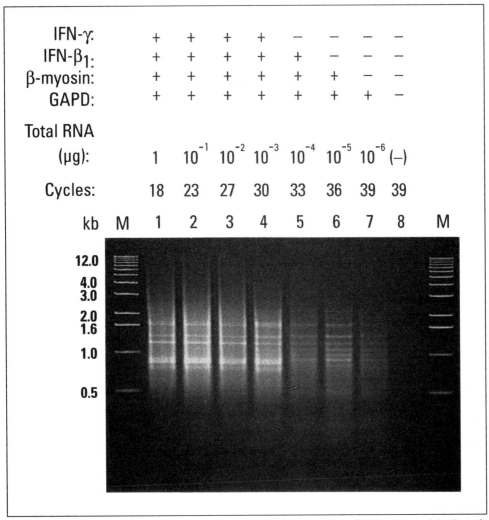

Figure 5. Relationship between the starting amounts of total RNA and quality of cDNA. Human skeletal muscle total RNA was serially diluted as indicated (lanes 1–7). Lane 8 is a negative control (no RNA). First-strand cDNA and PCR double-stranded cDNA were synthesized as described in the "Protocols" section for variable numbers of cycles (indicated). The quality of the cDNA was determined by both gel patterns and gene representation. For a gene representation test, SMART PCR cDNA in each lane was used as a template for further PCR amplification with 4 sets of gene-specific primers: interferon-γ (IFN-γ), interferon-β1 (IFN-β1), β-myosin and *GAPD*. PCR conditions were the same as in the "Protocols" section, except that 40 cycles were performed for IFN-γ and IFN-β1, and 30 cycles were performed for β-myosin and *GAPD*. Five microliters of the corresponding final PCR products were loaded on a 1.1% agarose–EtdBr gel, along with 1-kb DNA size markers (lanes M). "+" and "-" indicate the presence and absence of positive PCR signals on gels, respectively.

used to ensure the gene representation. We typically start with 1 µg of total RNA, which can easily be obtained from small amounts of cells and tissues.

Advantages and Limitations of the SMART Technology

The SMART PCR method has several improvements over other PCR-based methods (12). RT and template switching are catalyzed by a single enzyme, so no additional ligation and homopolymer tailing are needed. The coupled cDNA synthesis and template switching take only one hour. The process is completed in a single tube, eliminating multiple manipulations. The combination of SMART technology with long PCR generates double-stranded cDNA of up to 6 kb and gene representation approximately 90%–95%.

The primary drawback of the SMART PCR method is that the TS oligonucleotide occasionally primes the first-strand cDNA synthesis. When this occurs, the cDNA molecules contain the same anchor sequences at both ends. Sequence analysis of cDNA libraries constructed by the SMART method indicates that the frequency of bi-anchored cDNAs is about 2% (data not shown). For cDNA library construction, these artifactual clones can be eliminated by the use of directional cloning procedures. However, for rapid amplification of cDNA ends (RACE), these cDNAs generate a higher level of background as compared to the anchor ligation-based procedure (5). Thus, we do not recommend the SMART PCR method for RACE.

Applications of SMART PCR

The most important feature of SMART PCR is that large amounts of high-quality cDNA can be generated from small amounts of total RNA. This makes it possible to perform several gene cloning and analysis techniques that traditionally require several micrograms of poly(A+) RNA.

Construction of cDNA Libraries from Small Amounts of Total RNA

Typically, 5–10 µg of poly(A+) RNA are necessary to construct cDNA libraries comprehensive enough to obtain low-abundance cDNAs (2×10^6 clones). We have found that SMART PCR-generated cDNA can be used to generate high-quality cDNA libraries of this complexity (20,22).

cDNA Subtraction

cDNA subtraction is a valuable method for identification and cloning of differentially expressed genes (21). However, effective cDNA subtraction requires the use of high concentrations of cDNA during the hybridization step. At least several hundred nanograms of driver cDNA per microliter of hybridization buffer are necessary for both traditional subtraction procedures and the PCR-based suppression subtractive hybridization method (9). We have found that SMART PCR-generated cDNA can be used as starting material for the PCR-based suppression subtractive hybridization. Please refer to Diatchenko et al. (in this volume) for a detailed protocol.

"Virtual" Northern Blots

SMART PCR cDNA is also useful for researchers who wish to analyze transcript size and expression patterns by hybridization but lack sufficient poly(A^+) RNA or total RNA for Northern blots. "Virtual" Northern blots can be generated using SMART PCR-amplified cDNA instead of total or poly(A^+) RNA, and they often provide information similar to that provided by standard Northern blots (8).

REFERENCES

1. **Akowitz, A. and L. Manuelidis.** 1989. A novel cDNA/PCR strategy for efficient cloning of small amounts of undefined RNA. Gene *81*:295-306.
2. **Apte, A.N. and P.D. Siebert.** 1993. Anchor-ligated cDNA libraries: a technique for generating a cDNA library for the immediate cloning of the 5′ ends of mRNAs. BioTechniques *15*:890-893.
3. **Barnes, W.M.** 1994. PCR amplification of up to 35-kb DNA with high fidelity and high yield from lambda bacteriophage templates. Proc. Natl. Acad. Sci. USA *91*:2216-2220.
4. **Belyavsky, A., T. Vinogradova and K. Rajewsky.** 1989. PCR-based cDNA library construction: general cDNA libraries at the level of a few cells. Nucleic Acids Res. *17*:2919-2932.
5. **Chenchik, A., L. Diachenko, F. Moqadam, V. Tarabykin, S. Lukyanov and P.D. Siebert.** 1996. Full-length cDNA cloning and determination of mRNA 5′ and 3′ ends by amplification of adaptor-ligated cDNA. BioTechniques *21*:526-534.
6. **Chomczynski, P. and N. Sacchi.** 1987. Single-step method of RNA isolation by acid guanidinium thiocyanate-phenol-chloroform extraction. Anal. Biochem. *162*:156-159.
7. **Clark, J.M.** 1988. Novel non-templated nucleotide addition reactions catalyzed by procaryotic and eucaryotic DNA polymerases. Nucleic Acids Res. *16*:9677-9686.
8. **CLONTECH Laboratories.** 1996. CapFinder PCR cDNA synthesis kit: generation of high-quality cDNA for a wide spectrum of application. CLONTECHniques *XI(4)*:2-3.
9. **Diatchenko, L., Y.F. Lau, A.P. Campbell, A. Chenchik, F. Moqadam, B. Huang, S. Lukyanov, K. Lukyanov et al.** 1996. Suppression subtractive hybridization: a method for generating differentially regulated or tissue-specific cDNA probes and libraries. Proc. Natl. Acad. Sci. USA *93*:6025-6030.
10. **Domec, C., B. Garbay, M. Fournier and J. Bonnet.** 1990. cDNA library construction from small amounts of unfractionated RNA: association of cDNA synthesis with polymerase chain reaction amplification. Anal. Biochem. *188*:422-426.
11. **Frohman, M.A., M.K. Dush and G.R. Martin.** 1988. Rapid production of full-length cDNAs from rare transcripts: amplification using a single gene-specific oligonucleotide primer. Proc. Natl. Acad. Sci. USA *85*:8998-9002.
12. **Froussard, P.** 1992. A random-PCR method (rPCR) to construct whole cDNA library from low amounts of RNA. Nucleic Acids Res. *20*:2900.
13. **Gubler, U. and B.J. Hoffman.** 1983. A simple and very efficient method for generating cDNA libraries. Gene *25*:263-269.
14. **Hu, W.S. and H.M. Temin.** 1990. Retroviral recombination and reverse transcription. Science *250*:1227-1233.
15. **Kato, S., S. Sekine, S.W. Oh, N.S. Kim, Y. Umezawa, N. Abe, M. Yokoyama-Kobayashi and T. Aoki.** 1994. Construction of a human full-length cDNA bank. Gene *150*:243-250.
16. **Kellogg, D.E., I. Rybalkin, S. Chen, N. Mukhamedova, T. Vlasik, P.D. Siebert and A. Chenchik.** 1994. TaqStart Antibody™: "hot start" PCR facilitated by a neutralizing monoclonal antibody directed against *Taq* DNA polymerase. BioTechniques *16*:1134-1137.
17. **Kulpa, D., R. Topping and A. Telesnitsky.** 1997. Determination of the site of first strand transfer during Moloney murine leukemia virus reverse transcription and identification of strand transfer-associated reverse transcriptase errors. EMBO J. *16*:856-865.
18. **Maruyama, K. and S. Sugano.** 1994. Oligo-capping: a simple method to replace the cap structure of eukaryotic mRNAs with oligoribonucleotide. Gene *138*:171-174.
19. **Patel, P.H. and B.D. Preston.** 1994. Marked infidelity of human immunodeficiency virus type 1 reverse transcriptase at RNA and DNA template ends. Proc. Natl. Acad. Sci. USA *91*:549-553.
20. **Sambrook, J., E.F. Fritsch and T. Maniatis.** 1989. Molecular Cloning: A Laboratory Manual, 2nd ed. CSHL Press, Cold Spring Harbor, NY.
21. **Swendeman, S.L. and M.P. La Quaglia.** 1996. cDNA subtraction hybridization: a review and an application to neuroblastoma. Semin. Pediatr. Surg. *5*:149-154.
22. **Zhu, Y., A. Chenchik and P.D. Siebert.** 1996. Synthesis of high-quality cDNA from nanograms of total or poly(A^+) RNA with the CapFinder cDNA library construction kit. CLONTECHniques *XI(1)*:2-4.

23 Protein Production from PCR Products

Steven J. Winder
Institute of Cell and Molecular Biology, University of Edinburgh, Edinburgh, Scotland

OVERVIEW

Polymerase chain reaction (PCR) generation of cDNA for cloning into protein expression vectors is now the most efficient way of making even the largest constructs. Advances in proofreading polymerases and automated DNA sequencing technology mean that laborious subcloning through many vectors to generate the correct restriction sites compatible with a particular expression vector is a thing of the past. Rather than provide comprehensive details of the methodology of generating expression constructs by PCR and because of the diversity of expression systems, species used and problems with the expressed proteins, this chapter serves to provide some simple tips that can be universally applied to the design of protein expression constructs.

BACKGROUND

With the need for precise protein products for functional and structural studies or for antibody production and with the advent of high-fidelity proofreading DNA polymerases and automated DNA sequencing, PCR is now the method of choice for generating the majority of protein expression constructs. Protein expression in bacteria, yeast, amoebae, plants, cultured eukaryotic cells and reticulocyte lysates usually relies on the engineering of restriction sites into cDNA to generate constructs that will read in-frame with the initiating sequences in the particular expression vector. Although on rare occasions these might be found by chance in the particular cDNA or the construct design might take these into account, in the majority of cases it is necessary to design restriction sites de novo for incorporation into the expression construct. This can sometimes be achieved by cloning a particular cDNA through a series of cloning vectors with differing restriction sites to generate a novel site by blunt ending or filling in complementary overhangs, but in the vast majority of cases this is not possible or not desirable. With PCR, one can design a precise protein expression construct that might differ from the wild-type sequence only by the presence of the initiating methionine. With the emphasis of this volume on reverse transcription (RT)-PCR, it is worth noting that expression constructs can be generated directly by RT-PCR, but clearly the precise sequence of the target RNA would need to be known. We have successfully isolated and expressed cDNA to published sequences by direct RT-PCR from cDNA libraries and cloning into expression vectors, but for the purposes of this chapter we shall assume that a cDNA of known and verified sequence has already been isolated by RT-PCR or other methods.

Choice of Expression Vector

Certain aspects of expression vector choice will depend on the use to which the final protein product is to be put, e.g., antibody production, structural studies or biochemical or cell biological applications, or whether the protein needs to be isotopically or otherwise labeled in any way. Other factors that can affect vector choice are posttranslational modifications such as glycosylation and myristylation (not carried out in bacteria), whether the protein is an integral membrane protein or whether it is otherwise processed in some unique way. A comprehensive review of protein expression vectors is beyond the scope of this article; a more comprehensive treatise with detailed methods can be found elsewhere (1,2). Irrespective of the species in which the vector is to be used, there are basically two types of vector: ones in which the protein is expressed as a native molecule with no additional fusions or tags (frequently used for structural work in which extraneous amino acids might interfere) and ones in which the protein sequence of interest is fused either amino-terminally or carboxy-terminally, or both, to another protein. This other protein might act to drive high-level expression, as in the case of the commonly used glutathione S-transferase (GST) (7) or maltose binding protein (MBP) fusions in bacteria (3) and usually acts also as an affinity tag to aid in protein purification. Some vectors contain just such an affinity tag, as exemplified by the poly-histidine tag vectors; others contain tags that aid in subsequent identification of the protein when used in other experiments, such as luciferase or green fluorescent protein for direct visualization by fluorescence microscopy or small antigenic epitopes like the Myc tag for immunological detection or immunoprecipitation. The "sandwich" vectors, vectors encoding both amino- and carboxy-terminal fusions, can be particularly useful when it is essential that a full-length unproteolyzed protein be produced. Irrespective of which vector and organism are to be used for the protein expression, considerable care must be taken in both the design and generation of the expression construct, for which care in primer design and choice of coding sequence for expression are paramount.

Primer Design

As with most PCR-based strategies, primer design is a critical factor. As with PCR for other applications, the usual precautions against primer dimerization and secondary structure formation, and optimizing melting temperature and specificity should be taken into consideration. Several programs are available for this purpose. In addition, depending on the expression system to be used, care should be taken with codon usage because this can profoundly affect translation in some species. In *E.*

coli, certain rarely used arginine codons (AGA and AGG) and, to a lesser extent, other rarely used codons act as translational repressors and therefore reduce the amount of protein product (6). In *Dictyostelium*, certain codons are not used at all, making expression constructs produced from heterologous cDNA difficult, if not impossible, to express. In all cases, care should be taken to avoid low-frequency codons for the species in which the protein is to be expressed. If necessary, change the codon usage in the primer used to generate the PCR product. A comprehensive list of codon usage for a large number of species can be found in Wada et al. (9). Clearly, restriction enzymes appropriate to the expression vector to be used must be incorporated into both 5′ and 3′ primers with the addition of a stop codon prior to the restriction site in the 3′ primer, and the 5′ restriction site(s) must be in frame with the reading frame or initiation codon in the expression vector. Many commercial expression vectors incorporate stop codons in all 3 frames 3′ of the multiple cloning site. However, for precision and with the ease of generation by PCR, it is worthwhile to include one at the correct place in the 3′ primer. We routinely purify the initial PCR products, restriction digest and ligate directly into the expression vector. If this is the chosen route of cloning, then care must also be taken to include extra bases 5′ of the restriction site in the 5′ primer and 3′ of the restriction site in the 3′ primer. Many restriction enzymes will not cleave efficiently at the ends of DNA strands, particularly *Nde*I and *Nco*I, commonly used 5′ restriction sites in the pET series of bacterial nonfusion expression vectors (8). Both of these sites contain an initiating ATG sequence (*Nde*I: CAT**ATG** and *Nco*I: CC**ATG**G) often used in frame with the coding sequence to be expressed, and *Bam*HI can also be conveniently used to clone cDNA in frame (GGA TCC = Gly Ser). In the case of *Nde*I and *Nco*I, up to an extra 6 bases must be included to accommodate their cleavage requirements. Other common enzymes with more stringent requirements are *Hin*dIII, *Not*I, *Pst*I, *Sal*I and *Xho*I. Tables of cleavage close to the end of DNA fragments are included in the catalogs of most major suppliers of restriction enzymes. A schematic example of a pair of PCR primers for generation of a construct for cloning into a pET bacterial expression vector is shown in Figure 1. It is also possible to clone PCR products directly into cloning vectors using the A overhangs generated by certain DNA polymerases. To my knowledge however, none of these is universally compatible with expression vectors, and although in some cases there is some advantage in doing two cloning steps (one into the T-overhang vector and another into the expression vector), the only saving would be on the length of the PCR primers needed initially.

Expression Construct Design

As with the choice of expression vector depending on the ultimate use for the protein, the design of the expression construct itself will also reflect the use for the protein. Using PCR to generate expression constructs means that constructs can be made that precisely encode individual protein domains, structural elements or antigenic sequences, or contain specific mutations or alterations for functional analysis for expression as fusions or nonfusions. Clearly, a knowledge of protein structure or function is essential to successful construct design. In the case of protein production for immunization to raise antibodies, however, some of the work can be done by computer. Most protein or gene analysis software includes algorithms that will predict the most likely antigenic sequences, usually charged and hydrophilic sequences predicted to be surface features. If the aim is to generate an antibody to a novel protein, then these are a good place to start, whether expressed as a fusion or a nonfusion. Antibodies can be raised against any expressed protein, whether recovered as a soluble folded protein or, in the case of bacterially expressed protein, as insoluble inclusion bodies. This latter case can be an aid to purification, especially if antibodies against denatured protein are required. For structural or functional studies however, it is usually desirable to produce soluble protein at a high concentration, and in these cases, a knowledge of the structural or functional motifs is important to generate as nearly as possible a soluble folded product. Small differences in construct can produce profound differences in protein expression level, solubility and function. For example, a series of 3 constructs encoding the first 233, 246 and 254 amino acids of the amino-terminal actin binding domain of dystrophin, designed on sound structural grounds, produced insoluble and unrecoverable protein, highly soluble and apparently functional protein and soluble but nonfunctional protein, respectively (10). Even single amino acid substitutions can have equally dramatic effects on solubility and expression level. It is also worth noting that although bacterially expressed proteins are not N-terminally blocked by acetylation or other modification, they can be subject to N-terminal methionine excision (4). This should be taken into account if the amino-terminal methionine is crucial for protein activity. High-level protein expression, in bacteria in particular, is not a precise science.

Generating the Construct

The actual PCR of the expression construct and subsequent cloning steps should be routine and will not be described in detail here. It is sufficient to say that because of the good match between primers and cDNA template used to generate expression constructs, stringent PCR conditions can be used to reduce mismatches and other errors. With high-fidelity proofreading polymerases, however, there might be a temptation not to check the generated construct for errors once it is cloned into the expression vector. This is a false economy; the DNA sequencing of the generated expression construct is a small chore compared to the weeks of work that could ensue trying to optimize protein expression from a flawed expression construct that will never produce protein. Always sequence the construct.

```
5′  primer
a.                         M   T   N   S   E   K   I   L
b.          GGGTTT CAT ATG ACG AAC AGT GAG AAG ATC CTG
c.          overhang Nde I    optimised coding sequence

3′  primer
d.                     T   S   L   F   E   V   L
e.                    ACA TCT TTG TTT GAG GTG CTA TAG GTCGAC TACCCT
f.                        coding sequence          stop  Sal I  overhang
```

Figure 1. **Typical primer design for pET expression vector constructs.** Lines a. and d. represent the amino acid sequences at the amino and carboxy termini, respectively, of the protein to be expressed. Lines b. and e. are the actual primer sequences used but, in the case of the 5′ primer, optimized for *E. coli* codon usage. In the 5′ primer, the ATG in the *Nde*I site is in frame and is the initiation codon for the construct. Lines c. and f. are descriptions of the sequences in lines b. and e., respectively. The 3′ primer is shown in its sense orientation for clarity but must be reversed and complemented prior to synthesis.

Troubleshooting

If constructs have been carefully designed and protein production is still problematic, then there are several steps that can be taken to improve the situation without recourse to remaking the expression construct, although in some cases the latter might be inevitable. Most of what follows applies to protein expression in bacteria, although in some cases it will also apply to other systems. In my laboratory, we routinely express proteins as nonfusions for structural and functional studies; however, when proteins are insoluble, we turn to fusion vectors to try to generate soluble protein and then cleave off the fused protein at the specific encoded proteolysis site. As a matter of routine, we incorporate restriction sites into the original expression construct so that the construct can be easily switched among three of the most commonly used bacterial expression vectors (11,12). By incorporating an in-frame BamHI site 5′ of the NdeI site containing the initiating methionine in the 5′ primer and a SalI site 3′ of the stop codon in the 3′ primer, one can easily clone among the MBP fusion vector series pMAL, GST fusion vector series pGEX, and pSJW1, a pET-type nonfusion vector (8) with a modified multiple cloning site (11,12). The results of a test expression with a construct produced in this way and expressed in all three vectors is shown in Figure 2. In this case, the expression levels were highest in the nonfusion vector pSJW1. If soluble protein expression cannot be achieved by simply switching between vectors, several other factors can influence the solubility of expressed protein in E. coli. E. coli will easily synthesize protein at a range of temperatures that will support growth (10°–43°C), and in some cases this can influence the partitioning of protein between the soluble phase and inclusion bodies. If the protein being expressed (or RNA from which it is expressed) is under the control of a "leaky" promoter and is either incorrectly folded or toxic to the bacteria, it will be packaged into inclusion bodies and/or inhibit cell growth. The latter is a particular problem with the pET vectors under the control of the bacteriophage T7 RNA polymerase promoter, which is generally not tightly controlled in the standard host E. coli strain BL21(DE3). This can be remedied however by using the strains BL21(DE3)pLysS and BL21(DE3)pLysE, which carry a T7 lysozyme gene that acts as a strong inhibitor of the T7 promoter, thus stopping the leaky expression. It also aids in lysis and recovery of protein from the bacteria postinduction. If this is not sufficient and pET vectors are still the vector of choice, then with the

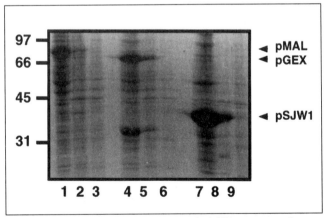

Figure 2. Expression of the same construct in 3 different expression vectors. A 230 amino acid construct from the carboxy terminus of the cytoskeletal protein utrophin was expressed in either pMAL (lanes 1–3), pGEX (lanes 4–6) or pSJW1 (lanes 7–9) as previously described (11,12). The 15% sodium dodecyl sulfate polyacrylamide gel shows equivalent quantities of postinduction E. coli-soluble fractions (lanes 1, 4 and 7), postinduction cells (lanes 2, 5 and 8) and preinduction cells (lanes 3, 6 and 9). Numbers on the left refer to the relative position of molecular weight markers (kDa), and annotated arrows on the right refer to the relative positions of the expressed proteins in the 3 different vectors.

expression vector in a cloning strain (e.g., *E. coli* HMS174 lacking T7 polymerase and therefore lacking the ability to transcribe from the T7 promoter), cells can be infected with bacteriophage CE6, which carries a T7 RNA polymerase gene, thus initiating transcription from the plasmid and protein expression but at the expense of phage development (8). As a matter of course for expression from bacterial vectors, we transform freshly made calcium-competent *E. coli* BL21(DE3) (this strain lacks the *lon* and *omp*T proteases, which can cause proteolysis problems during subsequent protein purification) with the expression construct, plate it out onto selective medium, grow it overnight and inoculate each liter of liquid culture medium directly with up to 50 colonies from the transformation. From a single transformation using 100 µL of calcium-competent cells and 1 µL of vector DNA, it is possible to spread 6 plates, from which the colonies can be used to inoculate up to 12 L of medium that will reach an optical density (OD) of 1 within approximately 5 h. Following a 2-h induction, cells can be harvested by centrifugation within the day, ready for subsequent lysis and protein production. This avoids problems associated with plasmid loss, overgrowth and relative dilution of the expressed protein by nonexpressing bacteria when growing large-scale cultures overnight or following other regimes. The level and length of induction can also affect both amount of protein and its partitioning in the cell. With most vectors, we generally use 0.5 mM isopropyl β-D-thiogalactopyranoside (IPTG) for 2 h, but maximum (soluble) protein expression can be reached before this time. As with all new expression constructs, it is wise to check protein expression levels at preinduction and different postinduction times, and if expression is too rapid, reduce the time, amount of inductant, temperature or combinations of the three. If, despite all attempts and the type of expression vector used, proteins still go to inclusion bodies, coexpression of bacteriocin release protein (5) has been successful in our hands in altering the ratio of soluble to insoluble protein in the bacteria or in recovering protein from the culture medium. For unknown reasons, however, despite taking all the above into consideration, some constructs still do not express. In these relatively rare cases, the only recourse is to redesign the construct and start again. For methods of recovering both soluble and insoluble protein from bacteria, see Reference 2.

SUMMARY

I do not want to suggest that protein expression from PCR products is not normally quite straightforward; the need to resort to some of the more obscure methodologies described above is fortunately seldom. With careful attention to choice of expression vector, construct and primer design and checking the integrity of the expression construct, it is possible with any number of systems to achieve high-level protein production.

ACKNOWLEDGMENTS

This work was supported by Muscular Dystrophy Group Grant No. RA3/116 and Wellcome Trust Grant No. 042180.

REFERENCES

1. Brent, R.E., P.F. Schendel, S. Tabor, A.R. Shatzman, M.S. Gross, M. Rosenberg, P. Riggs, E.R. Lavallie et al. 1995. Protein expression, p. 16.1-16.61. *In* Ausubel, F., R. Brent, R.E. Kingston, D.D. Moore,

J.G. Seidman, J.A. Smith and K. Struhl (Eds.). Short Protocols in Molecular Biology. John Wiley & Sons, New York.
2. **Glover, D.M. and B.D. Hames.** 1995. DNA Cloning 2: Expression Systems. IRL Press, Oxford.
3. **Guan, C., P. Li, P.D. Riggs and H. Inouye.** 1987. Vectors that facilitate the expression and purification of foreign peptides in *Escherichia coli* by fusion to maltose binding protein. Gene *67*:21-30.
4. **Hirel, P.-H., J.-M. Schmitter, P. Dessen, G. Fayat and S. Blanquet.** 1989. Extent of N-terminal methionine excision from *Escherichia coli* proteins is governed by the side-chain length of the penultimate amino acid. Proc. Natl. Acad. Sci. USA *86*:8247-8251.
5. **Hsuing, H.M., A. Cantrell, J. Luirink, B. Oudega, A.J. Veros and G.W. Becker.** 1989. Use of bacteriocin release protein in *E. coli* for excretion of human growth hormone into the culture medium. Bio/Technology *7*:267-271.
6. **Inouye, M. and G.-F.T. Chen.** 1990. Suppression of the negative effect of minor arginine codons on gene expression: preferential usage of minor codons within the first 25 codons of the *Escherichia coli* genes. Nucleic Acids Res. *18*:1465-1473.
7. **Smith, D.B. and K.S. Johnson.** 1988. Single-step purification of poly-peptides expressed in *Escherichia coli* as fusions with glutathione-S-transferase. Gene *67*:31-40.
8. **Studier, F.W., A.M. Rosenburg, J.J. Dunn and J.W. Dubendorff.** 1990. Use of T7 RNA polymerase to direct expression of cloned genes. Methods Enzymol. *185*:60-89.
9. **Wada, K., S. Aota, R. Tsuchiya, F. Ishibashi, T. Gojobori and T. Ikemura.** 1990. Codon usage tabulated from the GenBank genetic sequence data. Nucleic Acids Res. *18*(Suppl):2367-2411.
10. **Way, M., B. Pope, R.A. Cross, J. Kendrick-Jones and A.G. Weeds.** 1992. Expression of the N-terminal domain of dystrophin in *E. coli* and demonstration of binding to F-actin. FEBS Lett. *301*:243-245.
11. **Winder, S.J. and J. Kendrick-Jones.** 1995. Protein production in 3 different expression vectors from a single PCR product. Anal. Biochem. *231*:271-273.
12. **Winder, S.J. and J. Kendrick-Jones.** 1996. Protein production in 3 different expression vectors from a single PCR product (Vol 231, Pg 271, 1995). Anal. Biochem. *236*:190.

24 | In Vitro Cloning: A Method for PCR Amplification of Individual Unknown DNA Fragments Suitable for Direct Sequencing

Konstantin A. Lukyanov
Shemyakin and Ovchinnikov Institute of Bioorganic Chemistry, Russian Academy of Science, Moscow, Russia

OVERVIEW

We have developed a polymerase chain reaction (PCR)-based method called in vitro cloning that can be used instead of classical DNA cloning in bacteria and other in vivo systems. The in vitro cloning method permits molecule-by-molecule PCR amplification of complex unknown DNA mixtures. The products of in vitro cloning (in vitro clones) can be used for direct sequencing, hybridization analysis, RNA synthesis and other routine molecular biology procedures.

BACKGROUND

The two main goals of molecular cloning in bacteria, phages and other in vivo systems are: *(i)* production of a large amount of each DNA molecule type and *(ii)* creation of conditions for sequencing these molecules by introducing different sites for sequencing primer annealing at different ends of the unknown sequence.

PCR-based techniques have replaced in vivo cloning in a number of tasks in which the sequence of the molecule under investigation is known at least partially. Recent methodological achievements have nearly allowed creation of a technique for in vitro cloning of unknown DNA. Of special note are the following considerable steps forward: *(i)* partially or even completely unknown DNA can be amplified by attachment of artificial sequences for primer annealing (5,12,14); *(ii)* PCR products can be easily sequenced directly (11,13); *(iii)* the power of amplification is sufficient to process single molecules (4); and *(iv)* long and accurate PCR gives access to long DNA molecules and increases the fidelity of amplification (1).

So we could imagine a simple scheme for cloning by PCR. One should ligate an oligonucleotide adaptor to all DNA ends in the sample, dilute the sample to get the concentration at which one DNA molecule corresponds to the volume being taken for amplification and perform a number of PCRs with a primer that corresponds to the adaptor. However, in this case, amplification products would have the same flanking sequence at both ends and therefore could not be sequenced using these flanking sequences as primer annealing sites (two texts would be read simultaneously).

Recently, we have described the suppression PCR effect that strongly inhibits amplification of symmetrically flanked DNA molecules (2,9,12). Using the suppression PCR effect, we have developed an in vitro cloning technique that, though close to the ligation–dilution–amplification scheme mentioned above, ensures that the final amplification product is flanked by different adaptor sequences at different ends. Such fragments can be easily sequenced from both ends.

In vitro cloning is very convenient when no more than several dozen clones are needed. Potential applications include differential screening, subcloning of long DNA fragments and cloning of PCR products with low complexity.

PROTOCOLS

Materials and Reagents

Oligonucleotides

The following suppression adaptors (2) were found to be useful for in vitro cloning. This list contains two subsets: *(i)* adaptors 1–4 have one outer part (underlined) including the sequence of the T7 RNA polymerase promoter, and *(ii)* adaptors 5–7 have another outer part (double-underlined). The best combination is adaptors 1 and 2. Adaptors 3 and 4, 5 and 6, 5 and 7, and 6 and 7 also give good results. In these combinations, the adaptors used have the same outer part. This makes it possible to perform in vitro cloning in one PCR using a single PCR primer identical to either underlined or double-underlined sequences. Also, it is possible to combine the adaptors from the first and second subset (e.g., adaptors 1 and 5 and adaptors 1 and 7). However, in this case, two nested PCRs for each in vitro cloning are necessary (8).

Note: All oligonucleotides must be gel-purified.

Suppression Adaptors

- Adaptor 1:

    ——————— *Not*I
    ——————— *Srf*I/*Sma*I
    5'-GTAATACGACTCACTATAGGGCTCGAGCGGCCGCCCGGGCAGGT
                                     3'-CCCGTCCA

- Adaptor 2:

    ——————— *Eag*I/*Eae*I
    5'-GTAATACGACTCACTATAGGGCAGCGTGGTCGCGGCCGAGGT
                                     3'-CGGCTCCA

- Adaptor 3:

                                                                                       *Mlu*I, *Sal*I, *Srf*I

  5′-<u>GTAATACGACTCACTATAGGGC</u>ACGCGTGGTCGACGGCCCGGGCTGGT

                                                                   3′-CCCGACCA

- Adaptor 4:

  5′-<u>GTAATACGACTCACTATAGGGC</u>ACTCTCCGACCTCTCACCGAGCGGT

                                                                   3′-GGCTCGCCA

- Adaptor 5:

  5′-<u>TGTAGCGTGAAGACGACAGAA</u>AGGGCGTGGTGCGGAGGGCGGT

                                                                   3′-GCCTCCCGCCA

- Adaptor 6:

                                                                     *Sal*I, *Srf*I

                                                     *Mlu*I

  5′-<u>TGTAGCGTGAAGACGACAGAA</u>AGTCGACGCGTGCCCGGGCTGGT

                                                                     3′-CCCGACCA

- Adaptor 7:

                                                                       *Eag*I/*Eae*I

  5′-<u>TGTAGCGTGAAGACGACAGAA</u>AGCGTGGTCGCGGCCGAGGT

                                                                   3′-CGGCTCCA

PCR Primers

- PCR primer 1 (for in vitro cloning using adaptors 1–4): 5′-GTAATACGACTC-ACTATAGGGC
- PCR primer 2 (for in vitro cloning using adaptors 5–7): 5′-TGTAGCGTGAA-GACGACAGAA
- Nested PCR primer 1 (for adaptor 1): 5′-TCGAGCGGCCGCCCGGGCAGGT
- Nested PCR primer 2 (for adaptor 2): 5′-AGCGTGGTCGCGGCCGAGGT

General Reagents

- Sterile deionized H$_2$O
- Dilution buffer: 20 mM HEPES-HCl, pH 7.5, 50 mM NaCl, 0.2 mM EDTA
- TM buffer: 10 mM Tris-HCl, pH 7.5, 1 mM MgCl$_2$
- Mineral oil (Sigma Chemical, St. Louis, MO, USA)
- dNTP mixture (10 mM each)
- Ethidium bromide (0.1 mg/mL)
- 50× Advantage™ KlenTaq Polymerase Mix and 10× PCR buffer (CLONTECH Laboratories, Palo Alto, CA, USA)

  Note: The 50× mixture contains KlenTaq-1 DNA Polymerase (an exo⁻, N-terminal deletion of *Taq* DNA polymerase), a proofreading polymerase and TaqStart™ Antibody. This polymerase mixture is optimized for long and accurate PCR (1). The TaqStart Antibody provides automatic hot-start PCR (7). A hot-start PCR must be used in both protocols (see below), and TaqStart Antibody is the only convenient way to do a hot-start PCR in a 96-well plate.

- PCR primer 1
- Nested PCR primer 1
- AmpliCycle™ Sequencing Kit (Perkin-Elmer, Norwalk, CT, USA)

- Aerosol barrier tips
- 0.5-mL PCR tubes
- 96-well PCR plates with sealing films (MJ Research, Watertown, MA, USA)
- Perkin-Elmer DNA Thermal Cycler 480
- MJ Research PTC-100™ Thermal Cycler

Protocol 1: Differential Screening of the Suppression Subtractive Hybridization (SSH)-Subtracted Samples Using In Vitro Cloning

The following protocol is intended for users of the PCR-Select™ cDNA Subtraction Kit (CLONTECH). The user manual for this kit contains the detailed protocols of the subtraction procedure and differential screening of an in vivo-cloned subtracted library. Alternatively, we suggest performing in vitro cloning of the subtracted sample and subsequent screening of the in vitro clones.

Materials and Reagents

- PCR-Select cDNA Subtraction Kit
- Nested PCR primer 2
- QIAquick™ PCR Purification Kit (Qiagen, Chatsworth, CA, USA)
- *Rsa*I enzyme and 10× *Rsa*I buffer (New England Biolabs, Beverly, MA, USA)
- [α-^{32}P]dCTP
- DECAprime™ II DNA Labeling Kit (Ambion, Austin, TX, USA)
- CHROMA SPIN™-100 Columns (CLONTECH)
- Denaturation buffer: 0.4 M NaOH, 1.5 M NaCl
- Neutralizing buffer: 1 M Tris-HCl, pH 7.0, 1.5 M NaCl
- ExpressHyb™ Hybridization Solution (CLONTECH)
- 20× standard saline citrate (SSC)
- 10% sodium dodecyl sulfate (SDS)

Procedures

Preparation of the probes

It is possible to use the tester and driver cDNA as probes for differential screening. However, to increase the sensitivity of the screening, we recommend preparing the probes from cDNAs subtracted in opposite directions (plus- and minus-subtracted cDNAs).

1. Prepare a 50-μL PCR mixture:
   33 μL H$_2$O
   5 μL 10× PCR buffer
   1 μL dNTP mixture (10 mM)
   4 μL nested PCR primer 1 (10 μM)
   4 μL nested PCR primer 2 (10 μM)
   1 μL 50× Advantage KlenTaq Polymerase Mix
   Pour mixture into two 0.5-mL PCR tubes (24 mL in each tube).
2. Dilute 2 μL of secondary PCR products of plus and minus subtractions in 98 μL of water and put 1 μL of the dilution in 24 μL PCR mixture. Overlay with mineral oil.

3. Perform 10 PCR cycles using the Perkin-Elmer DNA Thermal Cycler 480 under the following conditions: denaturation at 94°C for 30 s, annealing at 68°C for 30 s and extension at 72°C for 1.5 min.
4. Purify both samples using the QIAquick PCR Purification Kit in accordance with the supplier's recommendations. Elute cDNA from the column with 50 µL of 1× *Rsa*I buffer. It is important to remove all ethanol from columns before elution.
5. Add 10 U *Rsa*I enzyme to each tube, mix and incubate at 37°C for 1.5 h.
6. Purify the samples using the QIAquick PCR Purification Kit. Elute cDNA with 50 µL of TM buffer (10 mM Tris-HCl, pH 7.5, 1 mM $MgCl_2$). Store at -20°C.
7. Label 2 µL of the cDNA samples obtained using [α-^{32}P]dCTP (50 µCi per probe) and random priming (e.g., using DECAprime II DNA Labeling Kit).
8. Purify probes using CHROMA SPIN-100 columns.

Estimation of quantity of the target cDNA molecules in the subtracted hybridization mixture

Use the plus-subtracted cDNA sample obtained after the second subtractive hybridization and add 200 µL of dilution buffer as an initial sample for subsequent dilution.

1. Prepare the sequential dilutions (100-, 300-, 900-, 2700-, 8100-, 24 300-, 72 900-fold) of the plus-subtractive hybridization mixture using dilution buffer (20 mM HEPES-HCl, pH 7.5, 50 mM NaCl, 0.2 mM EDTA). Change tips and mix solution well after each dilution.
2. Prepare a 200-µL PCR mixture:
   162 µL H_2O
   20 µL 10× PCR buffer
   4 µL dNTP mixture (10 mM)
   10 µL PCR primer 1 (10 µM)
   4 µL 50× Advantage KlenTaq Polymerase Mix
   Pour mixture into eight 0.5-mL PCR tubes (24 µL in each tube).
3. Add 1 µL of each dilution to tubes 1–7. Tube 8 is a negative control sample for checking the DNA contamination. Overlay with mineral oil.
4. Incubate the tubes in a thermal cycler at 72°C for 1 min to extend the adaptors.
5. Immediately commence 40 cycles of amplification under the following conditions: denaturation at 94°C for 30 s, annealing at 66°C for 30 s and extension at 72°C for 1.5 min.
6. Load 5 µL of each PCR product on a 2% agarose gel and run it. Each band in samples containing no smear corresponds to a single molecule taken at the start of the PCR. Select the dilution containing about 10 target cDNA molecules per microliter.

Preparation of in vitro clones suitable for differential screening

1. Prepare a 500-mL PCR mixture:
   395 µL H_2O

50 μL 10× PCR buffer
10 μL dNTP mixture (10 mM)
25 μL PCR primer 1 (10 μM)
10 μL ethidium bromide (0.1 mg/mL)
3 μL cDNA dilution (ca. 10 cDNA molecules/μL)
10 μL 50× Advantage KlenTaq Polymerase Mix

2. Pour 5 μL into each well of the 96-well plate using an 8-channel pipettor (without mineral oil). To avoid evaporation of the PCR mixture, carefully cover the plate with a sealing film.
3. Commence 43 PCR cycles using the following program for the PTC-100 Thermal Cycler:

   Step 1: 72°C for 1 min
   Step 2: 94°C for 10 s
   Step 3: 66°C for 30 s
   Step 4: 72°C for 1.5 min
   Step 5: go to Step 2, perform 42 times
   Step 6: end.

   Use heat lid and temperature calculated for plate with 10 μL PCR mixture in each well. If different type of thermal cycler is used, the cycling parameters must be optimized for that machine.
4. Put the plate on a UV lamp and select the wells containing PCR products (it is stained by ethidium bromide). Plate should contain 20–30 stained wells. If it contains many more or fewer positive wells, repeat Step 1 using respectively fewer or more cDNA molecules.

   Note: Do not expose the plate before PCR to UV light to avoid possible damage of DNA and primer.
5. Add 10 μL of water to each positive well, mix by pipetting and analyze 5 μL of each PCR product on a 2% agarose gel.
6. Select the wells containing one band only.

The PCR products obtained (in vitro clones) can be directly used for differential screening and sequencing.

Differential screening

1. Spot 2 μL from each selected well on two nylon filters in the same order and dry the filters.
2. Place filters in denaturation buffer (0.4 M NaOH, 1.5 M NaCl).
3. Place filters in neutralizing buffer (1 M Tris-HCl, pH 7.0, 1.5 M NaCl).
4. Dry filters.
5. Cross-link the DNA to the membranes using a UV linking device (Stratalinker® UV Crosslinker; Stratagene, La Jolla, CA, USA) at 120 mJ.
6. Hybridize filters in separate bottles with plus- and minus-subtracted cDNA probes in ExpressHyb hybridization solution overnight at 68°C.
7. Wash filters at 68°C in 2× SSC, 0.5% SDS solution 2 times and after that in 0.1× SSC, 0.5% SDS solution 2–3 times (20 min each step).

In Vitro Cloning: A Method for PCR Amplification

8. Put wet filters between two pieces of plastic film and expose to X-ray film using intensifying screen at -70°C.
9. Compare autoradiographs and select the dots that preferentially (more then 10-fold difference) hybridize with plus probe.

PCR products of interest (that gave differential signal) can be directly sequenced using nested PCR primer 1 or 2 by any protocol for sequencing of PCR product. We recommend using the AmpliCycle Sequencing Kit.

Protocol 2: In Vitro Generation of Nested Deletions for Sequencing Large Cloned DNA Fragments

Suppression adaptor 1 and PCR primer 1 are used in this protocol. The use of these oligonucleotides is compatible only with the vectors containing the sequence of PCR primer 1 (T7 RNA polymerase promoter and adjacent sequence). For instance, pBluescript®, pBS or Phagescript (all from Stratagene) vectors can be used.

Materials and Reagents

- DNase I (Sigma)
- DNA Polymerase I, Klenow Fragment (5 U/µL; New England Biolabs)
- T4 DNA Ligase (400 U/µL) and 10× buffer (New England Biolabs)
- Tris-HCl, pH 7.5 (0.5 M)
- Bovine serum albumin (BSA) (1 mg/mL)
- $MnCl_2$ (100 mM)
- EDTA, pH 8.0 (100 mM and 0.5 M solutions)
- dNTP mixture (1 mM each)
- NaOAc (3 M)
- Ethanol (96%)
- Neutral phenol
- Chloroform–isoamyl alcohol (24:1)

Procedures

DNase I digestion and adaptor ligation

1. Dilute DNase I up to 0.1 ng/µL (2.7×10^{-4} U/µL) using buffer (50 mM Tris-HCl, pH 7.5, 0.1 mg/mL BSA).
2. Mix the following reagents (on ice):
   64 µL H_2O
   10 µL Tris-HCl, pH 7.5 (0.5 M)
   10 µL BSA (1 mg/mL)
   10 µL $MnCl_2$ (100 mM)
   5 µL DNA (500 ng/µL)
   1 µL DNase I (0.1 ng/µL = 2.7×10^{-4} U/µL)
3. Incubate the tube at 16°C.
4. Take 20 µL of reaction mixture after 30 s and 1, 2, 4 and 10 min of incubation. To stop reaction, immediately place it in the tubes containing 20 µL of 100

mM EDTA.

5. Analyze 5 µL of each sample by electrophoresis in a 1.2% agarose gel. Select the sample that contains mostly the linear vector and almost no visible smear of degraded DNA (the use of more digested DNA results in the loss of long PCR products).
6. Perform phenol–chloroform extraction and ethanol precipitation of the selected sample.
    a. Add 10 µL of neutral phenol and vortex mix thoroughly.
    b. Add 50 µL of chloroform–isoamyl alcohol (24:1) and vortex mix thoroughly.
    c. Centrifuge the tube at $10\,000\times g$ for 5 min at room temperature.
    d. Carefully remove the top aqueous layer and place in a clean tube. Discard the interphase and lower phase.
    e. Add 1/10 vol of 3 M NaOAc (pH 4.8) and 2.5 vol of 96% ethanol.
    f. Vortex mix thoroughly and centrifuge the tube at $10\,000\times g$ for 15 min at room temperature.
    g. Remove the supernatant carefully and discard.
    h. Air dry the pellet for 5–15 min to evaporate residual ethanol.
    i. Dissolve the pellet in 8 µL of H_2O.
7. To ensure blunt ends, treat the DNA sample with Klenow enzyme:
   5 µL H_2O
   1 µL 10× ligase buffer
   1 µL dNTP mixture (1 mM)
   2 µL digested DNA (step 6i)
   1 µL DNA polymerase I, Klenow fragment (5 U/µL)
   Incubate for 15 min at 37°C.
8. Inactivate Klenow enzyme by heating at 70°C for 10 min.
9. To perform ligation of suppression adaptor 1, add to the reaction mixture the following reagents:
   4 µL H_2O
   1 µL 10× ligase buffer
   4 µL adaptor 1 (10 µM)
   1 µL T4 DNA ligase (400 U/µL)
   Incubate for 2 h at room temperature.
10. To terminate the reaction, add 1 µL 0.5 M EDTA and heat the tube at 70°C for 10 min.

Estimation of quantity of the target DNA molecules in the ligation mixture

1. Prepare the sequential dilutions (10^4-, 10^5-, 10^6-, 3×10^6-, 10^7-, 3×10^7-, 10^8-, 3×10^8-, 10^9-fold) of the ligation mixture using dilution buffer (20 mM HEPES-HCl, pH 7.5, 50 mM NaCl, 0.2 mM EDTA). Change tips and mix solution well after each dilution.
2. Prepare a 200-µL PCR mixture:
   162 µL H_2O

20 µL 10× PCR buffer
4 µL dNTP mixture (10 mM)
10 µL PCR primer 1 (10 µM)
4 µL 50× Advantage KlenTaq Polymerase Mix
Pour mixture into ten 0.5-mL PCR tubes (20 µL in each tube).

3. Add 1 µL of each dilution to tubes 1–7. Tube 8 is a negative control sample for checking the DNA contamination. Overlay with mineral oil.
4. Incubate the tubes in a thermal cycler at 72°C for 1 min to extend the adaptors.
5. Immediately commence 40 cycles of amplification under the following conditions: denaturation at 94°C for 30 s, annealing at 66°C for 30 s and extension at 72°C for 5 min.
6. Load 5 µL of each PCR product on a 2% agarose gel and run it. Each band in samples containing no smear corresponds to a single molecule taken at the start of PCR. Select the dilution containing about 10 target cDNA molecules per microliter.

Preparation of in vitro clones suitable for sequencing

1. Prepare a 500-µL PCR mixture:
   395 µL H$_2$O
   50 µL 10× PCR buffer
   10 µL dNTP mixture (10 mM)
   25 µL PCR primer 1 (10 µM)
   10 µL ethidium bromide (0.1 mg/mL)
   3 µL DNA dilution (ca. 10 DNA molecules/µL)
   10 µL 50× Advantage KlenTaq Polymerase Mix
2. Pour 5 µL in each well of the 96-well plate using an 8-channel pipettor (without mineral oil). To avoid evaporation of the PCR mixture, carefully cover the plate with a sealing film.
3. Commence 43 PCR cycles using the following program for the PTC-100 Thermal Cycler:
   Step 1: 72°C for 1 min
   Step 2: 94°C for 10 s
   Step 3: 66°C for 30 s
   Step 4: 72°C for 1.5 min
   Step 5: go to Step 2, perform 42 times
   Step 6: end.
   Use heat lid and temperature calculated for plate with 10 µL PCR mixture in each well. If a different type of thermal cycler is used, the cycling parameters must be optimized for that machine.
4. Put the plate on a UV lamp and select the wells containing PCR products (it is stained by ethidium bromide). The plate should contain 20–30 stained wells. If it contains many more or fewer positive wells, repeat Step 1 using respectively fewer or more cDNA molecules.

   Note: Do not expose the plate before PCR to UV light to avoid possible damage of DNA and primer.

5. Add 10 µL of water to each positive well, mix by pipetting and analyze 5 µL of each PCR product on a 1.2% agarose gel.
6. Select the wells that contain single bands of different length.

PCR products of interest can be directly sequenced using nested PCR primer 1 by any protocol for sequencing of PCR products. We recommend using the AmpliCycle Sequencing Kit. The sequence of the whole DNA insert can be composed of several sequences obtained by means of their overlapping.

RESULTS AND DISCUSSION

Principle of the Method

In the first step of our in vitro cloning procedure (Figure 1), a mixture of two different suppression adaptors (A and B) is ligated to double-stranded DNA fragments. Each adaptor is a pair of oligonucleotides of uneven length complementary to each other at the 3′ end of the longer oligonucleotide, giving one blunt end. However, only the longer oligonucleotide is ligated because 5′ ends of both oligonucleotides lack phosphate groups. The strand complementary to the ligated longer oligonucleotide is produced before the first PCR cycle at temperatures at which the smaller oligonucleotide already falls off its binding site, but longer DNA molecules still retain the double-stranded structure (in fact, it happens during the hot-start step of PCR). We consider such adaptors more useful than standard fully double-stranded ones because they are cheaper and do not require phosphorylation or precautions against multiple-adaptor ligation.

After attachment, two types of DNA fragments are generated statistically: *(i)* symmetrically flanked molecules bearing the same adaptors on both ends (AA or BB) and *(ii)* asymmetrically flanked molecules bearing different adaptors on the ends (AB or BA). Note that only asymmetrically flanked molecules (AB and BA) will be amplified in subsequent PCR with primers corresponding to the outer parts of the adaptors because after denaturation of the DNA strands, complementary ends of each strand of symmetrically flanked molecules readily anneal to each other (forming the pan-like structure), hiding the primer annealing site before the primer can bind to it (suppression PCR effect; References 2, 9, 12).

In the second step, to obtain each DNA molecule by itself, the ligation mixture is diluted to get the concentration when about one DNA molecule corresponds to the volume being taken for amplification. Then PCR with primers corresponding to the outer parts of the adaptors is performed. As a result, PCR products derived from individual molecules are obtained. We call such PCR products in vitro clones. In vitro clones can be directly sequenced because of the presence of different adaptor sequences at the ends of the DNA fragments.

We can draw the following analogies between in vitro and in vivo cloning. First, suppression adaptors act as vectors. Ligation of the adaptors gives the possibility to amplify the DNA fragment of interest and to sequence it. For convenience, the adaptors can carry T7 or SP6 RNA polymerase promoters and restriction sites. Second, dilution of up to a single DNA molecule per reaction fulfills the role of transformation of bacteria (in both processes, DNA molecules are separated from each other). Third, molecule-by-molecule PCR amplification corresponds to spreading and grow-

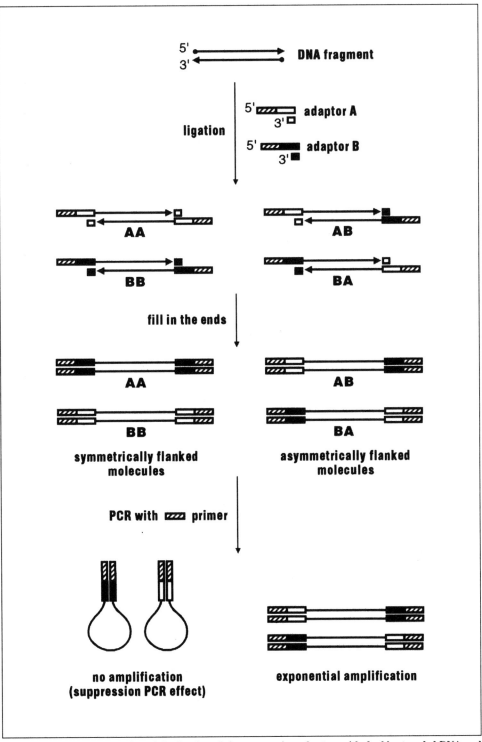

Figure 1. Schematic illustration of the ligation of the suppression adaptors with double-stranded DNA and subsequent PCR. Optimal combination of the adaptors having the identical outer parts is shown. Hatched box represents the outer parts of the adaptors as well as the PCR primer. Open and solid boxes represent the inner parts of adaptors A and B, respectively.

ing transformed bacterial cells on the plate.

The in vitro cloning method has several modifications, each of which can be used in all embodiments. First, the used pair of suppression adaptors might have completely different sequences from each other or might have identical outer or inner parts. The use of the adaptors with an identical outer part is preferential. Second, amplification can be performed in one PCR (40–45 cycles) or in two nested PCRs. Amplification of single molecules in one PCR was found to be reproducible if one primer corresponding to the identical outer parts of the adaptors is used. If two primers drive amplification, two nested PCRs are usually necessary. Third, it is possible to carry out in vitro cloning starting from one DNA molecule per reaction or from several DNA molecules having different lengths per reaction. In the latter case, after amplification, the different products in each tube can be discriminated using gel electrophoresis. The bands on the gel can be treated as individual clones; they can be screened and excised from the gel for separate reamplification and direct sequencing. Fourth, amplification can be performed in tubes or in 96-well plates. Fifth, for quick selection of compartments containing in vitro cloning products, the ethidium bromide can be added to the PCR mixture.

The in vitro cloning method permits molecule-by-molecule PCR amplification of complex DNA mixtures. The products of in vitro cloning can be used for direct sequencing, hybridization analysis, RNA synthesis and other routine molecular biology procedures. DNA sources include sets of DNA fragments that have the same or different sequences and lengths and that are suitable for PCR amplification (usually up to 10 kb). In vitro cloning may find use in a wide variety of situations instead of cloning in bacterial or phage systems.

Differential Screening by In Vitro Cloning

One of the most suitable applications of in vitro cloning is differential screening of subtracted libraries prepared by the suppression subtractive hybridization (SSH) method (3,6,9) (see also Diatchenko et al. in this book), because SSH produces molecules that are already flanked by suppression adaptors. So the procedure of differential screening using in vitro cloning looks quite simple and fast (Figure 2). First of all, one needs to perform PCR amplification of sequential (e.g., 3-fold) dilutions of the cDNA sample obtained after subtractive hybridization to estimate the number of molecules capable of amplification. The typical result of such amplification is as follows. After amplification in the first several dilutions, cDNA looks like a smear; in several middle dilutions, cDNA looks like different sets of bands or single bands; and the last dilutions are empty. Each band in the samples containing no smear corresponds to a single molecule taken at the start of PCR. After this estimation, one can generate the required number of in vitro clones. The simplest way of obtaining in vitro clones is PCR in a 96-well plate in the presence of ethidium bromide. After amplification (starting with 20–30 cDNA molecules per plate), the wells containing PCR products are stained by ethidium bromide while empty wells are black (Figure 2). The excess of the black wells guarantees that most of the colored wells contain single in vitro clones that can be subjected to differential screening. For this purpose, in vitro clones are spotted on two nylon membranes in the same order and hybridized with probes prepared from tester and driver cDNA and/or from cDNAs that had been subtracted in opposite directions (plus- and minus-subtracted cDNAs). The in vitro

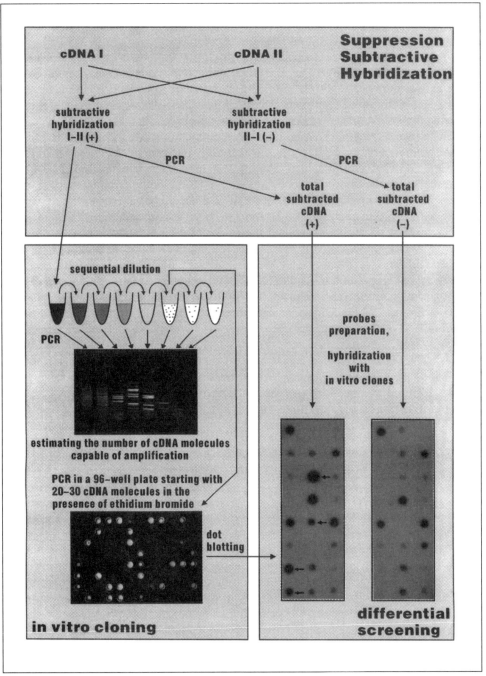

Figure 2. **Strategy for differential screening of SSH-subtracted cDNA using in vitro cloning.** SSH procedure: for the plus subtraction (+), cDNA I is used as a tester, and cDNA II is used as a driver. For the minus (-) subtraction, cDNA II is used as a tester, and cDNA I is used as a driver. The total subtracted cDNAs are prepared by PCR starting from 10^3–10^4 cDNA molecules. In vitro cloning procedure: the plus-subtractive hybridization mixture is diluted up to single cDNA molecules per particular volume. After that, the necessary number of in vitro clones is prepared by PCR in a 96-well plate. The presence of ethidium bromide in the PCR mixture allows to easily select the wells containing the PCR products (such wells are colored under UV illumination, while the empty wells are black). Differential screening procedure: the in vitro clones are spotted on two nylon membranes and hybridized with radioactively labeled plus and minus total subtracted cDNAs. The differentially hybridized in vitro clones are designated by arrows.

clones that preferentially hybridize with tester and/or plus probe represent the differentially expressed genes and can be directly sequenced by any method designed for sequencing of uncloned PCR products.

Differential expression of selected in vitro clones can be confirmed by RT-PCR with specific primers designed based on their sequences. Alternatively, it can be confirmed by Northern blot hybridization with initial RNA samples or by Southern blot hybridization with initial amplified cDNA samples as well as with plus- and minus-subtracted cDNA samples, using in vitro clones as probes.

There are several advantages of the in vitro cloning procedure as compared to the routine cloning methods. First, for in vitro cloning, one can use PCR only instead of time-consuming bacterial or phage cloning and subsequent purification of DNA or PCR amplification of inserts. Second, although in vitro cloning requires more than 40 PCR cycles, point mutations introduced by *Taq* DNA polymerase will be undetectable by direct sequencing of PCR products because of the random distribution of such mutations in each amplified DNA molecule. This eliminates the problem of low fidelity of amplified cDNA. At the same time, cloning of subtracted samples obtained by PCR amplification reveals the PCR substitutions. Third, in vitro cloning resolves the problem of competition in amplification and cloning between different sequences in complex cDNA samples by providing better representation of low-abundance sequences. For instance, the usual source of artifacts in conventional cloning of complex amplified cDNA is excess PCR cycles. After saturation of amplification, the cDNA in the PCR product becomes denatured. Among all the divergent molecules, only high-abundance cDNA species can reanneal into the form suitable for cloning. Low-abundance cDNA species are present mostly in single-stranded form and are eliminated from the cloned library. As a result, representation of the cloned library does not reflect the initial amplified sample. Finally, in vitro cloning never results in rearrangements (duplications, deletions, inversions etc.) of the DNA.

Other Applications of the In Vitro Cloning Method

Obviously, in vitro cloning can be convenient if no more than several dozen clones are needed. Below, we briefly describe several applications of in vitro cloning found to be useful in our practice.

The in vitro cloning method can be used to prepare suitable templates for all regions of a long DNA fragment to be sequenced. Schematic representation of this application is shown in Figure 3. As an example of a large DNA to be sequenced, the fragment cloned in a plasmid vector is shown. A set of nested fragments having different lengths suitable for long-range sequence determination can be generated by partial digestion with a nonspecific exonuclease, endonuclease or four-base cutting restriction endonuclease. Optimally, one can digest the initial DNA sample using DNase I (it creates double-stranded nicks in the presence of manganese chloride). The treatment should be quite slight (about one cleavage per molecule) to generate a pool of long fragments with different ends. After that, the product of digestion is ligated with one suppression adaptor, diluted up to single molecules per reaction and amplified using long-distance PCR (1) with one primer corresponding to the outer part of the adaptor and the other primer corresponding to a unique sequence of the vector DNA (optimally, these primers should be the same). Although after ligation all DNA fragments bear the adaptor on both ends, only sub-fragments flanked by a vector-specific primer and adaptor are subjected to amplification because of the suppression PCR effect. As a result, the set of overlapping PCR products of different lengths corresponding to individual molecules is collected. All such PCR fragments have the same sequence at the ends bearing vector-specific primer and different sequences at the ends bearing adaptor. Each PCR product can be partially sequenced using a primer corresponding to the inner part of the suppression adaptor. The

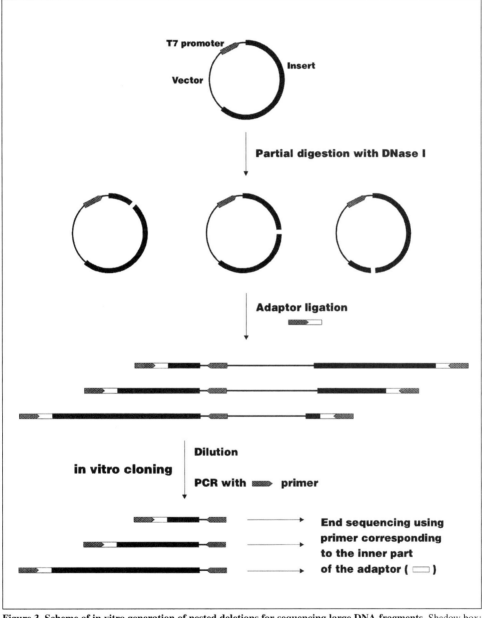

Figure 3. **Scheme of in vitro generation of nested deletions for sequencing large DNA fragments.** Shadow box: the T7 promoter (arrow shows the direction of RNA synthesis); open box: the inner part of the suppression adaptor; black box: insert DNA; thin line: vector DNA. Uncut molecules and over-digested molecules lacking vector's T7 promoter are not shown.

position of the obtained sequence on the whole DNA of interest depends on the site at which cleavage has occurred. The sequence of the whole DNA fragment can be composed of several sequences obtained by means of their overlapping.

In vitro cloning can also be used for subcloning of relatively long (usually more than 1 kb) DNA fragments. The long DNA fragment is digested by any standard procedure (e.g., with restriction endonucleases). The cleavage product is ligated to a pair of suppression adaptors in a mixture, diluted up to single molecules per volume taken for amplification and PCR-amplified using primer(s) corresponding to the outer parts of the adaptors.

The in vitro cloning method can also be used to purify the major PCR products of interest from the mixture containing some background PCR products.

In vitro cloning can be applied to analyze the products of PCR with degenerate primers designed on the basis of amino acid sequence (usually highly conservative sequences). Such a PCR product (even if it looks like a single band by gel electrophoresis) usually contains several DNA fragments having the same length but slightly different sequences (this heterogeneity is due to the presence of several members of a gene family). The PCR product should be ligated with a pair of suppression adaptors and subjected to in vitro cloning. Because all DNA molecules in this application have the same length, one should have in vitro clones only in a small part (e.g., 1/10) of all reactions to be sure that no more than one molecule is present in the reaction.

Troubleshooting

The biggest danger of in vitro cloning is contamination by DNA bearing the suppression adaptors. DNA contamination is a common problem for all PCR-based techniques (10), but in vitro cloning is very sensitive to it because of the extremely small quantity of target DNA. Often, the first several successful in vitro cloning experiments are the source of DNA contamination for subsequent ones. So we recommend to conform to the following rules at the very beginning: *(i)* use a separate work place for dilution of the DNA samples and preparing PCR mixtures. This place should be as far as possible from the places where PCR products are analyzed; *(ii)* cover the work place with a piece of filter paper and change it after each experiment; *(iii)* the pipettors, plasticware packs and all reagents that are used in initial steps of in vitro cloning (before PCR amplification) should not be used in other experiments; *(iv)* use the aerosol barrier tips only; and *(v)* use small aliquots of all reagents to avoid contamination of stock solutions.

When contamination occurs, it is not possible to dilute the target DNA up to zero; starting from certain dilution, all subsequent dilutions show the same DNA concentration. Also, after PCR in a 96-well plate, all (or almost all) wells contain PCR products, although the estimate was only 20–30 DNA molecules per plate.

Because of high probability of DNA contamination, we recommend having a negative control sample (without DNA) in each amplification of sequential DNA dilutions. In general, to ensure the absence of DNA contamination, the tubes or wells without PCR products should be present after in vitro cloning.

To eliminate DNA contamination, we recommend carrying out the following procedures, checking the presence of the contamination after each step: *(i)* change all reagents (including water and mineral oil) and plasticware packs; *(ii)* change or carefully wash pipettors; and *(iii)* change the suppression adaptors and primer used.

ACKNOWLEDGMENTS

I thank my coworkers Sergey Lukyanov, Mikhail Matz and Arkady Fradkov for their hard work in developing the in vitro cloning method. I am grateful to Luda Diatchenko for adaptor design and discussion of this work and also to Marina Tretyak for help in the manuscript preparation. This work was supported by CLONTECH Laboratories.

REFERENCES

1. **Barnes, W.M.** 1994. PCR amplification of up to 35-kb DNA with high fidelity and high yield from λ bacteriophage templates. Proc. Natl. Acad. Sci. USA *91*:2216-2220.
2. **Chenchik, A., S.A. Lukyanov, K.A. Lukyanov, N.G. Gurskaya, V.S. Tarabykin, E.D. Sverdlov, L. Diatchenko and P.D. Siebert**, Inventors; CLONTECH Laboratories, Assignee. 1996 Oct 15. Method for suppression DNA fragments amplification during PCR. US patent 5,565,340. 1996.
3. **Diatchenko, L., Y.-F.C. Lau, A.P. Campbell, A. Chenchik, F. Moqadam, B. Huang, S.A. Lukyanov, K.A. Lukyanov et al.** 1996. Suppression subtractive hybridization: a method for generating differentially regulated or tissue-specific cDNA probes and libraries. Proc. Natl. Acad. Sci. USA *93*:6025-6030.
4. **Ehlich, A., V. Martin, W. Muller and K. Rajewsky.** 1994. Analysis of the B-cell progenitor compartment at the level of single cells. Curr. Biol. *4*:573-583.
5. **Frohman, M.A., M.K. Dush and G.R. Martin.** 1988. Rapid production of full-length cDNAs from rare transcripts: amplification using a single gene-specific oligonucleotide primer. Proc. Natl. Acad. Sci. USA *85*:8998-9002.
6. **Gurskaya, N.G., L. Diatchenko, A. Chenchik, P.D. Siebert, G.L. Khaspekov, K.A. Lukyanov, L.L. Vagner, O.D. Ermolaeva, S.A. Lukyanov and E.D. Sverdlov.** 1996. The equalizing cDNA subtraction based on selective suppression of polymerase chain reaction: cloning of the Jurkat cells' transcripts induced by phytohemagglutinin and phorbol 12-myristate 13-acetate. Anal. Biochem. *240*:90-97.
7. **Kellogg, D.E., I. Rybalkin, S. Chen, N. Mukhamedova, T. Vlasik, P.D. Siebert and A. Chenchik.** 1994. TaqStart antibody™: "hot start" PCR facilitated by a neutralizing monoclonal antibody directed against *Taq* DNA polymerase. BioTechniques *16*:1134-1137.
8. **Lukyanov, K.A., M.V. Matz, E.A. Bogdanova, N.G. Gurskaya and S.A. Lukyanov.** 1996. Molecule by molecule PCR amplification of complex DNA mixtures for direct sequencing: an approach to *in vitro* cloning. Nucleic Acids Res. *24*:2194-2195.
9. **Lukyanov, S.A., N.G. Gurskaya, K.A. Lukyanov, V.S. Tarabykin and E.D. Sverdlov.** 1994. Highly efficient subtractive hybridisation of cDNA. Bioorg. Khim. *20*:701-704.
10. **Orrego, C.** 1990. Organizing a laboratory for PCR work, p. 447-454. In M.A. Innis, D.H. Gelfand, J.J. Sninsky and T.J. White (Eds.), PCR Protocols: A Guide to Methods and Applications. Academic Press, San Diego.
11. **Sarkar, G. and M.E. Bolander.** 1995. Semi exponential cycle sequencing. Nucleic Acids Res. *23*:1269-1270.
12. **Siebert, P.D., A. Chenchik, D.E. Kellogg, K.A. Lukyanov and S.A. Lukyanov.** 1995. An improved PCR method for walking in uncloned genomic DNA. Nucleic Acids Res. *23*:1087-1088.
13. **Trower, M.K., D. Burt, I.J. Purvis, C.W. Dykes and C. Christodoulou.** 1995. Fluorescent dye-primer cycle sequencing using unpurified PCR products as templates; development of a protocol amenable to high-throughput DNA sequencing. Nucleic Acids Res. *23*:2348-2349.
14. **Welsh, J., J.-P. Liu and A. Efstratiadis.** 1990. Cloning of PCR-amplified total cDNA: construction of a mouse oocyte cDNA library. Genet. Anal. Tech. Appl. *7*:5-17.

Index

A

Affinity capture   151, 259, 260, 262, 266, 268
Amplicons   129-132, 135, 136, 138-140, 193-195, 197-201
Anchored oligo(dT) primers   145, 147, 148, 152, 159-165, 168, 262, 306, 311-313
Angiotensin   72, 77-80, 82, 84-87
Archival specimens   3-6, 8, 12, 14, 16, 17
Automation   129, 151, 183, 209, 287
Avian myeloblastosis virus (AMV)   58, 59, 61, 87, 148

B

Band selection   172, 173, 176, 178, 179
Biopsy   3, 4, 17, 87, 104, 151, 168, 236, 305
Breast carcinoma   3, 4, 12, 14, 16

C

Capture probes   129, 131, 133-139, 272, 275-278, 281, 283-287
Cardiac fibroblasts   72, 77, 82, 85, 86
Cardiac myocytes   72, 86, 87
cDNA
   amplifiable single-stranded cDNA   272, 278, 286
   cDNA amplification   25, 113, 178, 218, 229, 274, 305, 306, 308, 309, 312, 313, 315, 318
   cDNA fragments   96, 194, 195, 198, 201, 205, 206, 219, 226, 228, 232-234, 239-248, 250-255, 259-264, 266-268, 274
   cDNA library construction   287, 305, 306, 309, 315, 318
   cDNA subtraction   215, 236, 305, 306, 318, 332
   enriched cDNA libraries   240, 244, 259-269
   full-length cDNAs   241, 242, 255, 257, 259-269, 287, 314
Cell-specific fingerprints   160
Competitor   72-74, 78-82, 85-88, 91, 92, 94, 96-100

Creating a ribonuclease-free environment   20

D

Differential display (DD)   30, 145-155, 159-168, 171-180, 183-190, 204, 239, 240, 242, 260
Differential gene expression   145, 159, 171-174, 177-180, 184, 190, 193-195, 201, 203, 204, 207-210, 213, 215, 216, 226, 231, 232, 234, 236, 239, 240, 242, 243, 255, 260, 266, 318, 342
Differential screening   145, 146, 172, 201, 203, 213, 215, 220, 228, 230, 232-234, 236, 239, 330, 332-334, 340, 341
Direct sequencing   287, 329, 340, 342
Display gel   239, 240, 242, 246, 248-251, 254, 255, 267
Dithiothreitol (DTT)   10, 11, 29-31, 37, 40, 50, 58-64, 94, 118, 136, 153, 154, 174, 184, 186, 196, 197, 217, 243, 274, 279, 294, 307
DNA polymerases
   Deep Vent™   149, 217, 307
   Pfu   149, 274, 283-285
   Pwo   284
   Taq   30, 58, 59, 73, 75, 78, 81, 83, 93, 95, 136, 149, 155, 161, 162, 166, 175, 176, 186, 197-200, 204, 206, 217, 224, 274, 276, 277, 282-285, 331, 342
   Tth   66, 148, 149, 295-297
   Vent®   28, 149, 217, 295, 296, 307
DNA walking   289-291, 294, 295, 297, 299
DNase treatment   19-21, 24, 26, 153, 174
Dopamine D_2 receptor (D2R)   129, 136, 137, 139, 140
Dot blotting   151, 207, 228, 229, 232, 233
Down-regulated gene products   242, 254, 259, 261

E

Efficiency   113-117, 119, 126
Elongation factor 1α (EF-1α)   72, 77, 78, 80, 82, 84-87

Enzyme-linked immunosorbent assay (ELISA) 97, 98, 100, 132, 135
Equations 113-117, 119, 122-125, 139
Escherichia coli 66, 196, 197, 207, 217, 260, 274, 283, 286, 324-326
Estrogen receptor (ER) 3, 4, 7, 11-16, 317
Expression constructs 321-326
Expression vectors 262, 286, 321-326

F

Formalin-fixed tumor samples 3-9, 11-13, 15-17
Frozen sections 57, 65
Fusion proteins 322-325

G

Gene capture 271-273, 276, 278, 284, 286, 287
Gene families 271, 272, 286, 344
Gene-specific oligonucleotides 10, 11
Glyceraldehyde-3-phosphate dehydrogenase (GAPD) 6, 7, 11-15, 103, 113, 115, 117, 119-126, 179, 216, 221, 222, 224-226, 232, 235, 316, 317
Guanidinium thiocyanate 19-21, 23, 25, 39-41, 43, 49-53, 57, 64-66, 118, 174, 196, 204, 307

H

Heparin 36, 38, 40, 43
Heterogeneous hybridization 129, 131, 132, 136
Homologous gene cloning strategy 271
Homologous genes 271, 272, 281, 284-286
Hybridization assays 129-133, 135, 136

I

In vitro cloning 329-333, 335, 337-345
Internal standards 72, 78, 79, 103, 114, 115, 126, 136, 137, 139

L

Linker-capture subtraction (LCS) 203-205, 207-211

M

Magnetic beads 99, 132, 136-138, 183, 185-187, 189, 190, 196, 261, 265, 268, 272, 277, 279-281, 285, 286
Magnetic separation 189, 272, 273, 277
Mathematical modeling 116, 123, 125, 127, 135, 216, 235
Microplates 129-141, 226, 229, 247, 249
Microscopic biological systems 172, 174, 178, 180
Moderate- to low-abundance transcripts 171, 172, 174, 177-180
Moloney murine leukemia virus (MMLV) 6-8, 73, 87, 104, 107, 118, 136, 148, 154, 174, 175, 184, 186, 190, 306, 307, 311, 312, 315
mRNA
    absolute quantification of mRNA 71, 78, 23, 83, 85, 87, 89, 91, 92, 100
    aldolase mRNA 57, 58, 60, 62-64
    apolipoprotein B (apoB) mRNA 44-46
    fatty acid synthase (FAS) mRNA 43, 44
    insulin-like growth factor I (IGF-1) mRNA 44, 45
    insulin-receptor mRNA 27, 28, 30, 32
    large mRNA 35, 42, 45
    low-abundance mRNA 32, 72, 171, 177-180, 234, 236
    mRNA capture 184, 189
    mRNA quantitation 72, 103, 104, 111, 136
    thyroglobulin (Tg) mRNA 41, 42
Multiplex competitive RT-PCR titration assay (MPTA) 71, 72, 74-89
Mung bean nuclease 203, 204, 206, 209

N

Nested deletions 335, 343
Nonfusion proteins 323-325

Nontemplated nucleotide addition 306, 311, 312
Normalization 7, 12, 14, 16, 60, 85, 87, 108, 110, 114, 117, 119, 123, 124, 178, 210, 213, 215, 234-236, 253, 254
Northern blotting 3, 42, 44-47, 88, 103, 113, 116, 121, 125-127, 146, 151, 152, 172, 173, 177, 179, 180, 201, 207, 208, 215, 233-236, 283, 285, 305, 306, 319, 342

O

Oligo(dT) 30, 33, 92, 94, 99, 104, 107, 108, 145-149, 152, 159-168, 172, 184-190, 194-197, 204, 205, 273, 274, 282-285, 305, 306, 310-313
Optimization of reverse transcription 86

P

Paraffin-embedded tumor samples 3-9, 11-13, 15-17
pET vectors 323-325
Phenol–chloroform extraction 22, 37, 41, 43, 47, 52, 53, 65, 79, 94, 153, 166, 195, 197, 198, 200, 219, 228, 243, 244, 253, 293, 294, 300, 307, 335, 336
Poly(A+) RNA 107, 108, 159-162, 164, 165, 168, 196, 204, 205, 213-215, 218-220, 226, 231, 233-235, 262, 273, 276, 280-282, 285, 305, 309, 314-316, 318, 319
Polymerase chain reaction (PCR)
    arbitrarily primed PCR (AP-PCR) 147, 159-169, 172, 239
    gene capture PCR (GC-PCR) 271-273, 278, 284, 286, 287
    hot-start PCR 61, 75, 83, 88, 186, 187, 217, 277, 282, 283, 296, 331, 338
    inverse PCR 78, 79, 289
    nested PCR 27, 28, 30-33, 216, 291, 300, 330-332, 335, 338, 340
    SMART™ PCR 215, 227, 231, 287, 305, 306, 311, 314-319
    suppression PCR 215, 236, 279, 291, 292, 299, 330, 338, 343
    touchdown PCR 292, 301

Preferential identification 171-173, 175, 177, 179, 181
Primer selection 174, 178
Prostaglandin $F_{2\alpha}$ (FP) receptor 91, 92, 96, 97, 99
Prostate cancer 160, 165-168, 204, 207, 208
Protein expression 321, 322, 324-326

Q

Quantitation 65, 72, 80, 103, 104, 109, 111 113, 115, 121, 129, 131-133, 134-141, 178, 198

R

Recombinase A (RecA) protein 259-264, 266-268
Reduction of complexity 239
Relatively rare transcripts 172, 180
Representational difference analysis (RDA) 193-201, 204, 209
Restriction digestion 76, 77, 84-87, 243, 244, 252, 300
Reverse transcription polymerase chain reaction (RT-PCR)
    competitive RT-PCR 71, 91-93, 95, 97-101, 103, 104, 108, 109, 113, 127, 136
    multiplex RT-PCR 71, 72, 74-89, 103-107, 109, 111
    quantitative RT-PCR 69, 72, 91-93, 98-100, 104, 152, 173, 177
    single-cell RT-PCR 27, 32, 33
Ribonuclease protection assay 88
RNA fingerprinting 147, 159, 160, 162, 163, 168, 172, 176, 179, 204, 239
RNA isolation 4, 5, 19-26, 35-37, 39, 41-45, 47, 49, 51, 53, 55, 57, 59, 61, 63-67, 77, 86, 150, 151, 153, 161, 165, 166, 231, 244, 279
    from solid tumors 19, 21, 23, 25, 26
    with ammonium sulfosalicylate 20, 22, 25, 26
    with guanidinium thiocyanate 21, 23, 25, 41
RNase 3, 6-8, 19, 20, 24, 28-30, 32, 33, 36-43, 53, 57-66, 72, 92, 104, 107,

118, 136, 148, 154, 161, 174, 177, 185, 196, 197, 217, 262, 273, 274, 279, 312

RNase inhibitor   37, 57-66, 72, 92, 104, 107, 118, 136, 174, 274, 279

RNase-A-Sepharose   57, 65

S

Selected primers   111, 171-173, 176-180

Sequencing large DNA fragments   335, 343

Sequential hybridization   272

Signal peptidase gene   160, 161, 166-168

Southern blotting   31, 173, 177, 179, 180, 273, 275, 281-285, 287, 291, 299, 342

Standard curve 72, 91-101, 133, 135

Subcloning   226, 321, 330, 344

Subtractive hybridization   171, 194, 199, 203, 206, 209, 213-217, 219-221, 223, 225, 227, 229, 231, 233, 235-237, 239-242, 244, 248-250, 252, 254, 255, 259-261, 267, 318, 332, 333, 340

Suppression subtractive hybridization (SSH)   213-217, 219, 221, 223-225, 227-229,

T

Template switching (TS) 310-313, 318
   TS oligonucleotide   305-307, 310-313, 315, 318
   TS reaction   305, 310, 311, 313, 315

Tissue sampling   19-21, 47, 48, 50, 51

Total RNA 5, 19, 39, 42, 44-46, 53, 57, 63-65, 82, 85-87, 89, 103, 104, 110, 111, 117, 121, 137, 140, 147, 159, 161-166, 168, 171, 173-175,177-180, 190, 196, 204, 209, 213, 215, 218, 226, 231, 232, 236, 272, 273, 276, 278-281, 284, 305, 307-309, 314-319

Transcriptome   171, 239, 259

Triple helix   259, 260, 262-264, 266, 268

Tumors   3-9, 11-17, 19-26, 38, 146, 160, 165-168, 204, 207, 208, 266, 299

U

Up-regulated gene products   242, 254, 259-261

V

Virtual Northern blot   215, 234-236, 305, 319